T0317672

Design and Development of Aircraft Systems

Aerospace Series

Helicopter Flight Dynamics: Including a Treatment of Tiltrotor Aircraft, 3rd Edition
Gareth D. Padfield CEng, PhD, FRAeS

Space Flight Dynamics, 2nd Edition
Craig A. Kluever

Performance of the Jet Transport Airplane: Analysis Methods, Flight Operations, and Regulations
Trevor M. Young

Small Unmanned Fixed-wing Aircraft Design: A Practical Approach
Andrew J. Keane, András Sóbester, James P. Scanlan

Advanced UAV Aerodynamics, Flight Stability and Control: Novel Concepts, Theory and Applications
Pascual Marqués, Andrea Da Ronch

Differential Game Theory with Applications to Missiles and Autonomous Systems Guidance
Farhan A. Faruqi

Introduction to Nonlinear Aeroelasticity
Grigorios Dimitriadis

Introduction to Aerospace Engineering with a Flight Test Perspective
Stephen Corda

Aircraft Control Allocation
Wayne Durham, Kenneth A. Bordignon, Roger Beck

Remotely Piloted Aircraft Systems: A Human Systems Integration Perspective
Nancy J. Cooke, Leah J. Rowe, Winston Bennett Jr., DeForest Q. Joralmon

Theory and Practice of Aircraft Performance
Ajoy Kumar Kundu, Mark A. Price, David Riordan

Adaptive Aeroservoelastic Control
Ashish Tewari

The Global Airline Industry, 2nd Edition
Peter Belobaba, Amedeo Odoni, Cynthia Barnhart

Modeling the Effect of Damage in Composite Structures: Simplified Approaches
Christos Kassapoglou

Introduction to Aircraft Aeroelasticity and Loads, 2nd Edition
Jan R. Wright, Jonathan Edward Cooper

Theoretical and Computational Aerodynamics
Tapan K. Sengupta

Aircraft Aerodynamic Design: Geometry and Optimization
András Sóbester, Alexander I J Forrester

Stability and Control of Aircraft Systems: Introduction to Classical Feedback Control
Roy Langton

Aerospace Propulsion
T.W. Lee

Civil Avionics Systems, 2nd Edition
Ian Moir, Allan Seabridge, Malcolm Jukes

Aircraft Flight Dynamics and Control
Wayne Durham

Modelling and Managing Airport Performance
Konstantinos Zografos, Giovanni Andreatta, Amedeo Odoni

Advanced Aircraft Design: Conceptual Design, Analysis and Optimization of Subsonic Civil Airplanes
Egbert Torenbeek

Design and Analysis of Composite Structures: With Applications to Aerospace Structures, 2nd Edition
Christos Kassapoglou

Aircraft Systems Integration of Air-Launched Weapons
Keith A. Rigby

Understanding Aerodynamics: Arguing from the Real Physics
Doug McLean

Design and Development of Aircraft Systems, 2nd Edition
Ian Moir, Allan Seabridge

Aircraft Design: A Systems Engineering Approach
Mohammad H. Sadraey

Introduction to UAV Systems, 4th Edition
Paul Fahlstrom, Thomas Gleason

Theory of Lift: Introductory Computational Aerodynamics in MATLAB/Octave
G.D. McBain

Sense and Avoid in UAS: Research and Applications
Plamen Angelov

Morphing Aerospace Vehicles and Structures
John Valasek

Spacecraft Systems Engineering, 4th Edition
Peter Fortescue, Graham Swinerd, John Stark

Unmanned Aircraft Systems: UAVS Design, Development and Deployment
Reg Austin

Gas Turbine Propulsion Systems
Bernie MacIsaac, Roy Langton

Aircraft Systems: Mechanical, Electrical, and Avionics Subsystems Integration, 3rd Edition
Ian Moir, Allan Seabridge

Basic Helicopter Aerodynamics, 3rd Edition
John M. Seddon, Simon Newman

System Health Management: with Aerospace Applications
Stephen B. Johnson, Thomas Gormley, Seth Kessler, Charles Mott, Ann Patterson-Hine, Karl Reichard, Philip Scandura Jr.

Advanced Control of Aircraft, Spacecraft and Rockets
Ashish Tewari

Air Travel and Health: A Systems Perspective
Allan Seabridge, Shirley Morgan

Principles of Flight for Pilots
Peter J. Swatton

Handbook of Space Technology
Wilfried Ley, Klaus Wittmann, Willi Hallmann

Cooperative Path Planning of Unmanned Aerial Vehicles
Antonios Tsourdos, Brian White, Madhavan Shanmugavel

Design and Analysis of Composite Structures: With Applications to Aerospace Structures
Christos Kassapoglou

Introduction to Antenna Placement and Installation
Thereza Macnamara

Principles of Flight Simulation
David Allerton

Aircraft Fuel Systems
Roy Langton, Chuck Clark, Martin Hewitt, Lonnie Richards

Computational Modelling and Simulation of Aircraft and the Environment, Volume 1: Platform Kinematics and Synthetic Environment
Dominic J. Diston

Aircraft Performance Theory and Practice for Pilots, 2nd Edition
Peter J. Swatton

Military Avionics Systems
Ian Moir, Allan Seabridge, Malcolm Jukes

Aircraft Conceptual Design Synthesis
Denis Howe

Design and Development of Aircraft Systems

Third Edition

Allan Seabridge and Ian Moir

This edition first published 2020

Edition History
American Institute of Aeronautics & Astronautics (1e, 2004); John Wiley & Sons Ltd (2e, 2012)

Registered Offices
John Wiley & Sons, Inc., 111 River Street, Hoboken, NJ 07030, USA
John Wiley & Sons Ltd, The Atrium, Southern Gate, Chichester, West Sussex, PO19 8SQ, UK

Editorial Office
The Atrium, Southern Gate, Chichester, West Sussex, PO19 8SQ, UK

For details of our global editorial offices, customer services, and more information about Wiley products visit us at www.wiley.com.

Wiley also publishes its books in a variety of electronic formats and by print-on-demand. Some content that appears in standard print versions of this book may not be available in other formats.

Library of Congress Cataloging-in-Publication Data

Names: Seabridge, Allan, author. | Moir, Ian, author.
Title: Design and development of aircraft systems / Allan Seabridge and Ian Moir.
Description: Third edition. | Hoboken, N.J., USA : Wiley, 2020. | Series: Aerospace series | Includes bibliographical references.
Identifiers: LCCN 2019039099 (print) | LCCN 2019039100 (ebook) | ISBN 9781119611509 (hardback) | ISBN 9781119611554 (adobe pdf) | ISBN 9781119611516 (epub)
Subjects: LCSH: Airplanes–Design and construction. | Aeronautics–Systems engineering.
Classification: LCC TL671.2 .S39 2020 (print) | LCC TL671.2 (ebook) | DDC 629.135–dc23
LC record available at https://lccn.loc.gov/2019039099
LC ebook record available at https://lccn.loc.gov/2019039100

Cover Design: Wiley
Cover Images: Section view of aircraft brake components © Steve Mann/Shutterstock, Brake Assembly of aircraft © Standard store88/Shutterstock

Set in 9.5/12.5pt STIXTwoText by SPi Global, Pondicherry, India

Printed and bound by CPI Group (UK) Ltd, Croydon, CR0 4YY

10 9 8 7 6 5 4 3 2 1

Contents

About the Authors

Allan Seabridge was until 2006 the Chief Flight Systems Engineer at BAE Systems at Warton in Lancashire in the UK. In over 45 years in the aerospace industry his work has included the opportunity to work on a wide range of BAE Systems projects, including Canberra, Jaguar, Tornado, EAP, Typhoon, and Nimrod, and the opportunity to act as reviewer for Hawk, Typhoon, and the Joint Strike Fighter, as well being involved in project management, research and development, and business development. In addition, Allan has been involved in the development of a range of flight and avionics systems on a wide range of fast jets, training aircraft, and ground and maritime surveillance projects. From experience in BAE Systems with systems engineering education he is keen to encourage a further understanding of integrated engineering systems. An interest in engineering education continues since retirement with the design and delivery of systems and engineering courses at a number of UK universities at undergraduate and postgraduate level, including the Universities of Bristol, Cranfield, Lancaster, Loughborough, Manchester, and the West of England. Allan has been involved at Cranfield University for many years and been an external examiner for the MSc course in Aerospace Vehicle Design for a three-year period.

Allan has co-authored a number of books in the Aerospace Series with Ian Moir, all published by John Wiley. He is currently a member of the BAE Systems Heritage Department at Warton and is fully involved in their activities, working closely with a colleague to produce a project history book published by the Heritage Group: *EAP: The Experimental Aircraft Programme* by Allan Seabridge and Leon Skorzcewski.

Ian Moir after 20 years in the Royal Air Force as an engineering officer, went on to Smiths Industries in the UK where he was involved in a number of advanced projects. Since retiring from Smiths he spent some time as a highly respected consultant. Ian has a broad and detailed experience working in aircraft avionics systems in both military and civil aircraft. From the RAF Tornado and Army Apache helicopter to the Boeing 777 Electrical Load Management System, Ian's work kept him at the forefront of new system developments and integrated systems in the areas of more-electric technology and system implementations. After more than 50 years of experience in Aerospace Ian has now retired.

Series Preface

The field of aerospace is multi-disciplinary and wide ranging, covering a large variety of products, disciplines, and domains, not merely in engineering but in many related supporting activities. These combine to enable the aerospace industry to produce innovative and technologically advanced vehicles. The wealth of knowledge and experience that has been gained by expert practitioners in the various aerospace fields needs to be passed onto others working in the industry and also researchers, teachers, and the student body in universities.

The *Aerospace Series* aims to be a practical, topical, and relevant series of books aimed at people working in the aerospace industry, including engineering professionals and operators, engineers in academia, and allied professions such as commercial and legal executives. The range of topics is intended to be wide ranging, covering design and development, manufacture, operation and support of aircraft, as well as topics such as infrastructure operations and current advances in research and technology.

Modern aircraft are a good example of a complex high-value 'systems of systems' where the systems addressing all aspects of operation, for instance, the flight control, engine control, electrical power, and hydraulic systems, need to be designed to function effectively both individually and also as part of the overall system. Further complications arise as these interacting sub-systems often have conflicting requirements, will be in service for many years, and, of course, aircraft are intended to be operated by a human pilot. The safety and efficient operation of all aircraft depends upon the success of the system design.

This is the third edition of *Design and Development of Aircraft Systems* and provides an excellent introduction to aircraft systems and the systems development process, with a focus on students studying in the areas of aerospace or systems who have the aim of being employed in the aviation industry or related areas. The material covered in previous editions has been expanded and includes a new chapter on Integration and Complexity. The book is a fine complement to the other Aircraft Systems-related books in the *Aerospace Series.*

Jonathan Cooper and Peter Belobaba

Acknowledgements

> *There is no invention that does not possess a history, none that does not build on, or learn from or owe a debt to the work of others.*
>
> Joseph Swan, 1828–1914.

From 'Swan, 1924' by Sean O'Brien, *Litmus: Short Stories from Modern Science*, Ra Page 9ed.), Comma Press, 2011.

This work is the culmination of many years of work in the field of military and civil aircraft systems engineering. My work experience has been enriched by the opportunity to work with a number of universities at undergraduate and postgraduate level to develop and add to degree courses, where the delegates unwittingly became critics and guinea pigs for my subject matter. Discussions during the courses with academics and students have broadened my knowledge considerably. In particular I would like to mention the Universities of Manchester, Loughborough, Cranfield, Bristol, the West of England, and Lancaster for their MSc and short courses attended by students and engineers from industry.

At Cranfield special thanks must go to Dr Craig Lawson, Dr Huamin Jia, and Professor Shijun Guo for inviting me to participate in their MSc modules in air vehicle design and short courses in aircraft systems design. Their international students have been most attentive and have made significant contributions to my knowledge. Dr Craig Lawson has also contributed an important section in Chapter 12 on the estimation of fuel penalties as part of the trade-off process and has provided me with much information and advice.

I want to acknowledge the BAE Systems Heritage group members at Warton for their advice, comments, photographs, and words which they contributed freely, and often unknowingly. The content of Figure 5.23 is reproduced with kind permission of one member, Brian Weller, from his informative and well written book *A history of the fly-by-wire Jaguar* (2018, published by BAE Systems Heritage department, ISBN 978-0-9 573 755-5-0), an excellent bit of history. Dennis Morley and Leon Skorczewski provided images from their extensive private collections of aircraft and component photographs. In addition to images they both provided much useful information. Jim Banks gave me a lot of help and examples from the design perspective, including a scheme included in Figure 3.10. Steve McDowell advised me on some of the legal aspects of my work and ensured that the work was suitable for publication.

At BAE Systems Warton Chris Preston gave me a lot of advice on the CAD process and images. David Coates was very helpful in advising me from the public relations perspective and guiding the process of obtaining permissions. Simon Leigh from Legal Department obtained the group sign-off from BAE Systems for permission to use their images. I would like to thank BAE Systems for their permission to use images in the following figures: 3.8, 3.10, 3.14, 3.16, 3.19, 7.6, 7.11, 7.12, 7.17. These images greatly enriched the explanations used in the text.

Figure 7.11 contains an image reproduced with kind permission of Merlin Flight Simulation Ltd, UK. Many thanks to Marion Neal of Merlin Flight Simulation for her ever helpful and cheerful response.

Many thanks are due to Ian Moir for all our past collaboration which contributed so much to this edition. This third edition of *Design and Development of Aircraft Systems* has been prepared without Ian Moir who has decided to retire finally after many previous attempts. He has been a colleague and a friend for more than 40 years and I really missed his input to this work. We did not always agree except on one thing: whatever we were working on was going to be the best that we could make it. The intention was always to get things right and to contribute to the learning of other engineers in the aerospace industry. I hope we succeeded.

I have received considerable help from the staff at Wiley, especially Anne Hunt and Eric Willner as well as their proof readers, copy editors, and publishing and production staff.

June 2019

Allan Seabridge
Dent, Cumbria, UK

Glossary of Terms

A4A	Airlines for America
ABS	Automatic braking system
AC	Alternating current
AC	Airworthiness circular: document offering advice on specific aircraft operations
ACARS	ARINC communications and reporting system
ACMP	AC-driven motor pump
ADC	Air data computer
ADP	Air-driven pump
ADIRS	Air data and inertial reference system
ADF	Automatic direction finding
ADM	Air data module
ADR	Accident data recording
ADS-B	Automatic dependent surveillance – broadcast (see IFF)
AFDS	Auto-pilot and flight director system
AFDX	Avionics fast-switched Ethernet
AHARS	Attitude heading and reference system
AIMS	Aircraft information management system (Boeing)
Al	Aluminium
ALU	Arithmetic logic unit
AMP	Air motor-driven pump
APU	Auxiliary power unit
ARINC	Air Radio Inc (US)
ARINC 400 Series	Series of ARINC specifications providing a design foundation for avionic equipment
ARINC 404	Early ARINC standard relating to the packaging of avionic equipment

ARINC 429	Widely used Civil Aviation data bus standard
ARINC 500 Series	Series of ARINC specifications relating to the design of analogue avionic equipment
ARINC 578	ARINC standard relating to the design of VHF omni-range (VOR)
ARINC 579	ARINC standard relating to the design of instrument landing systems (ILSs)
ARINC 600	Later ARINC standard relating to the packaging of avionic equipment
ARINC 600 Series	Series of ARINC specifications relating to enabling technologies for avionic equipment
ARINC 629	ARINC standard relating to a 2 Mbit/s digital data bus
ARINC 664	ARINC standard relating to an aircraft full multiplex (AFDX) digital data bus
ARINC 700 Series	Series of ARINC specifications relating to the design of digital avionic equipment
ARINC 708	ARINC standard relating to the design of weather radar
ARINC 755	ARINC standard relating to the design of multi-mode receivers (MMR)
ARP	Aerospace recommended practice (SAE)
ASIC	Application-specific integrated circuit
ATA	Air Transport Association
ATC	Air traffic control
ATI	Air transport instrument, a means of specifying the size of aircraft instruments
ATM	Air transport management
AWG	American wire gauge
Backwards compatibility	The ability of systems to be compatible with earlier developments/ configurations
BART	Bay Area Rapid Transport (San Francisco)
BC	Bus controller (MIL-STD-1553B data bus)
BCAR	British Civil Airworthiness Requirement
BIT	Built-in test
BMS	Business management system
CAD	Computer-aided design
CADMID	UK MoD procurement process
CAIV	Cost as an independent variable

CANBus	Automotive data bus
CB	Circuit breaker
CDR	Critical design review
CDU	Control and display unit
CFC	Chloro-fluoro-carbon compounds
CFD	Computational fluid dynamics
CFIT	Controlled flight into terrain
CG, cg	Centre of gravity
CNI	Communications, navigation, identification
Cold Soak	Prolonged exposure to cold temperatures
Com	Command channel
COMINT	Communications intelligence
COTS	Commercial off the shelf
CPIOM	Common processor input/output module
CPM	Common processing module
CPU	Central processing unit
CRM	Crew resource management
CS	Certification specification
CSG	Computer symbol generator
Cu	Copper
CVR	Cockpit voice recorder
DC	Direct current
DCMP	DC motor-driven pump
Def Stan	Defence standard
DME	Distance measuring equipment
DMC	Display management computer
DoD	Department of Defense (US)
DOORS	A requirements management tool
Downey cycle	Procurement model once used by the UK MoD
DRL	Data requirements list
DVI	Direct voice input
EASA	European Aviation Safety Administration
ECAM	Electronic check-out and maintenance (Airbus)

ECS	Environmental control system
EDP	Engine-driven pump
EDR	Engineering design requirements
EEC	Electronic engine controller
EFIS	Electronic flight instrument system
EICAS	Engine indication and crew alerting system
ELMS	Electrical load management system
EMC	Electro-magnetic compatibility
EMH	Electro-magnetic health
EMI	Electro-magnetic interference
EOS	Electro-optical system
EPB	External power breaker
ESM	Electronic support measures
ETOPS	Extended twin operations
EUROCAE	European Organisation for Civil Aviation Equipment
FAA	Federal Aviation Administration (US)
FADEC	Full authority digital engine control
FAR	Federal airworthiness requirement
FAV	First article verification
FBW	Fly-by-wire
FCS	Flight control system
FCU	Flight control unit
FL	Flight level
FMECA	Failure mode and criticality analysis
FMQGC	Fuel management and quantity gauging computer
FMS	Flight management system
FOB	Fuel on board
Forwards compatibility	The ability of systems to be compatible with future developments/configurations
FRR	Final readiness review
Full duplex	A data bus that passes data in a bi-directional manner
G&C	Guidance and control
GATM	Global air traffic management

GCB	Generator control breaker
GCU	Generator control unit
GHz	10^9 Hertz (gigaHertz)
GPS	Global positioning system
GPWS	Ground proximity warning system (see also TAWS)
GUI	Graphical user interface
gpm	Gallons per minute
Half Duplex	A data bus that passes data in a uni-directional manner
HALT	Hardware accelerated life test
HF	High frequency
HIRF	High-intensity radio frequency
HMI	Human–machine interface
HOTAS	Hands on throttle and stick
Hot soak	Prolonged exposure to high temperatures
HP	Horse power
HUD	Head-up display
IAS	Indicated airspeed
IC	Integrated circuit
ICD	Interface control document
IDG	Integrated drive generator
IEEE 1498	High-speed data bus
IFE	In-flight entertainment
IFF/SSR	Information friend or foe/secondary surveillance radar (see ADS-B)
ILS	Instrument landing system – an approach aid used for guiding the aircraft on a final approach to landing
IMA	Integrated modular architecture
IMINT	Image intelligence
INCOSE	International Council On Systems Engineering
INS	Inertial navigation system
I/O	Input/output
IPT	Integrated product team
IR	Infra-red
IRS	Inertial reference system

ISIS	Integrated standby instrument system
IT	Information technology
ITAR	International Traffic in Arms Regulations
JAA	Joint Aviation Authorities (Europe) (see EASA)
JAR	Joint airworthiness requirement
JASC	Joint aircraft system/component (FAA)
kbit	10^3 bit (kilobit)
LCD	Liquid crystal display
LCN	Load classification number (runway)
LED	Light-emitting diode
LfE	Learning from experience
LoC	Lines of code
LOX	Liquid oxygen
LP	Low pressure
LRI	Line replaceable item
LRU	Line replaceable unit
LVDT	Linear variable differential transformer
Mach	The speed of an aircraft in relation to the speed of sound
MAD	Magnetic anomaly detector
MAU	Modular avionics unit
Mbit	10^6 bit (megabit)
MCDU	Multi-function control and display unit
MCU	Modular concept unit
MEA	More electric aircraft
MHz	10^6 Hertz (megaHertz)
MIL-HBK	*Military Handbook*: a US military publication
MIL-STD-1553B	Widely used military data bus standard
MLS	Microwave landing system: an advanced approach aid used for guiding the aircraft on a final approach to landing
MMEL	Master minimum equipment list
MMR	Multi-mode receiver: a receiver containing GPS, ILS, and MLS receivers

MoD	Ministry of Defence (UK)
Mode S	A communication system used to exchange flight data between adjacent aircraft and air traffic control
Mon	Monitor channel
MPCDU	Multi-purpose control and display unit
MPP	Master programme plan
NASA	National Aeronautics & Space Agency (US)
NATO	North Atlantic Treaty Organisation
Nav Aids	Navigation aids
ND	Navigation display
NDA	Non-disclosure agreement
NRC	Non-recurring costs
NTSB	National Transportation Safety Board
OAT	Outside air temperature
OBOGS	On-board oxygen generating system
OOD	Object-oriented design
PBS	Product breakdown structure
PC	Personal computer
PDR	Preliminary design review
PFD	Primary flight display
PHM	Prognostics and health management
PRR	Production readiness review
psi	Pounds per square inch
PTU	Power transfer unit
Quadrax	A four-wire duel half duplex data bus connection arrangement that enables data to passed each way thereby effectively achieving bi-directional data transfers (favoured by Airbus)
QMS	Quality management system
RAM	Random access memory
RASP	Recognised air surface picture
RAT	Ram air turbine

R&D	Research and development
RDC	Remote data concentrator
RF	Radio frequency
RFI	Request for information
RFP	Request for proposal
RIO	Remote I/O
ROM	Read only memory
RT	Remote terminal (MIL-STD-1553B data bus)
RTCA	Radio Technical Committee Association (US)
RVDT	Rotary variable differential transformer
SAE	Society of Automotive Engineers (US)
SAHRS	Secondary attitude and heading reference system
SARS	Severe acute respiratory syndrome
SATCOM	Satellite communications
SBAC	Society of British Aerospace Companies (UK)
SDD	System design document
SDR	System design review
sfc	Specific fuel consumption
SIGINT	Signals intelligence
SOW	Statement of work
SPC	Statistical process control
SRR	System requirements review
SSA	System safety analysis
SSPC	Solid state power controller
SSR	Software specification review
Stanag	Standardisation agreement (NATO)
SysML	System modelling language
System of systems	A systems embracing a collection of other systems
S/UTP	Shrouded unshielded twisted pair
TAS	True airspeed
TAWS	Terrain avoidance warning system
TCAS	Traffic collision avoidance system
TCP	Tri-cresyl phosphate

TRR Test readiness review

TRU Transformer rectifier unit

TV Television

Twinax A two-wire half-duplex data bus connection that allows unidirectional data transfers (favoured by Boeing)

UAV Unmanned air vehicle

UK United Kingdom

UML Unified modelling language

US, USA United States (of America)

USMS Utility systems management system

UTP Unshielded twisted pair

UV Ultra-violet

VHF Very high frequency

VMS Vehicle management system

VOC Volatile oil compound

VOR VHF Omni-Range: a commonly used navigation beacon in civil aerospace

WBS Work breakdown structure

1

Introduction

1.1 General

In three companion books in the Aerospace Series – *Aircraft Systems*, *Civil Avionic Systems*, and *Military Avionics* (Moir and Seabridge 2006, 2008, 2013) – the authors described the technical aspect of systems for military and commercial aircraft use, in essence the engineering of systems and system products. Other books in the series described the technical aspects of various systems, for example fuel systems (Langton et al. 2009) and display systems (Jukes 2004). However, we did not dwell on the mechanism by which such systems are designed and developed, although the process of systems development is a most important aspect that contributes to the consistency, quality, and robustness of design.

The first edition of this book tried to make amends for this omission and described the design and development process and the lifecycle of typical aircraft systems. Since its initial publication the material in the book has been used in a number of postgraduate courses and continuing professional development short courses for aerospace systems engineers, and has been developed to suit the engineering audience in response to questions received and discussions held during course delivery.

The second edition continued in this vein, widening its scope a little to offer subjects to people in the same industries who did not specialise in engineering but needed to have some knowledge of how engineers worked, for example procurement, contracts, and support.

This third edition has been produced to continue this introduction to aircraft systems and the systems development process for students studying systems or aerospace subjects and wishing to enter the aircraft industry or related industries and for organisations sponsoring these people. The content is intended to be of interest to people intending to join or already working in:

- organisations directly involved in the design, development, and manufacture of manned and unmanned fixed-wing and rotary-wing aircraft, both military and commercial
- systems and equipment supply companies involved in providing services, sub-systems, equipment, and components to the manufacturers of aviation products
- organisations involved in the repair, maintenance, and overhaul of aircraft for their own use or on behalf of commercial or military operators

Design and Development of Aircraft Systems, Third Edition. Allan Seabridge and Ian Moir.

- commercial airlines and armed forces operating their own or leased aircraft on a daily basis
- organisations involved in the training of personnel to work on aircraft.

The book is also aimed at educational establishments involved in the teaching of systems engineering, aerospace engineering or specialist branches of the topic such as avionics or equipment engineering at high school, university undergraduate or postgraduate level. It is also suitable for short courses intended for the professional development of industry professionals and practitioners.

These are the sort of people who will be found in the broad range of stakeholders in complex aerospace projects. Figure 1.1 gives an example of the aviation system and some of the people and groups affected by the system or directly affecting the system. This diagram has been developed to illustrate the stakeholders in the development of an aircraft solution to meet environmental considerations. A specific project will have its own specific set of stakeholders.

Each of these stakeholders will have a different perspective of the design and development process, and each is capable of exerting an influence on the process. For those directly involved it is vital that the design process is visible to all parties so that they can coordinate their contributions for maximum benefit to the project. A clear and well-documented process is essential to allow the stakeholders to visualise the design and development path as a framework in which to discuss their different perspectives. This can be used to establish boundaries, to air differences of opinion, and to arbitrate on differences of technical, commercial or legal understanding.

It is worth noting that since the first edition of this book there have been significant changes in business practice in the aerospace industry. Previously the development of

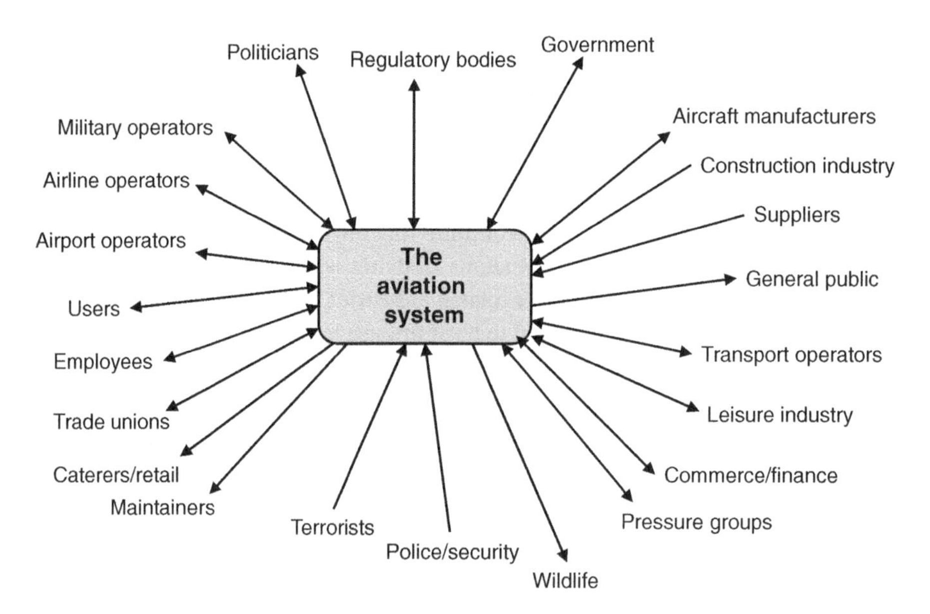

Figure 1.1 Stakeholders in the aviation system.

aircraft had been mainly in the hands of prime contractors appointed by the customer, with a supply chain competing for individual equipment and components. In modern aircraft development the first-tier suppliers compete at the system level and in many cases the supplier teams work on-site at the prime contractor's base. In many cases of international collaboration this usually means a number of prime contractor partner bases in different countries. In this situation the supplier and the prime engineering teams develop equipment and component specifications together as integrated product teams (IPTs). The system supplier is now typically responsible for system-level and component-level performance; in many cases the supplier also responsible for direct maintenance costs associated with their system. This change in business practices demands that the supplier base becomes 'systems smart' and this book should provide a valuable insight for the business community to fulfil this need effectively (Langton et al. 1999).

The principles established are equally applicable to other platforms, such as surface and sub-surface naval vessels, commercial marine vessels, and land vehicles. The aerospace industry is almost unique, given the nature of an aircraft, in having to address high integrity and availability, weight, volume, power consumption, cost, and performance issues. The conflict of competing system drivers often makes trade-offs more acute when attempting to achieve the optimum balance of meeting the customer's requirements and achieving an affordable product. There are also differences between commercial and military solutions that may demand a subtly different interpretation of the process and the standards that apply. The emergence of unmanned air vehicles broadens the system concept to incorporate ground stations for remotely piloted vehicles. The striving for autonomous unmanned vehicles will lead to more innovative approaches to design and will require more rigour in the certification of systems. Nevertheless, the process described in this book should be applicable, albeit with suitable tailoring.

Although the text is formed around examples that are mainly aeronautical platform-based the reader may also apply them to other high-value systems such as ground-based radar, communications, security systems, maritime and space vehicle-based systems, or even manufacturing or industrial applications.

What makes all these platforms and systems similar is that they are all complex, high-value products comprising many interacting sub-systems, and they are intended to be used by a human operator. They also share a common characteristic of having long operational lifecycles, often in excess of 25 years, usually with long gestation and development time-scales, and the need for operator and maintenance training and full-life in-service support. Such time-scales demand a rigorous, controlled, and consistent development process that can be used to maintain an understanding of the standard or configuration of the platform throughout its life in order to support repair, maintenance, and update programmes.

1.2 Systems Development

There are many valuable lessons to be learned from the field of systems engineering. The author believes that much of the theory and practice of systems engineering can be applied to the engineering of hardware- and software-based systems for use in aircraft. It is a broad field of practice that covers the behaviour of systems across a wide range of subjects, including organisational, operational, political, commercial, economic, human, and educational systems. The concept of systems and systems engineering operates at many different levels in many different types of organisation. Much of the early analysis of systems behaviour was concerned with organisational or management issues – the so-called 'soft' systems. This work led to an understanding of the interactions of communications, people, processes, and flows of information within complex organisations. (Checkland 1972; Lockett and Spear 1983).

An important outcome from this work was the emergence of 'systems thinking'. This term encompasses the ability to take a holistic or a total systems view of the development or analysis of any system. The key to this activity is the ability to take into account all influences or factors which may affect the behaviour of a system. This is accomplished by viewing the system as existing in an environment in which certain factors of importance to the understanding of the system are present. In this book the concept of a single environment has been extended to encompass layers or shells of environments that allow people in an organisation to take their own viewpoint, and to examine aspects of prime importance to themselves. In this way it is possible to examine a system from the top down and to allow individuals such as politicians, marketing, accountants, engineers, and manufacturing and support staff to critically examine and develop their own particular requirements (Burge 2019).

Another important property of systems is that they can be broken down into subsystems, almost indefinitely. Thus Figure 1.2 shows how a system can be considered as a system of systems which is a grouping of several sub-systems, which may not require detailed definition at the level at which the system is being examined. The owners of the

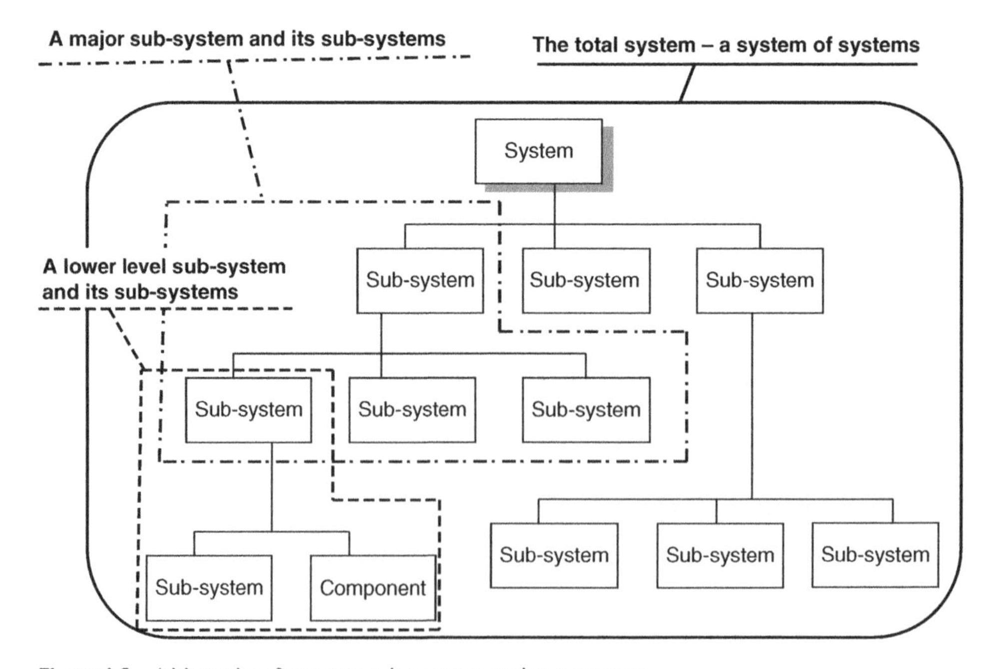

Figure 1.2 A hierarchy of systems, sub-systems, and components.

sub-systems, however, will regard their sub-system as being the system of prime importance and may choose to break it down into further sub-systems. This top-down sub-division, or decomposition, can take place from an abstract concept of a system right down to its hardware and software components. This hierarchy of systems, in which the top-level systems are important and exert an influence on lower-level systems, is the manner in which most complex systems are analysed and implemented. It is the way in which the key systems and systems architectural principles stated at the highest levels of system definition are preserved throughout the implementation and into the product.

For aircraft systems the ultimate and most elemental building blocks for a system are the components, physical components such as pumps, valves, sensors, effectors, etc., that determine the hardware characteristics of the system, or alternatively the software applications or modules that contribute to the overall system performance. The human, in the form of the pilot, crew member, passenger or maintainer, is also a vital part of the system.

The decision on how far to keep decomposing a system into sub-systems depends on the complexity of the system and the ability to view the functions and interfaces as a whole. At some stage it may become necessary to construct a boundary around a system in order to specify it to an external supplier for further analysis and design. An example of this is the definition of a sensor sub-system that will be more effectively developed and manufactured by a specialist supplier, maybe as a single item of equipment.

Such a breakdown of systems into sub-systems, and yet further sub-systems and components reinforces another important aspect of systems and their interconnections. The outputs from a system can form inputs to other systems. Indeed a system may produce an output that is fed back to its own input as feedback. Feedback loops are not confined to one stage of a system as feedback may occur over several concatenated or interconnected systems in order to produce system condition status or stability. Feedback may also be implemented using a data bus and multiplexed processing units which means that data latency must be taken into account. To enable this to happen effectively in a hard system, the system interfaces must be defined to ensure compatibility – that a system output is accepted and understood as an input so that it can be acted upon. This requires that interfaces are well defined and rigorously controlled throughout the development of the system.

It should also be noted that there have been significant changes in the aircraft supplier industry resulting in mergers and acquisitions leading to large organisations with aspirations extending their business to tender for larger systems contracts. The mergers have increased the capability of suppliers to the extent that this is a feasible and sensible proposition. At the same time some major prime contractors have focussed their sights on the major system of system management contracts, concentrating their capabilities on management of design, design of specialist integration tasks, final assembly, and qualification of the product.

The 'top-down' development of individual systems as practiced in many line management organisations is shown in Figure 1.3 at point A.

This is the development path with which most engineers are familiar for all aircraft systems, avionics systems, and mission systems treated as individual systems. However, there is often a need for something more than this straightforward development route. Point B on Figure 1.3 illustrates a case where certain systems are interconnected to form a synergistic integrated function, in other words a function is performed that is more than

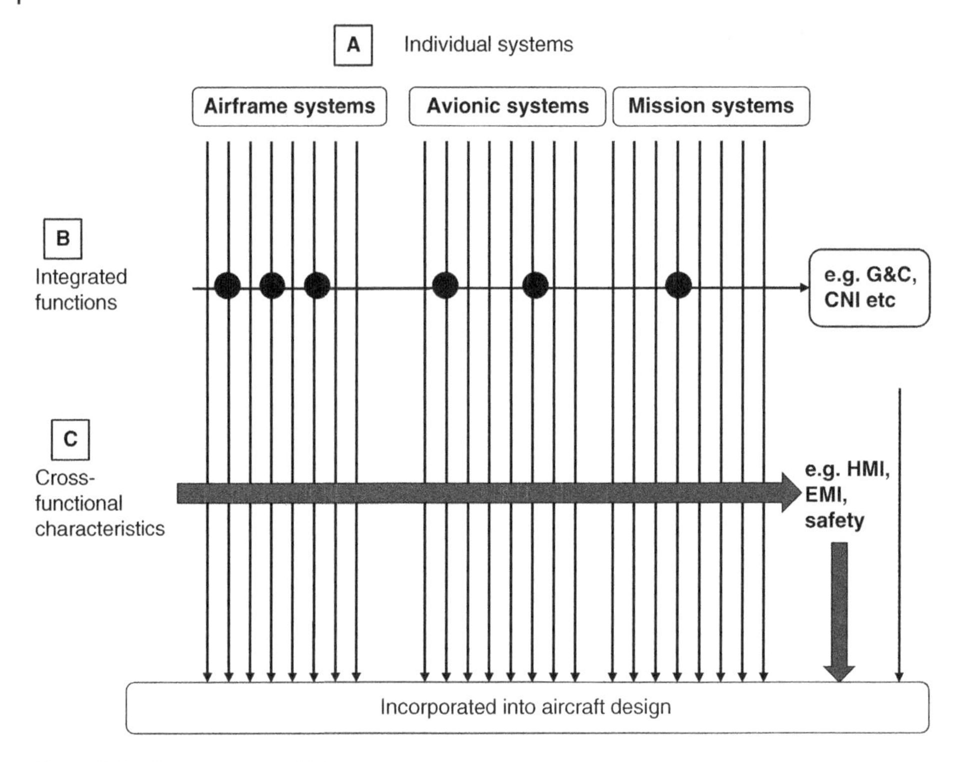

Figure 1.3 Some aspects of integration.

the sum of the individual system functions. An example of such a function is that of guidance and control (G&C) as an integration of the functions of flight control, hydraulics, automatic flight control, and fuel systems (see Chapter 6 for more detail). Also shown in this diagram is the integration of communications, navigation, and identification (CNI) systems.

Point C in Figure 1.3 illustrates an alternative view of integration, that of a design aspect that applies equally to all systems as a common discipline. Examples of this are safety, the human machine interface (HMI), electromagnetic health (EMH), and maintainability. These disciplines are governed centrally, usually by the chief engineer's office, and their impact on the individual systems will be gathered together to form a statement of design for the complete product.

The systems concepts described above can be used in aircraft systems engineering. They can be used to develop, from an understanding of a customer's top-level system requirements, a particular type of aircraft to perform a specific role and, after several successive analyses, or decompositions, can lead to an implementation of a product. The top-level system may be related to a need for national defence or for a transportation system which can be expressed in terms of people, communication, and processes, and eventually is expressed as a combination of various hardware products.

Such a top-level system is one that is conceived by many customers as representing their highest level of operational need. The role of systems engineering and systems integration is to ensure that the resulting combination of products can be shown to meet the overall requirements posed from this top level. The requirements set at the top level must flow down to the lowest level of product in a clearly traceable and testable manner so that the integrity – or fitness for purpose – of the product can be demonstrated to the customer and to regulatory bodies governing adherence to mandatory national and international regulations.

Systems thinking encompasses a process for the development of a system. This has been defined by Checkland (1972) and is based on a methodology defined by Hall (1962). Despite the age of this methodology its roots can be seen in many methods in use today. The methodology is:

- problem definition – essentially the definition of a need
- choice of objectives – a definition of physical needs and of the value system within which they must be met
- systems synthesis – the creation of possible alternative systems
- systems analysis – analysis of the hypothetical systems in the light of different interpretations of the objectives
- system selection – the selection of the most promising alternative
- system development – up to the prototype stage
- current engineering – system realisation beyond the prototype stage and including monitoring, modifying, and feeding back information to design.

For consistency across different projects it is usual for organisations to use a 'formal' process, either one defined by an industry standard or by their own developed process. Figure 1.4 shows two examples of how a process can be deployed in the design and development environment.

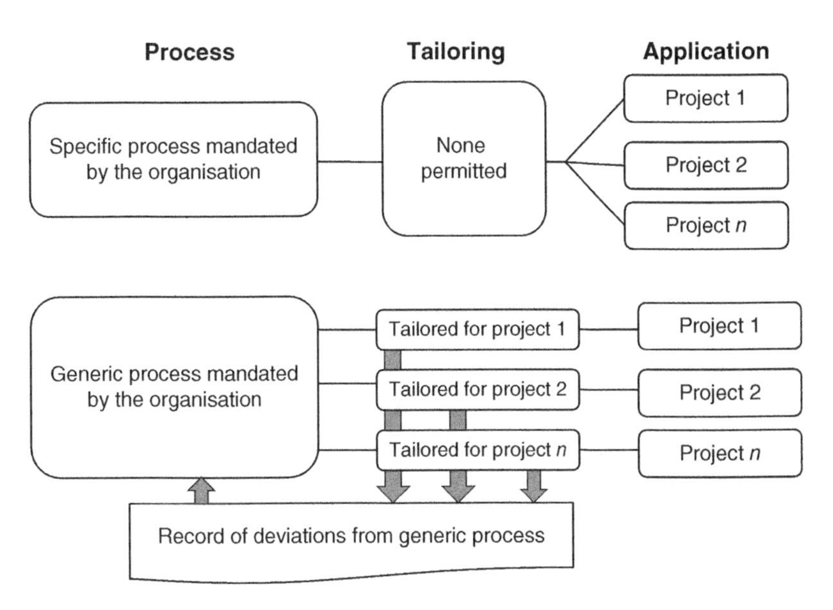

Figure 1.4 Process deployment examples.

Some organisations will mandate a project across all of their projects and will not tolerate any deviations at all. Other organisations will tolerate deviations or 'tailoring' of the process provided there is a sound reason for so doing. Such reasons include consideration of industrial partners in a joint project, taking account of their customer's preferences or tailoring the project to suit a project technology.

This book will aim to show how the process works for aircraft systems by taking a generic view of the process and providing specific examples. The intention is to promote a holistic view of systems in a world of increasing complexity.

1.3 Skills

No matter how good the systems engineering process, it can only succeed by the application of the skills of individuals and teams, and successful interactions between multi-disciplinary organisations. People are an essential element of the system, whether in its design and implementation or as its operators and users. Many skills are applied in the design, development, and manufacture of a successful system. It is important to recognise the need for skills and experience as well as the need for training to develop and maintain the skill base. This will ensure that skills do not become 'stale', and that individuals and teams are continuously aware of emerging techniques, technologies, methods, and tools that may enable or promote effective new systems, as well as ensuring that legacy skills are maintained to support products with long in-service lives.

Within a particular project the people or stakeholders in the organisation will differ from those shown in Figure 1.1, more likely being similar to those shown in Figure 1.5.

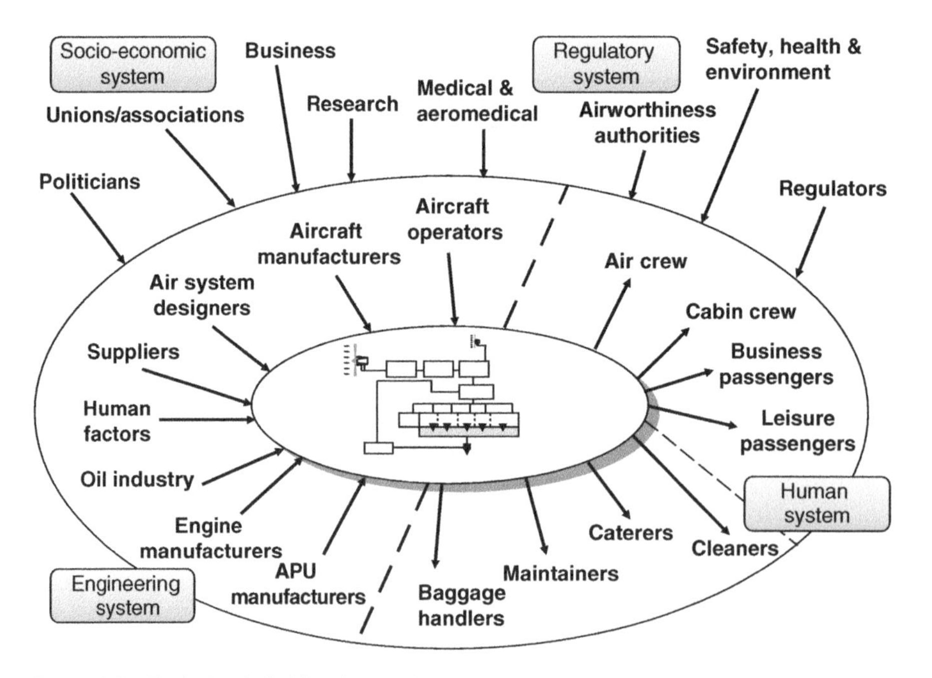

Figure 1.5 Typical stakeholders in a project.

Each chapter of this book will include a brief description of the typical skills that are particularly relevant to that part of the process being described. It must be recognised that skills can be taught but that experience can only be gained by working in the field and achieving levels of attainment. A particular skill that is difficult to describe, and that is usually acquired only with experience, is that of engineering judgement.

Skill and experience are an essential part of the capability of a systems engineering team and, together with the process and support tools, form the basis of sound systems engineering. The cognitive and personality characteristics of systems engineers (Frank 2000) must be appreciated by managers in order for them to build successful teams for the present and maintain capability for the future (Goodlass and Seabridge 2003).

1.4 Human Aspects

1.4.1 Introduction

One inescapable fact about aircraft is that they are normally designed to accommodate one or more humans and Figure 1.5 shows that the stakeholder population includes a number of such people. Whilst much of this book is concerned with the design and development of hardware to operate in a particular environment, the occupants of the aircraft necessarily occupy the same environment, although they are offered some protection by the airframe and the systems. These occupants must be taken into account in the design of the aircraft and those systems that affect them in some way.

It is worth examining the atmosphere in which various types of aircraft operate: the atmospheric environment on the ground and in the air, and the environment which the aircraft generates in its pressure cabin. The natural environment is a complex interaction of chemicals and electro-magnetic radiation. Only a portion of this atmosphere is used by aircraft, commonly to around 40 000 ft, although the ever-closer presence of the space-tourist industry will extend that beyond 360 000 ft. The most usual zones occupied by aircraft of various types are shown in Figure 1.6; also noted is FL260 (26 000 ft) which has been taken as a reference point for ozone reduction systems to start up and also as a point at which cosmic radiation monitoring should begin.

There are a number of factors originating in the environment of an aircraft that can have an impact on the long-term health of aircrew, cabin crew, and passengers as a result of prolonged or habitual exposure. They may arise as a result of poor design, but more often than not they are a fact of life, a result of the physics that arises from the operation of a high-speed machine and of the environment in which it operates. The machine can be considered as the workplace for aircrew and cabin crew, in which long-term exposure and damage to health may be inevitable unless action is taken to reduce the exposure to specific hazards. In considering the machine in this way – as the 'office' or workplace – it is no different to the ground-based office or factory in which many humans go to work on a daily basis. Legislation exists to protect them and their employers must respect the law or pay the consequences.

The occupants of a commercial aircraft are confined in a relatively small volume. Occupants include pilots, flight engineer, relief pilot, cabin crew, and passengers. Some of

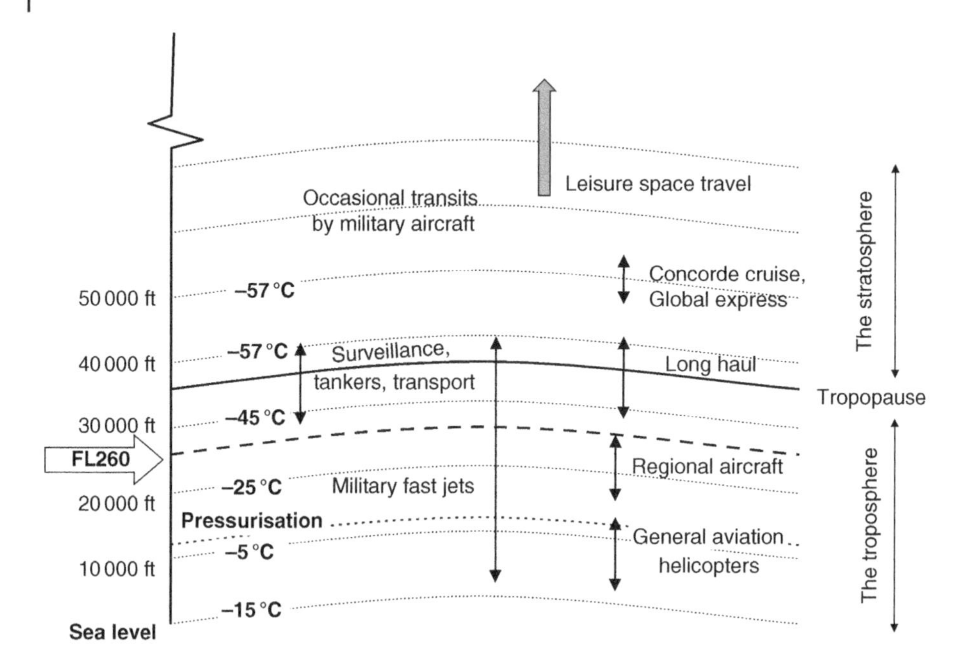

Figure 1.6 The atmosphere and air travel.

them may be anxious about flying; many may suffer from poor health or be predisposed to certain kinds of ailment from their genetic make-up or from recent illness. They breath re-circulated air obtained from the aircraft engines, they are exposed to noise, vibration, and motion for long periods of time, and at the same time the environment is subjecting the aircraft to solar and cosmic radiation. Under these circumstances it is not surprising that a few of the many millions of passengers complain about feeling unwell after flying or that frequent flyers claim to suffer from some kind of occupational effect.

Typical conditions that may be experienced on an aircraft are shown in Figure 1.7. Those conditions within the dotted shape are part of the aircraft interior environment and can be controlled; those outside the dotted line exist in the environment and are beyond the control of the designer. Not all these conditions apply to all types of aircraft, nor do they have a significant impact on non-frequent flyers, but they may have an impact on crews who are subjected to the conditions in the course of their job. They are, however, worthy of research by systems engineers designing safe systems and these conditions should be recorded with an acknowledgement that they have been considered in the design.

1.4.2 Design Considerations

The engineering teams designing different types of aircraft aim to meet customer specifications for performance and logistic support under demanding environmental and operational conditions. Many of these conditions are also inflicted on the aircrew, together with other conditions of operation resulting from inhabiting and operating a complex military machine, most often in peace time. Aircrew, cabin crew, and passengers of commercial

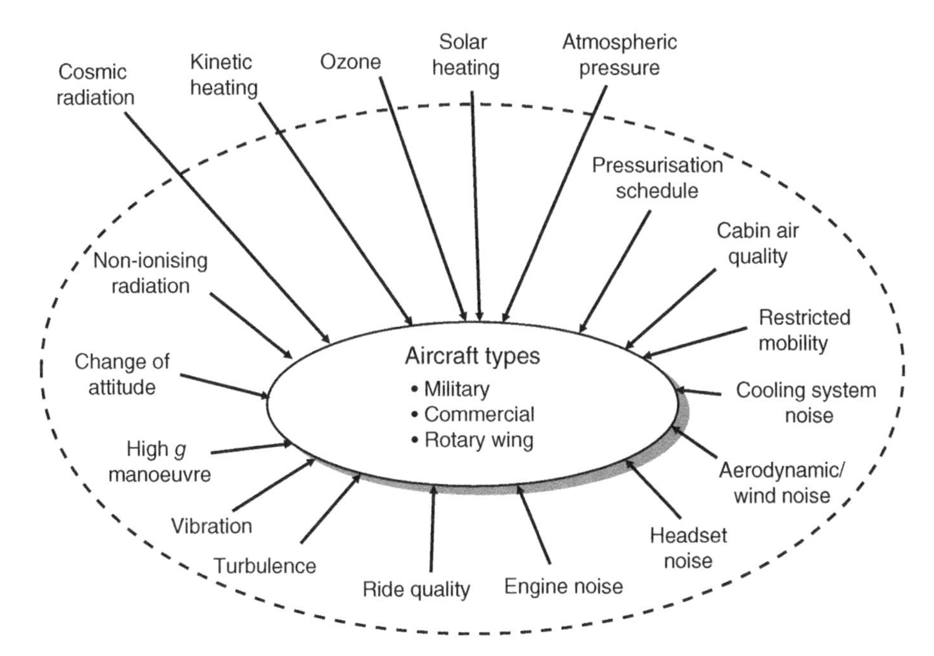

Figure 1.7 Sources of conditions that may be injurious to health.

aircraft also experience the same combination of conditions, albeit at less extreme limits. Singly, or in combination, these conditions can have an impact on the physical well-being of aircrew and passengers which may be apparent immediately or may only emerge after a long period of flying. In some cases, the effects may appear after flying employment has terminated for those whose career has been in aviation.

The good systems engineer needs to be mindful of these effects and the conditions most likely to cause them. This will enable the design of the aircraft to incorporate some alleviating aspects wherever possible, and most certainly to ensure that users of their products are aware of the risks and their duty of care to aircrew in their employment. At the same time organisations should be carrying out or supporting research to understand the issues and the risk that they pose.

The systems engineer must be sure to acquire knowledge and experience, and apply it in the engineering design of the aircraft and its systems:

- knowledge of legislation and its impact on design and operation
- awareness of research in the relevant field
- awareness of how to merge engineering and aero-medical or physiological aspects.

Mindful may not actually be the correct word. The designer and manufacturer of military aircraft has a responsibility, a duty of care to its own test pilots, and to its customers, to ensure that long-term use of the product does not pose a hazard to aircrew health. Similarly, the operator of commercial aircraft must take responsibility for the users of its aircraft, therefore there is a moral as well as a legal duty of care to users of the product. The operators of commercial aircraft have a similar duty of care to their aircrew, cabin crew, and passengers.

Legislation is continuously being revised to cope with differing workplace environments to protect workers. The responsible manufacturer of aircraft and responsible operators do their utmost to reduce the risk, but workplace legislation often advances faster than the design lifecycle of major aircraft products, which means that there is often a difference between in-service products and legislation.

Aircraft provide a dynamic environment that is the daily working environment – 'the office' – for aircrew and cabin crew. Some aspects of this environment are particularly harsh, especially for military aircrew. Prolonged exposure to these conditions may lead to long-term damage to health unless something is done to reduce the risk. This may be by design of the aircraft and its environment or by control of flying hours.

1.4.3 Legislation

Legislation exists to protect workers in their workplace. This is often interpreted as the protection of office and factory workers, and their working environment is well-regulated and governed. The workplace for aircrew is the cabin or cockpit of their aircraft and this is a dynamic environment that is less easy to regulate, but nevertheless contributes to their health.

The Health and Safety at Work Act 1974 in the UK outlines some general conditions that must be met by law to safeguard the health of people in their place of work. All employees have a statutory duty to observe the Act and to demonstrate that they do so. In addition, there are regulations that govern the exposure of workers to specific threats to their health, e.g. noise, vibration, ionising radiation, etc. All nations will have similar legislation in place.

The Act contains requirements that should be used as guidelines for any aircraft design. It puts the onus on product designers to undertake or promote the necessary research to discover and, so far as is reasonably practicable, to eliminate or minimise risks to health or safety to which the design or article may give rise. Furthermore, in the event of any legal action, the designer must prove that they have taken the necessary steps to eliminate risk, or that there was no practicable or viable alternative.

There is a possibility that aircraft operators may be subject to litigation by their aircrew claiming damage to health or impact on their career as a result of using manufactured products in their work. The customer may then claim that the aircraft manufacturer has made a contribution to this condition as a result of limitations in the design of the product.

1.4.4 Summary of Legal Threats

In the event that products supplied by an aircraft manufacturer are found to be unsafe or harmful to those operating them, there are several potential sanctions that may apply:

- Criminal prosecution is likely if it can be shown that the manufacturer is in breach of relevant legislation, particularly health and safety legislation but also a whole series of safety-related legislation. The penalties that the aircraft manufacturer would suffer will generally be fines, ranging from relatively small amounts to significant sums of money.

There is also the possibility of corporate manslaughter charges arising out of negligence leading to the death of individuals. Whilst at present this is unlikely, changes in the law have been mooted and could lead to prison sentences for company directors.
- Civil lawsuits are a possibility if individuals suffer injury as a result of the use of products and if that injury can be shown to have been caused by a defect within the product then aircraft manufacturers face the prospect of being sued by that individual for damages. The amount payable will depend upon the severity of the injury but in any event if held liable the aircraft manufacturer will be required to pay the legal costs of all parties involved. Customers may also take legal action against the manufacturer if the products that are supplied or maintained are defective. Again, the penalty for this will be damages and significant legal costs.
- Customer/public relations will suffer if the aircraft manufacturer receives a reputation for supplying products that are inherently unsafe and lead to users suffering harm; customers are less likely to purchase from them with the consequent effect on profits and shareholder value.
- Manufacturers need to ensure that the products they supply are as safe as possible given all the circumstances and that they continue to evaluate and minimise risk wherever possible. Failure to do this can have significant and far-reaching consequences.

1.4.5 Conclusions

It is clear that there are a number of different phenomena to which aircraft inhabitants are subject, knowingly if they are employed to operate the aircraft and un-knowingly if they are paying passengers. Operators are subject to these phenomena for long periods of time simply because they fly more often, whereas the leisure and business traveller will be only infrequently exposed. It is also clear that some inhabitants will be subjected to a number of these phenomena simultaneously, i.e. noise, vibration, g-manoeuvres, and hard landings all in the same flight. It is, however, common to see research reports and newspaper accounts applied to individual subjects. It is important to look at the integrated system – the machine, the human, and the combined effects of various phenomena.

The wise systems engineer will try to resolve this issue by taking into account all aspects of design of the vehicle. It should be noted, however, that organisations often divide their engineering teams into functional responsibilities and that makes it difficult to take an integrated viewpoint.

The staff of a company designing and releasing to service an aircraft need to be aware of the implications of the impact of their design on the inhabitants of the aircraft. All staff should be aware of legislation and should keep pace with any changes that occur. The company should publish procedures and processes that ensure that engineers are given guidance on where to look for standards, how to apply them, and how to deal with any deviations. Training should be made available to ensure that engineering staff are fully briefed on contemporary legislation.

Tracking and understanding legislation is an essential task for all aircraft companies. The requirements of the immediate customer and prospective customers must be included in the design standards. The requirements of certification agencies and government agencies mandating on health and safety must also be understood. Figure 1.8 illustrates an example

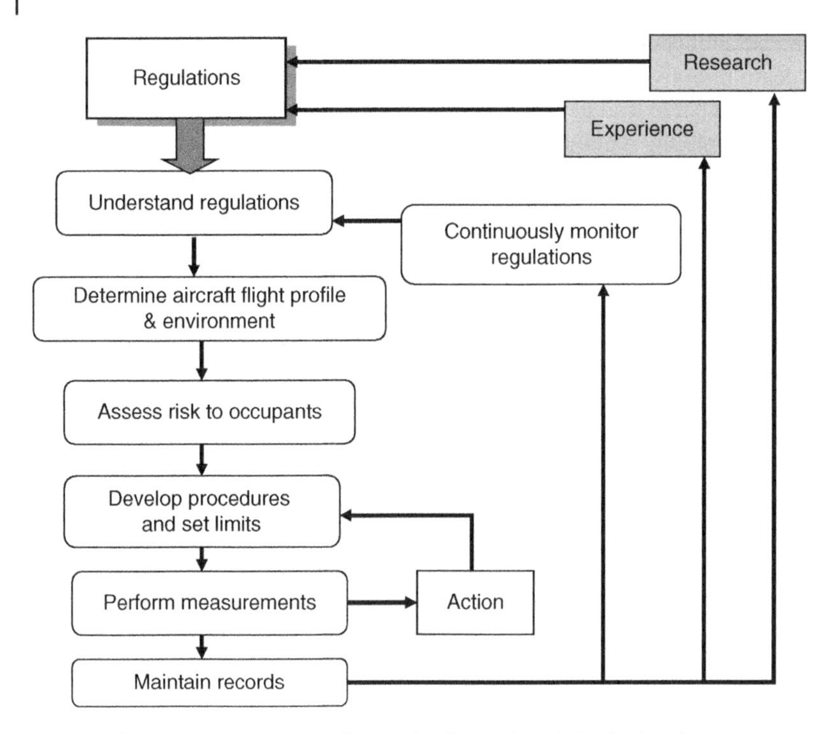

Figure 1.8 An example process for monitoring and applying legislation.

process for monitoring and applying legislation, and making a contribution to a body of knowledge that can be used for future updates of regulations.

1.5 Overview

The intention of this book is to provide a basic understanding of the principles of practical systems engineering, not to justify or recommend specific processes or tools. Examples will be used to illustrate the principles, but it is important to note that there is not one single 'right' approach to an engineering process, nor need there be. As long as there is consistency of approach in the partners in a project, and as long as the process works, then that is the correct approach for that project. This understanding will be particularly useful to engineers designing systems or equipment, and will provide essential background information for engineers or technicians using or maintaining the systems.

What this book aspires to do is to create an open-minded approach, so that systems engineers feel comfortable that the process they have chosen will produce a safe and successful result. It will also serve to introduce people to the language, jargon, and terms used in industry.

Exercises have been included at the end of most chapters to encourage readers to develop their reading of the chapter. There are no answers given; in many cases there are no 'right answers', but doing the work, alone or in groups, will help to develop the skills of understanding a system and developing it to a firm solution. Many references and suggestions for further reading have been provided to assist in this process, and the internet serves as a source of further information.

Chapter 2 addresses the general nature of an aircraft system and leads to a definition of such systems in the context of a physical application. Some characteristics of systems and their environments are introduced to encourage the reader to adopt a behavioural skill of broad systems thinking when addressing the analysis and design of systems. This description includes the associated ground systems, such as those required for support and logistics organisations to analyse fault and prognostic information, as well as the systems required to operate and analyse the information collected by unmanned air systems for real-time operations.

Chapter 3 examines a typical product lifecycle and describes example processes used in each phase of the lifecycle from concept through to retirement of the product. A view of people skills is also given to illustrate that the process of developing a successful product is a combination of processes and people. Consideration is also given to the extended development and operational lifecycle common to many aviation projects and the conflicts with rapidly moving technology in other sectors.

Chapter 4 describes how the influences of design drivers or factors in the system environment are exerted on the design process and how they affect the technical and economic feasibility of systems solutions. This illustrates the multi-disciplinary nature of systems engineering. These drivers will have a different influences in different industries, and may even change between projects or phases of a project. There is a need, therefore, to constantly examine the design drivers and prioritise them to ensure an appropriate response.

Chapter 5 looks briefly at system architectures and block diagrams to give a high-level view of systems design. This high-level view is used to show how simple block architectures can be used to gain an understanding of complex systems and their behaviours. Such simple architectures are used as a stimulus for communication between stakeholders. There is also a discussion of the complexity of modern architectures with functions and data in the system being shared and transmitted by various data bus systems and relayed to the crew on multifunction displays. The levels of complexity being encountered cast some doubt on the reality of exhaustively testing systems and on the understanding of systems status by the crew in major failure conditions.

Chapter 6 addresses systems integration as the discipline of combining systems in terms of functions performed, data produced, data used, systems interactions, and the HMI, leading to the production of a system that is fit for purpose. Integration is a most important topic as there is an increasing trend towards 'tighter' integration, especially as technology offers greater computing and storage power. There is a risk that the introduction of non-deterministic techniques in software languages and in data bus scheduling may lead to non-linear systems tending to behaviours that are unexpected and maybe even chaotic.

Chapter 7 describes methods of modelling used in the product lifecycle. Modelling a system is a quantitative description of the behaviour of a system to predict performance over a range of operating conditions at all phases of the lifecycle. At little cost modelling enables the system to be analysed under differing conditions that would be extremely time-consuming, sometimes impossible, to emulate in hardware. Modelling is used in various ways throughout the product lifecycle to perform trade-offs of different solutions. It is a quick and effective way of examining complex solutions before committing to design. Models can also be used to examine system performance for prediction and qualification, providing evidence to support qualification of the product long before the functional product is available.

Chapter 8 introduces some practical considerations based on experience in the industry in the areas of communication and criticism, both essential aspects of the open-minded systems approach. The considerations are not simply technical, but also address people and communication issues on the basis that complex projects undertaken by complex organisations demand clear and unambiguous communications in order to be successful.

Chapter 9 outlines the issues associated with the subject of configuration control and shows how the key system attributes must be maintained in a compatible manner. In this way forward and backward compatibility may be maintained between successive system or product development iterations, easing development and support costs. This control is essential in a product where many sub-systems will develop at different rates and it is inevitable that differing design standards will co-exist in the lifecycle. Also included is a discussion on control of information in the information age where ownership of multiple devices and access to many ways of creating and exchanging information may compromise information integrity.

Chapter 10 addresses an example of aircraft systems, showing how key aircraft systems all contribute to the total aircraft functionality and also interact with one another. Specific system examples are given.

Chapter 11 examines some issues of integration and complexity, and their potential impact on flight safety. Developments leading to increased integration, automation, and complexity in modern aircraft are described followed by a study of literature and reports of some unexplained events that may be due to complexity in modern systems. The issue of complexity in modern architectures that leads to decreasing visibility of design and functional performance and the difficulty, maybe of impossibility, of exhaustive testing of complex systems are discussed. A view is taken of the potential impact of all this on flight safety. Finally, the chapter looks at the possibility of complex systems integrated with an expanding air transport management system and with aircraft systems remaining powered up continuously for many days, which may lead to chaotic behaviour.

Chapter 12 presents the key characteristics of all aircraft systems in an abbreviated tabular form. The intention is to provide a brief summary of what each system is and to provide references to source material for further detailed descriptions.

A section provides a short process to assist engineers who need to examine their system further for the purpose of quantifying aspects of mass, power demand, dissipation, and fuel penalties. The tables contain an entry to enable students to identify the key components that need to be considered to do this. This is often done to provide a model of individual systems, or even a whole project, to enable trade-off studies to be conducted to evaluate different proposals.

Chapter 13 summarises the content of the book and provides a table of the systems covered in the book along with key integration and interfacing aspects. Also included are references to textbooks providing more detailed system descriptions. Each systems description in the tables contains information to enable students to 'size' a system for project work, for which typical parameters are mass, power demand, dissipation, and installation factors.

Exercises

1 Re-draw Figure 1.1 from the perspective of your own sector of the aviation system. Now take a personal view of the stakeholders and identify them by name to form a personal contact list for your sector.

2 Carry out this process using Figure 1.5 and apply it to the project on which you work.

3 Consider Figure 1.3. From your own experience can you think of an example of a project composed of a number of technical strands in the line management that would benefit from a cross-functional or integrating viewpoint? What benefits would result from this?

References

Burge, S. (2019). *Tutorial on How to 'Systems Think' INCOSE*. www.incoseuk.org.uk. [Accessed March 2019].

Checkland, P.B. (1972). Towards a systems based methodology for real world problem solving. *Journal of Systems Engineering* 3: 87–116.

Frank, M. (2000). Cognitive and personality characteristics of successful systems engineers. INCOSE 10th International Symposium.

Goodlass, Sue and Seabridge, Allan. (2003). Engineering tomorrow's systems engineers today. INCOSE 13th International Symposium

Hall, A.D. (1962). *A Methodology for Systems Engineering*. Van Nostrand.

Jukes, M. (2004). *Aircraft Display Systems*. Wiley.

Langton, R., Clark, C., Hewitt, M., and Richards, L. (2009). *Aircraft Fuel Systems*. Wiley.

Langton, R., Jones, G., O'Connor, S., and DiBella, P. (1999). Collaborative methods applied to the development and certification of complex aircraft systems. INCOSE Symposium, Brighton, UK.

Lockett, M. and Spear, R. (eds.) (1983). *Organisations and Systems*. Open University Press.

Moir, I. and Seabridge, A. (2006). *Military Avionic Systems*. Wiley.

Moir, I. and Seabridge, A. (2008). *Aircraft Systems*, 3e. Wiley.

Moir, I. and Seabridge, A. (2013). *Civil Avionic Systems*, 2e. Wiley.

Further Reading

Ackoff, R.L. (1977). Towards a system of system concepts. In: *Systems Behaviour* (eds. J. Beishon and G. Peters). Open University Press.

Buede, D.M. (2009). *The Engineering Design of Systems: Models and Methods*. Wiley.

Jenkins, G.M. (1977). The Systems Approach. In: *Systems Behaviour* (eds. J. Beishon and G. Peters). Open University Press.

Longworth, J.H. (2005). *Triplane to Typhoon*. Lancashire County Council.

Maier, M.W. and Rechtin, E. (2002). *The Art of Systems Architecting*, 3e. CRC Press.

Mynott, C. (2012). *Lean Product Development*. IET.

2

The Aircraft Systems

2.1 Introduction

A typical aircraft is equipped with a set of interacting systems that are combined to enable the aircraft to perform a particular role or set of roles. The systems that provide primary power and sources of energy have been described by the authors in detail (Moir and Seabridge 2008) as have the avionics systems that enable the aircraft to operate safely in controlled airspace (Moir and Seabridge 2013) and the military avionic systems or mission systems that enable a military aircraft to perform its role (Moir and Seabridge 2006). Each of the systems will have its own particular design requirements, and its own constraints and design drivers; some systems will stand alone, others will be integrated with one or more systems. They all have to be combined to provide the complete aircraft with the capability to perform its role.

The systems of an aircraft must also be designed to meet stringent design targets such as low mass, low power consumption, high performance, high accuracy, high integrity, high availability, and low cost, and must meet stringent safety targets. Some of these aims are conflicting; all of them are challenging to meet. In addition to the basic aims there are a number of 'design drivers', which are described in Chapter 4. This chapter will briefly describe the characteristics of the systems to illustrate the diversity of system implementations and design considerations.

2.2 Definitions

The term 'system' is used in many organisations: political, academic, commercial, educational, industrial, military, and technical. It is often encountered in day-to-day parlance and each user probably has in mind a particular understanding of the term. In this book the word 'system' applies to the various combinations of components and control units that perform a useful function in the operation of an aircraft, often with human interaction.

There are numerous definitions of a system in use in the engineering and technical communities. A dictionary definition (*Collins Dictionary and Thesaurus*) is as follows: *An assembly of electronic, electrical or mechanical components with interdependent functions, usually forming a self-contained unit.*

Design and Development of Aircraft Systems, Third Edition. Allan Seabridge and Ian Moir.

The MIL-HBK-338B *Electronic Reliability Design Handbook* (US Department of Defense 1998) uses a wider and more explicit definition: *A composite of equipment and skills and techniques capable of performing or supporting an operational role. A complete system includes all equipment, related facilities, materials, software, services and personnel required for its operation and support to the degree that it can be considered self-sufficient in its intended operational environment.*

This definition introduces the involvement of people and their skills as an integral part of a system. People are involved in the system, both in the definition of the original requirement, and also as the user of the system throughout its lifetime in service. The definition also includes facilities and services that may be provided as part of the system, or as the entire system. The totality of these elements is often summarised as a 'capability' to deliver and operate a system and many organisations use this terminology.

The Open University has long used another definition (Jenkins 1977):

- *A system is an assembly of parts, components, processes or functions connected together in an organised way.*
- *The parts, components, processes or functions do something.*
- *The parts, components, processes or functions are affected by being in the system and are changed if they leave it, i.e. the whole is greater than the sum of the parts.*
- *The particular assembly has been identified as being of special interest.*

In this definition the concept of synergy is important – the system consists of mutually interdependent elements that combine to form a useful, functional whole.

There are many other definitions used by authors and institutions, all equally useful to demonstrate particular points or to offer explanations. They are all equally valid in their own context and there need not be a dogmatically applied definition. An amalgamation of the key points of these definitions is shown in Figure 2.1 as a pictorial summary, which will

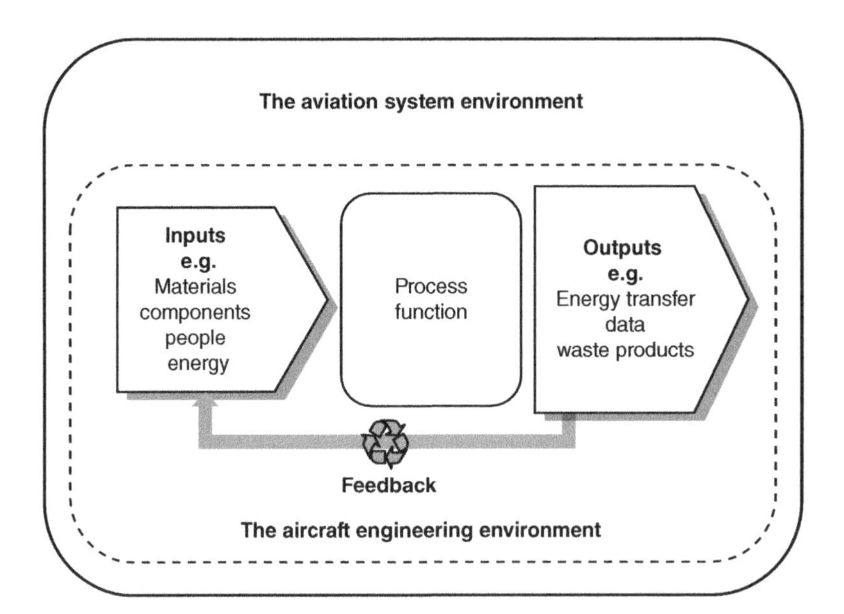

Figure 2.1 The aircraft engineering environment.

be used as the generic form of a system for the remainder of this book. This generic form will be developed later in this chapter to illustrate a generic model of a typical aircraft system which will form a basis for further explanations of aircraft systems and systems engineering processes.

Figure 2.1 shows the aircraft systems engineering environment which contains the systems described in this chapter within the aviation system environment (described in Chapter 1) that exert influences on the systems design from all of the stakeholders. This environment will be different for different projects. Ideally, it should be identified and developed fully at the start of a project to determine the key stakeholders and what influence they have on the total system and the individual sub-systems. Ideally, the stakeholders in the aviation systems should be prioritised in terms of their importance and their influence on the system design.

2.3 Everyday Examples of Systems

The word 'system' is often used loosely in everyday speech by people to describe large amorphous 'things' or corporations. These are complex things that defy a simple description. Examples include:

- natural systems such as the eco-system or solar system
- the National Health Service
- the building and construction industry
- integrated transportation systems
- manufacturing systems
- public utilities.

Similar national and international organisational systems can be seen in the aviation environment, where they exert influences and pressures on the design of projects. They include:

- regulatory systems
- air transport management (ATM)
- air traffic control
- international aircraft companies, such as Airbus and International Aircraft Engines, who supply V2500 engines, that are international consortia whose engineering, manufacturing, and support activities integrate a world-wide group of suppliers
- aircraft systems integration: the integration of the landing gear system on the A380 aircraft requires Airbus (UK) to manage the engineering effort of several engineering suppliers from a number of different countries from Europe and the USA.

Most of these examples exist because they have evolved over many years, rather than having been designed and developed as a system. Nevertheless, most of them today can be considered as having the classical properties of a system in the form of inputs, a process or function, outputs, and controlling feedback.

Systems such as the public utilities, e.g. electricity, gas and water, telecommunications, postal services, and transport, have a highly visible public aspect behind which lies a massive infrastructure. For example, behind the electrical socket outlet to which domestic and industrial appliances are connected is a structure that includes:

- generation of power from raw material energy sources – oil, gas, coal or hydro-electric
- distribution and transformation of power to users
- ordering, use, and disposal of raw materials
- metering and billing of consumers
- appliance manufacture
- high street appliance showrooms and outlets
- employment
- health and safety, and environment concerns
- repairs and maintenance
- research and development
- sales and marketing
- public relations
- legal services.

This list gives some idea of the complexity and diversity of the functions that characterise a large system.

Most of the systems listed above are examples of organisations which can be visualised at the top level as a collection of sub-systems dominated by people and processes. If the organisation is examined in detail successive layers of sub-systems become visible, becoming more mechanised to include machines and hardware components used by people to do their jobs. At the lower levels of some organisations things are designed and manufactured as a product or components. Raw materials and energy are converted to useful outputs. It is at this level that motor vehicles and aircraft emerge as familiar system-based products.

Looking at systems in this way produces a view of systems as hierarchies of sub-systems. At each level of the hierarchy there are sub-systems that conform to the general characteristics of a system and are subject to the same pressures as the top-level organisation, but they can also be seen to be autonomous in their own right.

In Figure 2.2 the military aircraft is considered as a complex set of interacting systems – a highly complex product consisting of systems, sub-systems, and components. However, it is only a small part of the output of the aircraft industry organisation that produced it – this also includes commercial aircraft, light aircraft, and rotary wing aircraft. At the top of the hierarchy is the system of the defence of the realm that includes government, all armed forces, conscripts, regulators, planners, etc.

The requirements that emerge from the operational scenarios required to defend the realm must be flowed down to all products to ensure that the entirety of the weapons and logistics to support them is in place. In Figure 2.2 the major system may be an air, land, sea or intelligence system; in this particular example the air system has been developed. Verification that these requirements have been met is flowed up and compared with the original requirement. If they do not match at any point then corrective action must be taken or a limitation in effectiveness must be acknowledged, in other words the system won't do what was intended of it, but something can be done to achieve something close to the desired outcome.

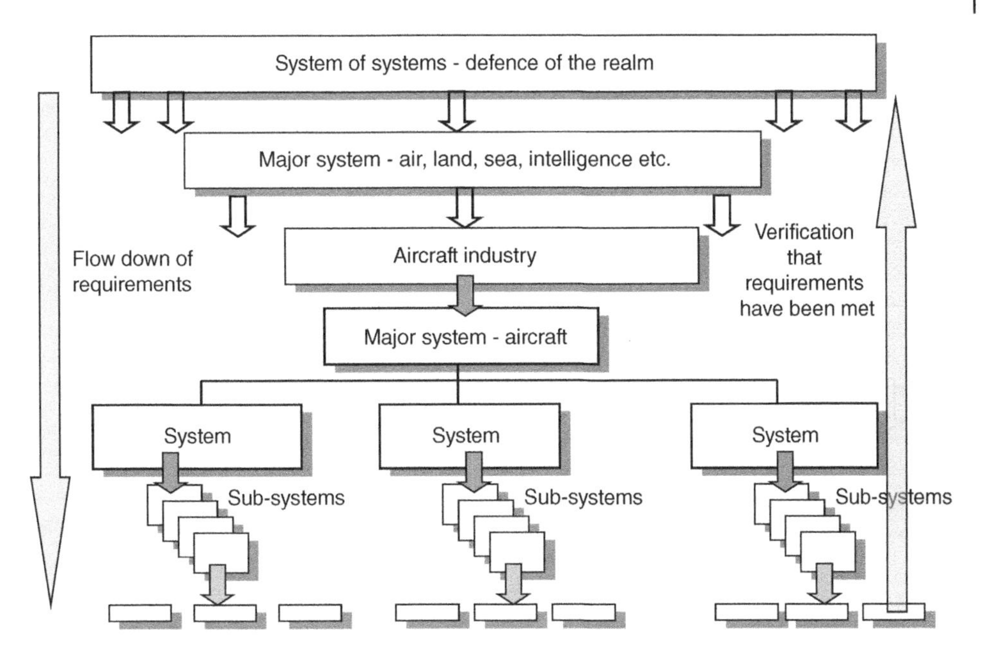

Figure 2.2 A hierarchy of systems requirements.

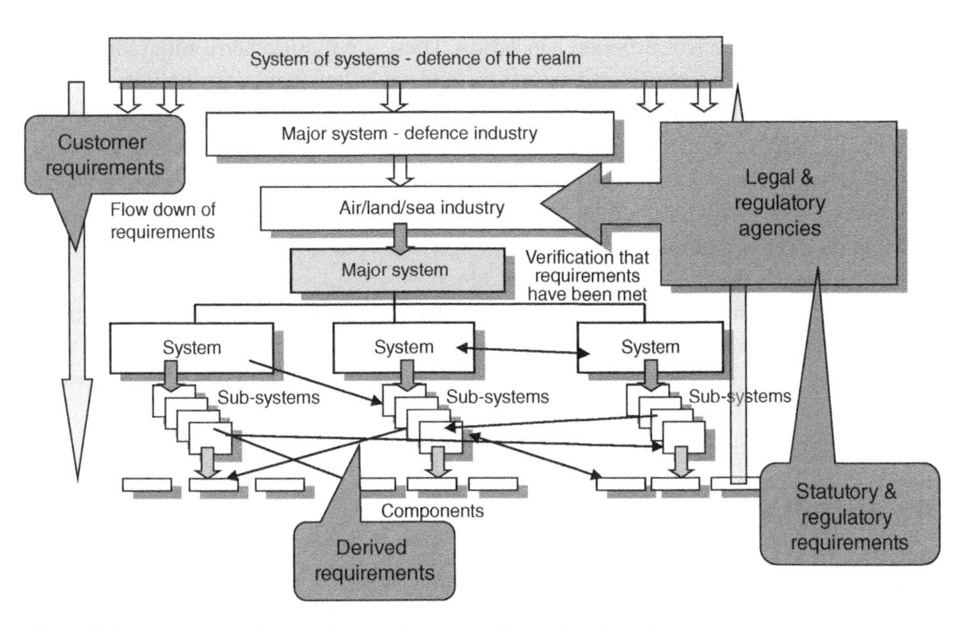

Figure 2.3 An extended view of the systems requirements hierarchy.

This flow down of requirements can be extended further to show that they need not necessarily emerge from the customer's requirement, as illustrated in Figure 2.3.

There is a key input of regulatory requirements from standards and directly from regulatory bodies. These regulatory inputs may be determined by flight safety, health and safety

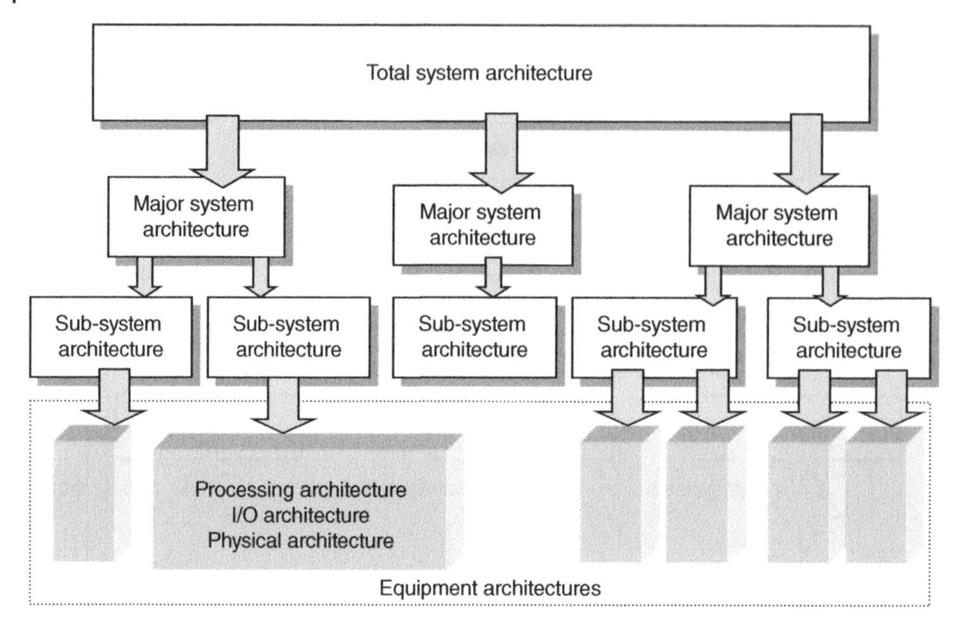

Figure 2.4 A product view of systems.

or by the need to meet environmental legislation. These are requirements often formed by international bodies and made mandatory through aviation regulatory bodies. A most important aspect of the figure, which has a strong influence on the design of systems, is shown by the derived requirements. These emerge as individual systems develop and relationships between systems begin to emerge as interfaces, sharing of functions, and physical interactions. Great skill is required by engineers to identify, define, and record these requirements. This is where the majority of design takes place in aerospace organisations and the result is the design data set for the product.

The resulting product view is shown in Figure 2.4, in which an overall system architecture consists of a number of individual sub-systems and sub-sub-systems completing with components. The architectural principles described in the uppermost system level must be obeyed.

2.4 Aircraft Systems of Interest

The Air Transport Association (ATA) chapter numbering system has been used for many years to provide a common referencing system for all civil aircraft. The system is controlled and published by the ATA, an airline association based in the USA for over 75 years that harmonises the requirements of the US air transport system. This organisation has recently been renamed Airlines for America (A4A), although its Charter remains much the same. The key pillars of the organisation are safety, engineering and maintenance, and flight operations and air traffic management.

The ATA chaptering system provides a unified referencing system whereby aircraft systems share common identifiers irrespective of aircraft type, e.g. Chapter 24 represents the aircraft electrical system whether the aircraft is a B747 or a small business jet. For the air transport engineering community this referencing system provides a consistent framework for aircraft technical documents and maintenance manuals.

A simplified version of the ATA referencing system is portrayed in Figure 2.5. Classic 'avionics' systems such as auto-flight, communications, recording and indicating (displays), and navigation are shown in segment 9.2. The highly integrated nature of modern air transport aircraft means that some or many ATA chapter functions may inter-react to provide a top-level aircraft function. An example of the levels of integration necessary to provide the mission management function of a typical aircraft operating in our skies today is given Chapter 10, Section 10.6.

The ATA 100 system was used until the early 2000s but has now been superseded by Joint Aircraft System/Component (JASC) four-digit codes (FAA 2002). This system provides a similar role to ATA but is now universally used for modern aircraft. For users of legacy aircraft and documentation systems the ATA referencing system has been retained in this edition.

Many of the systems described above are, in fact, collections of sub-systems, which, in combination, perform as a single system. Each individual sub-system will be designed and mechanised in different ways to perform its function, although there may be some over-riding rules for design laid down by the main system organisation.

The modern aircraft is also a system. The modern military aircraft is a collection of interdependent sub-systems designed for a specific role. The modern civil aircraft is similarly a

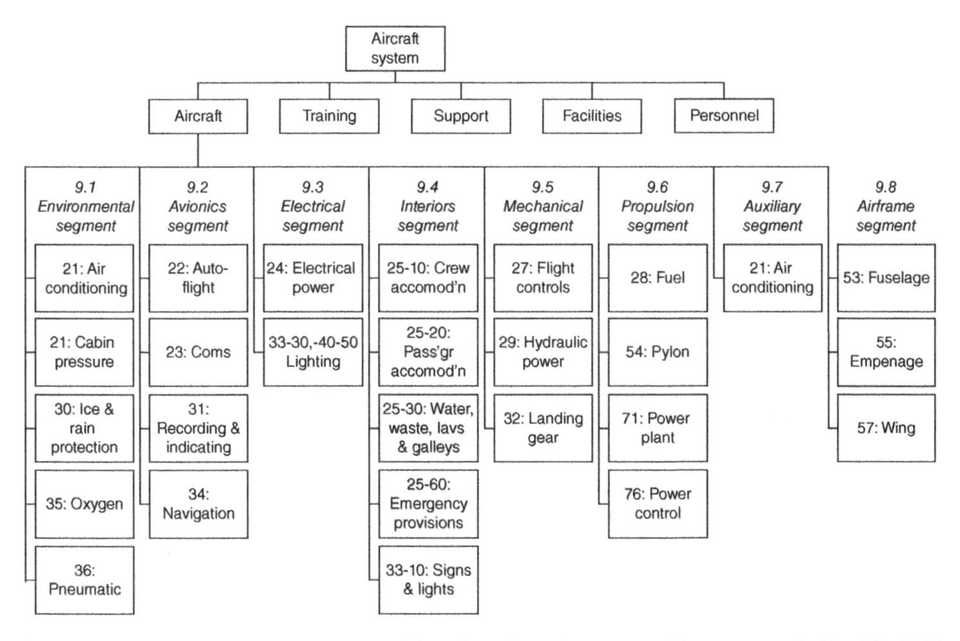

Figure 2.5 Simplified representation of the ATA referencing system. Now superseded by FAA JASC Codes (FAA 2002).

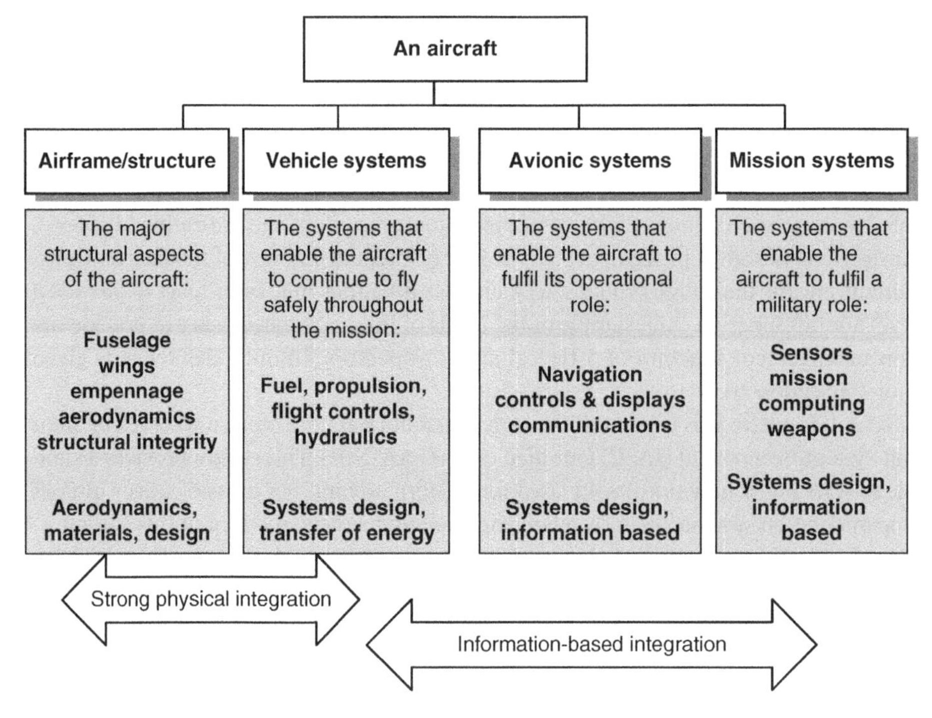

Figure 2.6 The aircraft as a set of sub-systems.

collection of sub-systems, many of them identical in their principles of operation to the military aircraft, although there are some that are significantly different. These sub-systems are designed to perform a specific individual task and are combined to form the whole aircraft in which the combination of the individual systems clearly performs a role that is greater than the sum of the individual parts. In other words, the sub-systems are acting in a synergistic manner.

The sub-systems can all be considered generically, and the remainder of this book will attempt to do that, whilst identifying any differences between the two types of aircraft.

Figure 2.6 shows the aircraft as a set of sub-systems that are common to most commercial and military types. These sub-systems map onto the domains in which many engineers are educated or into which they develop in their careers. Many prime contractor or aircraft manufacturer organisations are structured in this way.

The sub-systems show some interesting integration characteristics which must be taken into account in the overall system design. The vehicle systems show some very strong physical interactions with the airframe or structure. This arises because systems such as propulsion and fuel are very much a part of the structure. In a commercial aircraft the engines are usually incorporated in pods suspended from the wings and the thrust loads must be accounted for in the design of the wing; in a military fast jet the engines with their intakes and jet pipes must be incorporated into the structure. The fuel system tanks are similarly incorporated into the structure, especially the wing tanks. Heat and loads generated

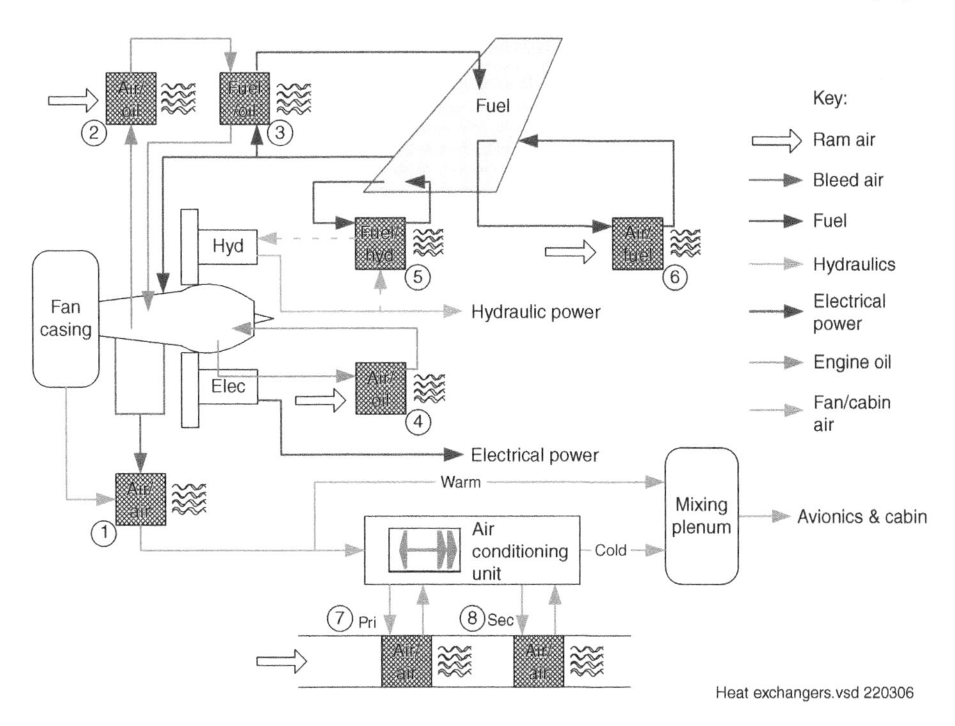

Figure 2.7 An example of system interactions (Moir and Seabridge 2008).

in many of the vehicle systems are translated into the structure. An example of such interaction is illustrated in Figure 2.7 and an explanation follows.

This example of significant interaction between systems shows how various systems operate together to reject waste heat from the aircraft. Heat is generated when fluids are compressed and also by energy conversion processes that are not totally efficient. Figure 2.7 depicts the interaction of several major systems, this time within the context of a civil aircraft. The diagram illustrates how a total of eight heat exchangers across a range of systems use the aircraft fuel and ambient ram air as heat sinks into which waste heat may be dumped. Starting with the engine the process is as follows:

1) Air extracted from the engine fan casing is used to cool bleed air tapped off the intermediate or high-pressure compressor (depending on engine type).
2) Air is used to cool engine oil in a primary oil cooler heat exchanger.
3) Fuel is used to cool engine oil in a secondary oil cooler heat exchanger.
4) The electrical integrated drive generator (IDG) oil is cooled by air.
5) The hydraulic return line fluid is cooled by fuel before being returned to the reservoir.
6) Aircraft fuel is cooled by an air/fuel heat exchanger.
7) Ram air is used in primary heat exchangers within the air-conditioning pack to cool entry bleed air prior to entering the secondary heat exchangers.
8) Secondary heat exchangers further cool the air down to temperatures suitable for mixing with warm air prior to delivery to the cabin.

The avionics and mission systems are mainly based on information structures and although there are demands for installation and low drag, much of the integration is based on data bus networks.

These systems and sub-systems can be further broken down into individual sub-sub-systems as described below.

2.4.1 Airframe Systems

The airframe can be viewed as a system since it is a complex and integrated set of structural components that supports the mass of systems and passengers, and carries loads and stresses throughout the structure. The airframe is designed and constructed as a set of sub-systems that are integrated to form the whole structure. This book will not describe the airframe any further, but will concentrate on the remainder of the aircraft systems, those that provide the airframe with the capability to perform its role.

2.4.2 Vehicle Systems

Aircraft vehicle systems are also known as general systems or utility systems. Many of these systems are common to both civil and military aircraft; they are a mixture of systems with very different characteristics. Some are high speed, closed loop, high integrity control such as flight controls; others are real-time data gathering and processing with some process control functions such as the fuel system. Yet others are simple logical processing such as undercarriage sequencing.

What they have in common is that they all affect flight safety in some way – in other words a failure to operate correctly may seriously hazard the aircraft, crew or passengers.

The functions of many of these systems are performed by software-based control units, either individual units or an integrated processing system such as a vehicle systems management system. This means that the software must be designed to appropriate levels of robustness. The following list provides a summary of the main purpose of each vehicle system.

- *Propulsion system*: provides the primary source of thrust and motive power via pilot demands, and electronic and hydro-mechanical fuel controls. This system provides the thrust and energy for flight, and also the motive power for the generation of electrical, hydraulic, and pneumatic systems.
- *Fuel system*: provides a source of energy for the propulsion system, which consists of tanks, a quantity measuring system, pumps, valves, non-return valves, and pipes to transfer fuel from tank to tank and to the engines. The fuel system is also used for the centre of gravity control and is the recipient of thermal energy from other systems as a result of its use in heat exchangers.
- *Electrical power generation and distribution system*: generates AC and DC power from the engine-connected generators and batteries, and distributes the power to all connected equipment, whilst protecting the electrical bus-bars and the electrical wiring harnesses from connected faults.
- *Hydraulic power generation and distribution system*: generates hydraulic power from engine-driven pumps and distributes hydraulic power to all connected systems. The

hydraulic supply must be ripple free and constant pressure under all demand conditions, provided by clean hydraulic fluid, and monitored by detect and isolate leaks. Much of the heat dissipated in the system will be transferred into the fuel system by fuel-cooled oil cooling heat exchangers.

- *Secondary power system*: provides a source of electrical, hydraulic, and cooling power for aircraft on the ground, and provides a form of energy to start the engines.
- *Emergency power generation system*: provides energy to allow safe recovery of the aircraft in the event of a major power loss.
- *Flight control system*: converts pilot demands or demands from guidance systems into control surface movements to control the aircraft attitude.
- *Landing gear system*: ensures that the aircraft is able to land safely at all loads and on designated runway surfaces. This includes the sequencing of all associated doors, and leg and wheel assemblies to fit in the landing gear bay.
- *Brakes/anti-skid system*: provides a safe form of braking without loss of adhesion under a wide range of landing speeds and loads.
- *Steering system*: provides a means of steering the aircraft under its own power or whilst being towed.
- *Environmental control system*: provides air of an appropriate temperature and humidity to ensure a safe and comfortable environment for crew, passengers, and avionic equipment.
- *Fire protection system*: monitors all bays where there is a potential hazard of fire, smoke or overheat, to warn the crew and to provide a means of extinguishing fire.
- *Ice protection system*: monitors external ambient conditions to detect icing conditions and prevent the formation of ice or to remove ice.
- *External lighting system*: ensures that the aircraft is visible to other operators and ensures runway/taxiway visibility during ground movements.
- *Probe heating system*: ensures that the pitot, static, attitude, and temperature probes on the external skin of the aircraft are kept free of ice.
- *Vehicle systems management system*: provides an integrated processing and communication system for interfacing with system components, performing built-in tests, performing control functions, providing power demands to actuators and effectors, and communicating with the cockpit displays.

Military aircraft also require the following systems:

- *Crew escape system*: provides a means of assisted escape for aircrew.
- *Canopy jettison or fragmentation system*: provides a means of removing the canopy from the aircraft or breaking the canopy material to provide a means of exit for escaping aircrew.
- *Biological and chemical protection system*: protects the crew from the toxic effects of chemical or biological contamination.
- *Arrestor mechanism*: provides a means of stopping the aircraft on a carrier deck or at the end of a runway.
- *In-flight refuelling system*: allows the aircraft to obtain fuel from a tanker aircraft.
- *Helicopter deck lock system*: secures helicopters to a carrier deck.

Commercial aircraft and large military aircraft require the following systems specifically for their use:

- *Galley system*: allows meals to be prepared and cooked for passengers.
- *Passenger evacuation system*: allows safe evacuation of passengers.
- *Entertainment system*: provides audio and visual entertainment for passengers.
- *Telecommunications system*: allow passengers to make telephone calls and send emails in flight.
- *Toilet and wastewater system*: provides hygienic management of toilets and water waste.
- Gaseous oxygen system: for passenger use in case of depressurisation.
- *Cabin and emergency lighting system*: provides general lighting for the cabin and galley, reading lights, exit lighting, and emergency lights to provide a visual path to the exit.

2.4.3 Interface Characteristics of Vehicle Systems

In order to control these systems interfaces must be designed to meet a wide range of sensors and actuator types. The input examples listed below have a diversity of type, range, source impedance, and slewing rate:

Input examples

- Relay or switch — Discrete 28 or 0 V
- Fuel gauge probe — Capacitance
- Fuel density — Fuel properties sensor
- Fuel properties — Permittivity sensor
- Rotational speed — Pulse probe (tachometer)
- Linear position — Linear variable differential transformer
- Rotary position — Shaft encoder, rotary variable differential transformer, synchro
- Actuator position — Potentiometer or variable differential transformer
- Temperature — Thermistor or platinum resistance
- Pressure — Barometric or piezo-electric
- Current (AC) — Current transformer
- Current (DC) — Hall effect sensor
- Level sensing — Thermistor
- Proximity — Proximity switch sensor

Output examples

- Valve commands — 28 or 0 V discrete
- DC motor — DC power drive
- Actuator drive — Low voltage analogue
- Actuator servo — Low current servo drive
- Fuel pump — High current drive
- Warning lamps — Lamp load filament or LED
- High power loads — Electrical contactor (up to 400 amps/phase)

2.4.4 Avionics Systems

Avionic systems are common to both civil and military aircraft. Not all aircraft types, however, will be fitted with the complete set listed below. The age and role of the aircraft will determine the exact suite of systems. The majority of the systems collect, process, transfer, and respond to data. Any energy transfer is usually performed by a command to a vehicle system. An example of this is a change to aircraft attitude demanded by the flight management system, which will be performed by the auto-pilot and flight control systems. The following list provides a summary of the main purpose of each avionic system.

- *Display and control systems*: provide the crew with information and warnings with which to operate the aircraft.
- *Communications system*: provides a means of communication between the aircraft and air traffic control and other aircraft.
- *Navigation system*: provides a world-wide, high accuracy navigation capability.
- *Flight management system*: provides a means of entering flight plans and allowing automatic operation of the aircraft in accordance with the plans.
- *Automated landing system*: provides the capability to make an automatic approach and landing under poor visibility conditions using an instrument landing system (ILS), microwave landing system (MLS) or global positioning system (GPS).
- *Weather radar*: provides information on weather conditions ahead of the aircraft, both precipitation and turbulence.
- *Interrogation friend or foe (IFF)/secondary surveillance radar (SSR)/transponder system*: provides information on the aircraft identification and height to air traffic control and provides identification information to other systems such as the traffic collision avoidance system (TCAS).
- *TCAS*: reduces the risk of collision with other aircraft in airport terminal areas.
- *Ground proximity warning system (GPWS)/terrain avoidance warning system (TAWS)*: reduce the risk of aircraft flying into the ground or into high ground.
- *Distance measuring equipment (DME)*: provides a measure of distance from a known beacon.
- *Automatic direction finding (ADF) system*: provides a bearing from a known beacon.
- *Radar altimeter*: provides an absolute reading of height above the ground or sea.
- *Air data measurement system*: provides information to other systems on altitude, air speed, outside air temperature, and Mach number.
- *Accident data recorder*: continuously records specified aircraft parameters for use in analysis of serious incidents.
- *Cockpit voice recorder*: continuously records specified aircrew speech for use in analysis of serious incidents. May also include video recording.
- *Internal lighting*: provides a balanced lighting solution on the flight deck for all panels and displays.

2.4.5 Interface Characteristics of Vehicle and Avionics Systems

Although both aircraft vehicle and avionics systems make extensive use of modern digital technology, processors and data buses, the ways in which these mechanisms are exploited are quite different. The differences between the tasks that vehicle systems and avionics systems perform lead to considerable variations, as described below.

2.4.5.1 Vehicle Systems

Vehicle systems have the following characteristics:

- not data intensive – signal types varied and multiple
- generally low data rates and iteration rates (some exceptions)
- lower data resolution – usually 8-bit and occasionally 12-bit resolution
- lower memory and throughput
- display intensive on an as-requested basis
- physically highly input/output (I/O) and wiring intensive.

2.4.5.2 Avionics Systems

Avionics systems have the following attributes:

- data and information intensive
- high data and iteration rates
- typically 32-bit floating point arithmetic manipulation
- high memory and throughput requirements
- display intensive
- not physical I/O intensive – minimal I/O wiring

2.4.6 Mission Systems

The military aircraft requires a range of sensors and computing to enable the crew to carry out designated missions. The mission systems gain information about the outside world from active and passive sensors and process this information to form intelligence. This is used by the crew, sometimes in conjunction with remote analysts on the ground, to make decisions that may involve attack. These decisions may, therefore, result in the release of weapons of defensive aids, an action which requires a particular set of safety and integrity design considerations:

- *Attack or surveillance radar*: provides information on hostile and friendly targets.
- *Electro-optical sensors*: provide a passive surveillance of targets.
- *Electronic support measures (ESM)*: a passive system that provides emitter information, range, and bearing of hostile transmitters.
- *Magnetic anomaly detector (MAD)*: confirms the presence of large metallic objects under the sea surface (submarines) prior to attack.
- *Acoustic sensors*: provide a means of detecting and tracking the passage of underwater objects.
- *Mission computing*: collates sensor information and provides a fused data picture to the cockpit or mission crew stations.
- *Defensive aids*: provide a means of detecting missile attack and deploying countermeasures.
- *Weapons system*: arms, directs, and releases weapons from the aircraft weapon stations.
- *Communications*: using a variety of different line-of-sight, high-frequency or satellite communications systems.

- *Station keeping*: provides a means of safely maintaining formation in conditions where station keeping lights are not permitted.
- *Electronic warfare systems*: detect and identify enemy emitters, collect and record traffic, and if necessary provide a means of jamming transmissions.
- *Cameras*: record weapon effects and provide a high-resolution image of the ground for intelligence purposes.
- *Head-up displays*: provide the crew with primary aircraft information and weapon aiming information.
- *Helmet-mounted displays*: provide primary flight information and weapon information to the crew, whilst allowing freedom of movement of the head.
- *Data link*: provides transmission and receipt of messages under secure communications using data rather than voice.

2.4.7 Interface Characteristics of Mission Systems

As well as extensive use of digital data technology as already outlined, mission systems utilise a wide range of electronic sensors covering up to ten decades of the electromagnetic spectrum ranging from 100 kHz (1×10^5 Hz) up to 1000 THz (1×10^{15} Hz). This covers those areas of the electromagnetic spectrum in which communications, radar, and electro-optic (EO) equipment operate. This is a highly complex topic and readers are referred to the *Military Avionics Systems* book Moir and Seabridge (2006) for further information.

2.5 Ground Systems

It is important to recognise that airborne systems will interact with a set of ground-based system as illustrated in Figure 2.8.

- *Flight test*: In the test phase of aircraft development there will be a need to collect information from the aircraft system for ground analysis. The results of the analysis inform the designers of the systems on the verification of their system design and will be used as part of the evidence of safe and correct operation. Data is collected from direct connection to system wiring or from the aircraft data bus network. It is stored in removable data media for removal after flight or is transmitted to a ground station by telemetry.
- *Health monitoring*: It is common for the health of airframe, engine, and aircraft systems to be monitored continuously to record observable failures, but more commonly to collect data to identify trends toward degraded performance so that more intelligent decisions can be made about equipment removal. Systems are in use to gather on-board data and relay it to ground facilities, such as engine health monitoring, structural health, and useage monitoring, and prognostics/diagnostics systems are found in many types of aircraft. Data is relayed to the ground by data link and the Air Radio Incorporated communications and reporting system (ACARS).
- *Accident investigation*: Data is collected from direct connections to aircraft systems and from aircraft data bus networks on a continuous basis in order to assist in determining the cause of accidents. The data is commonly stored in an accident data recorder designed

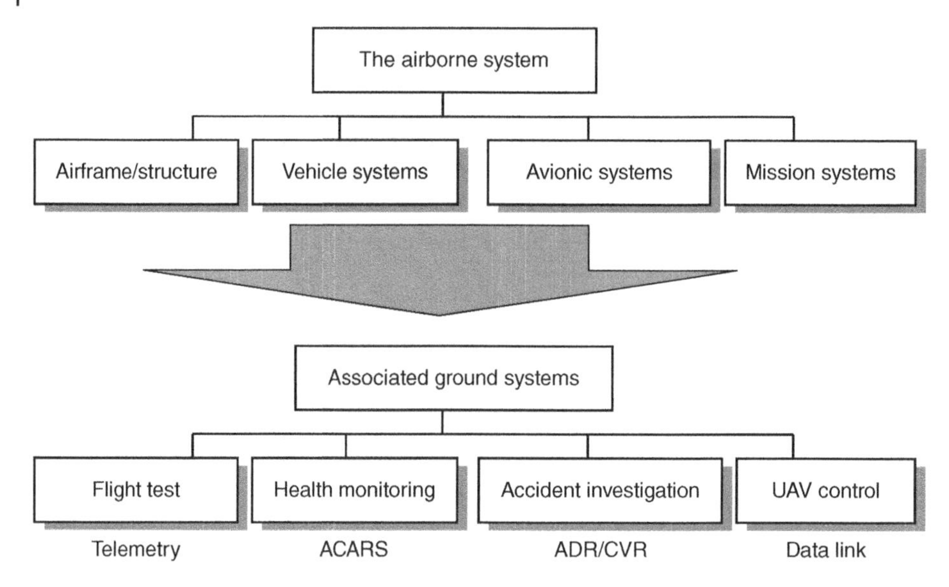

Figure 2.8 The integration of airborne and ground systems. ACARS, ARINC communications and reporting system; ADR/CVR, accident data recorder/cockpit voice recorder.

to withstand the rigours of crash, fire, and submersion in sea water. Systems data is complemented by cockpit voice recordings and video recording.

- *Unmanned air vehicle (UAV) control*: Unmanned air systems are used to collect information and to conduct military action, usually under the control of a human command structure. Even in the event of such vehicles acquiring more autonomy there will still be a need for information to be gathered on the ground for analysis and for commands to be sent to the vehicle. This will require the vehicle to be designed with telemetry and with communication paths to download information and upload commands.

2.6 Generic System Definition

An aircraft will be equipped with various combinations of the systems listed above according to its particular role. Some of the systems will be integral to the aircraft, others will be carried as role equipment in pallets or wing-mounted pods. The majority of these engineering systems are similar in their format. A generic aircraft system is shown in Figure 2.9 to illustrate the main attributes of any system.

- *Inputs* consist of combinations of the following:
  - Demand (or command) is a conscious input to the system to demand a deliberate response. The demand may be from an operator or from another system. Typically, the demand will result from the operator moving a selection mechanism, e.g. throttle levers, a switch, control column, steering wheel or tiller. Modern techniques have allowed demands to be obtained from direct voice input (DVI) or by cursor control devices such as a mouse or tracker ball. For unmanned or remotely piloted aircraft the demands will be from a ground facility using a data link.

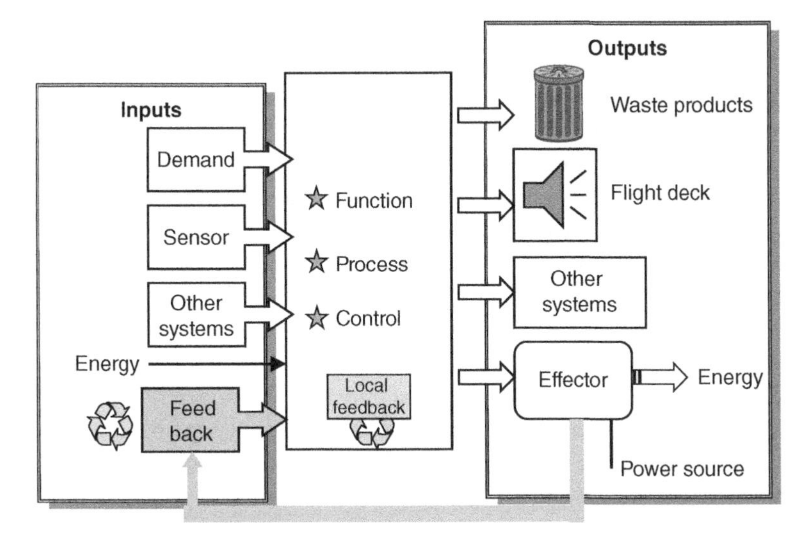

Figure 2.9 A generic aircraft system block diagram.

- Sensor inputs are provided to modify the behaviour of the system or to provide information to enable the function or process to be performed. Typically, this data is derived from sensors or measuring devices that monitor the system performance or environmental parameters such as speed, angular or rotary displacement, rates of change, pressure, temperature etc. in analogue or digital form.
- Other systems may provide information as determined by the requirements of the function or process to be performed. This data may be provided in analogue, discrete or digital format.
- Feedback is obtained from measuring devices or sensors in the output devices to allow control to be exercised for reasons of stability of the output.
- Energy is provided to enable the system to operate. This is usually in the form of alternating current or direct current from the electrical supply. This usually needs to be conditioned by the system to ensure that it is the correct voltage and free from transients or noise to ensure correct operation.
- A necessary process or function which may be performed by intellectual, physical, mechanical, electrical, electronic, fluidic or software driven means. The process can be performed by people or by natural or biological events, or by a machine or by a combination of person and machine. The latter combination is that most often encountered in aerospace and industrial systems, and it contains large sections of human machine interface challenges.

- *Outputs* consist of combinations of the following:
  - Effectors are devices that convert electrical energy into movement, rotary, linear or angular, often using another medium such as hydraulic oil or air at high pressure, although high-voltage electrical devices are becoming more common. These effectors are more commonly known as actuators, and act via mechanisms to move surfaces such as flying control surfaces, doors, landing gear, brake callipers etc.

- Other systems may require data or commands as inputs in order to complete their process. This may be in the form of analogue, discrete or digital data.
- Crew compartment indications and warnings make the crew are aware of the correct and incorrect operation of the system.
- Waste products are produced by the system as a result of the energy transformation or as a result of the operation of the system. Typical waste products are acoustic noise, electrical noise or interference, heat or vibration. All of these products can have a detrimental effect on other systems, or they can be a reason for other systems to exist. For example, the heat rejected by a system needs to be diverted to and dissipated by another system, usually a cooling system. Waste products can seriously affect the performance of the vehicle if they are not carefully considered during the design phase. The heat energy can also be exploited by devices that transform it into electrical energy.
- Feedback is used to enable a system to determine that its output command has reached a desired state in the desired time-scales and that the desired state is stable. Feedback appears as an input to the system and is derived from a measuring device that monitors the output of the system.
- External influences are exerted on the system and its components by the outside world and by other systems. Such influences must be clearly understood, and their impact on the design of the system and its performance must be taken into account during the design phase.

There are factors that influence the generic model shown in Figure 2.5 that make it less than ideal for some system implementations. Safety, integrity, availability, mission success, and customer perception are factors that influence the design of a system. Consideration of these factors can result in the introduction of redundancy of sensors, control process, and output devices in order to tolerate failures, whilst maintaining some degree of safe operation. This integrity of the basic control mechanism must be reflected through the entire system, including sources of power and the provision of information to the crew. In other words, the system must be safe from end to end. An example of redundancy is shown in Figure 2.10, which shows a dual-redundant system.

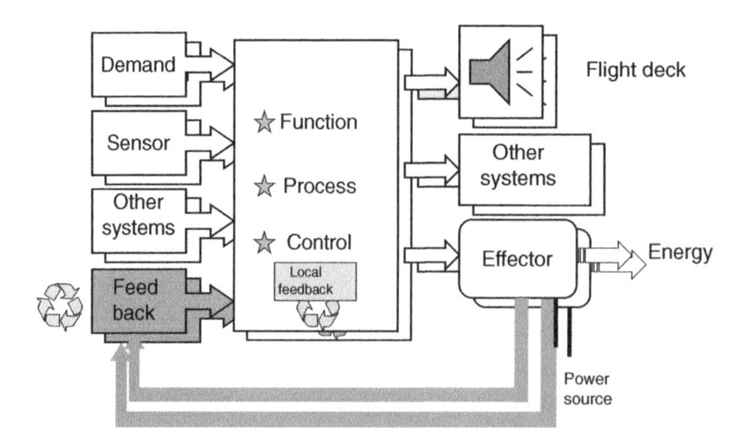

Figure 2.10 A dual-redundant aircraft system block diagram.

In this example all inputs, functions, and outputs are duplicated and are carefully separated to avoid faults or failures being propagated from one system to the other, known as a common mode failure. This philosophy can be extended to further levels of multiple redundancy; triple and quadruple are common in high integrity systems design.

It will be shown in Chapter 10 that an aircraft comprises a number of different systems with varying levels of redundancy that all make a contribution to meeting the necessary integrity and availability goals for the whole product.

Exercises

1 Search for more definitions of 'system'. Select what you think is the most relevant definition for your own work or study and adopt it.

2 Discuss your choice with colleagues to see how it differs from their understanding. If there is a difference try to understand if it matters or not.

3 Examine the lists of systems and input/output devices. Are they complete and what alternatives are likely to emerge that will influence future systems design?

References

FAA (2002). *FAA Joint Aircraft System/Component code tables and definitions*. Issue Feb 2002. Google JASC Codes PDF [Accessed March 2019].

Jenkins, G.M. (1977). The systems approach. In: *Systems Behaviour* (eds. J. Beishon and G. Peters). Milton Keynes: Open University Press.

Moir, I. and Seabridge, A. (2006). *Military Avionic Systems*. Wiley.

Moir, I. and Seabridge, A. (2008). *Aircraft Systems*, 3e. Wiley.

Moir, I. and Seabridge, A. (2013). *Civil Avionic Systems*, 2e. Wiley.

US Department of Defense (1998). *Handbook of Electronic Reliability Design*. MIL-HBK-338B. US Department of Defense.

Further Reading

Jukes, M. (2003). *Aircraft Display Systems*. Wiley.

See also references in Chapter 12.

3

The Design and Development Process

3.1 Introduction

Chapter 2 introduced an understanding of the numerous and disparate aircraft systems that need to be designed and developed into an integrated system solution to ensure that the aircraft is equipped to perform its stated tasks. To develop such a system from a customer's requirement through to implementation requires a discipline that will enable people to apply their skills and experience in a rigorous and consistent manner. It is important to recognise that the product moves through a number of stages in a lifecycle that cover initial concept, design and development, and in-service operation by a customer until the product is no longer required. In the case of an aircraft this entire lifecycle is generally about 25 years; with mid-life updates and refurbishment the life may exceed 50 years, as is the case with some aircraft in service today. Even the initial development phases before product design is sufficiently mature to commit to production are now longer than some technology life spans – in other words 'new' technology may be obsolete before it can even be used, yet alone stay in service for 25 years.

Inevitably in such a prolonged lifecycle there will be issues of currency of technology, obsolescence, changing requirements, application of different skills and processes, and changing legislation. There is a need for a disciplined approach to design and development in order to manage all these aspects. This chapter will look at best practice in related fields of engineering and describe a lifecycle process.

Throughout this prolonged lifecycle the skill set of the people involved will also change. Initially the skills will be those of understanding operational requirements and producing concepts to meet those requirements. To convert the concepts into a hard product requires engineers from a number of domains such as power generation, flight controls, radar, cockpit displays – those skills matching the systems described in Chapter 2. This chapter describes an example product lifecycle and the role of the engineers in that lifecycle.

3.2 Definitions

There are some important lessons and excellent practice to be gained from the field of systems engineering. There are a number of principles and practices that have much in common with established engineering processes, as others have observed: *It [systems*

Design and Development of Aircraft Systems, Third Edition. Allan Seabridge and Ian Moir.

engineering] is not a new discipline, since its history is deeply rooted in good industrial design practice (Eisner 2002). Some definitions from this field will be used to highlight good practice and promote cross-fertilisation and will also provide references for those who wish to read more about systems engineering.

As with the definition of a system in the previous chapter, there are numerous definitions of systems engineering. Different learned bodies and institutions, as well as practitioners of systems engineering, have formed their own understanding of the term. The International Council on Systems Engineering (INCOSE) uses the definition:

> *Systems Engineering is an interdisciplinary approach and means to enable the realisation of successful systems. It focuses on defining customer needs and required functionality early in the development cycle, documenting requirements, then proceeding with design synthesis and system validation while considering the complete problem:*
>
> - *Operations*
> - *Performance*
> - *Test*
> - *Manufacturing*
> - *Cost and schedule*
> - *Training and Support*
> - *Disposal*
>
> *Systems Engineering integrates all the disciplines and specialty groups into a team effort forming a structural development process that proceeds from concept to production to operation.*

The United States Department of Defense uses the following definition:

> *It [Systems Engineering] involves design and management of a total system which includes hardware and software, as well as other system life-cycle elements. The systems engineering process is a structured, disciplined and documented technical effort through which systems products and processes are simultaneously defined, developed and integrated. Systems Engineering is most effectively implemented as part of an overall integrated product and process development effort using multi-disciplinary teamwork.* (Consult US DoD Systems Engineering for contemporary definitions and an explanation of the SE process.)

NASA (Shisko 1995) describes systems engineering as *A robust approach to the design, creation and operation of systems*, and adds the following:

1) *Identification and qualification of goals*
2) *Creation of alternative system design concepts*
3) *Performance of design trades*
4) *Selection and implementation of the best design*
5) *Verification that the design is properly built and integrated*
6) *Post-implementation assessment of how well the system meets its stated goals.*

The key point that emerges from these definitions is that engineers engaged in the design and development of systems need a process that:

- encompasses the entire lifecycle of a product or system
- takes into account the interests and needs of a wide range of interested parties or stakeholders
- covers a wide range of topics and domains in a multi-disciplinary process
- takes into account the project and design drivers that influence the system solution
- allows for the understanding and management of complexity in a repeatable and consistent manner.

Underlying these definitions is the assumption that the approach to the design and implementation of a system must be disciplined and structured in order to bring together a number of elements of hardware and software into an integrated whole that does something. This structured approach is something that is inherent in the 'custom and practice' of engineering or problem solving. Its formalisation into a process means that it can be applied repeatably with continuous improvement.

The engineer developing a system must take into account a number of factors in the system environment that influence the outcome of his work. These factors (or design drivers) are considerations in trade-offs that must be made to arrive at a balanced system solution that meets the demands of the customer and the business. The design drivers are examined in detail in Chapter 4.

The design and development process is a combination of a process and people with appropriate skills to conduct the task. The process can be applied at all stages of the product lifecycle. What is more important is that all stages of the lifecycle are considered at the initial stages of the approach – in other words a whole lifecycle approach is taken. The following phase descriptions will provide an insight into the process and wide range of personal, technical, and managerial skills required. People are an integral part of the process, whether as developers of the system or as users. It is vital that people issues are considered throughout the lifecycle (for further information see Hall 1962; Checkland 1972; Jenkins 1972).

3.3 The Product Lifecycle

Figure 3.1 shows a typical aircraft product lifecycle from concept through to disposal at the end of the product's useful life. A historical example of an experimental project being developed through the entire lifecycle from concept to retirement can be found in Seabridge and Skorczewski (2016).

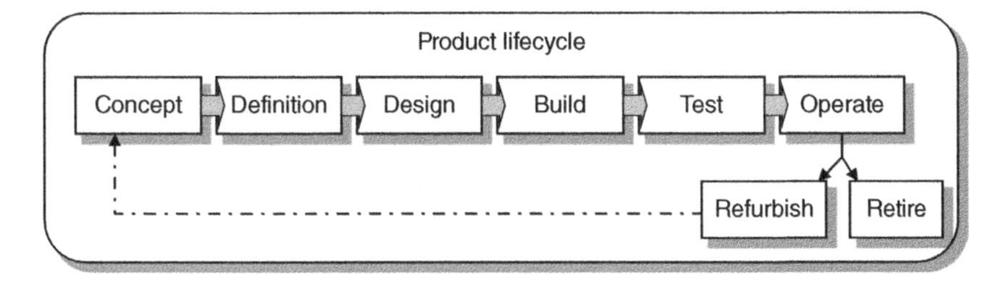

Figure 3.1 A typical aircraft product lifecycle.

Individual product lifecycles will differ from this but it is a sufficiently good model to illustrate the role of engineering in the design and implementation of a systems-based product. The lifecycle resembles the procurement lifecycle used by customers.

Within each step of the lifecycle there is an engineering process that is followed to ensure that the output from each stage is of the required quality. This process, or sequence of engineering activities, is a formalised representation of an intuitive process of engineering. Most system engineering organisations have adopted a process that is documented and used to ensure repeatability and high quality of work, and also to ensure that engineers working on different sites are using the same process. Individual organisations have developed their own specific processes and methods of imposing the process and governing its use. Hence, in the following descriptions the reference to process is intended to be generic or to provide examples only.

It is important to note that in practice the phases of the lifecycle are not necessarily strictly sequential. There is often an overlap, or concurrency, of work in the various phases. For this reason it is essential that there is good communication amongst all parties to ensure that work progresses in accordance with clearly understood interfaces. This understanding is essential to avoid errors or misunderstandings arising in the design process. An example of the cost of late detection of errors is discussed below.

Quite apart from concurrency the model shown is misleading in that it implies that all the stages of the lifecycle are of equal duration and this is not the case in practice. Figure 3.2 shows a more realistic situation and also gives some experience of typical durations encountered with contemporary aircraft.

Figure 3.2 shows that some development projects for many complex products (not only aircraft) may take from 10 to 20 years from concept to entry into service. It also shows that once in service, many types are still being used beyond what was once thought to be a sensible time frame. There are types in service today in which the original design has exceeded 50 years in service, sometimes in the same role but often modified from its original

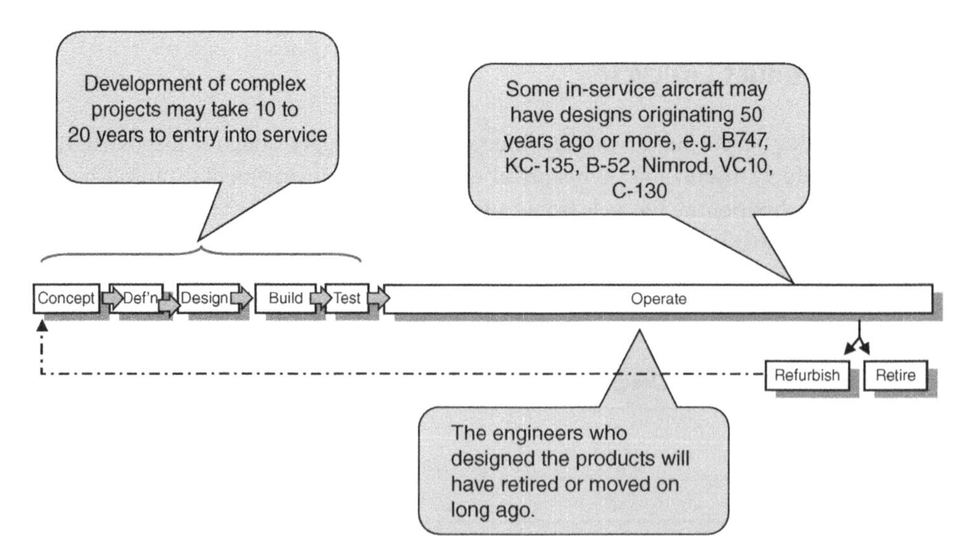

Figure 3.2 Some examples of lifecycle durations.

role to perform new tasks. For example, many commercial passenger-carrying aircraft are modified to carry freight and some commercial airframe types are modified for military roles such as troop transport, surveillance platforms, and air-to-air re-fuelling tankers. It is also clear that such time frames are beyond the working life of the engineers involved in the original design (see Chapter 4 for more on this topic).

There have been attempts to reduce the duration of some of these phases and this work still goes on. Terms such as 'rapid engineering', 'lean manufacturing', and 'concurrency' have been used, and in fact the techniques are in general use (Hartmann et al. 2004; Mynott 2011). The use of artificial intelligence and augmented reality has been proposed and use has been made of auto-code generation of software, largely successfully. Attention is now switching to computer-aided generation of system architectures. All of these mechanisms have a measure of risk that may only manifest itself in the test and operate phase, where the cost of rectification of errors is high. Care must be taken when considering the introduction of new design techniques, and it must always be remembered that the performance of many computer-aided techniques is only as good as the algorithms that define them.

Figure 3.3 shows some external influences on this extended lifecycle, most of which arise because of business pressures which demand that the priorities of suppliers and customers will change over such long timescales.

The first thing to note in this figure is that the lifecycle of supplier equipment is shorter than the aircraft lifecycle. Suppliers have multiple customers and are driven by the demands of competition and the need to continually develop their products and employ new technology to make their products more attractive. This means that decisions made early in the aircraft lifecycle may lead to the selection of obsolescent products, whereas decisions delayed until later can lead to programme delay. Thus obsolescence, which once made its presence felt during the operate phase, is now a threat at the initial stages of the development lifecycle.

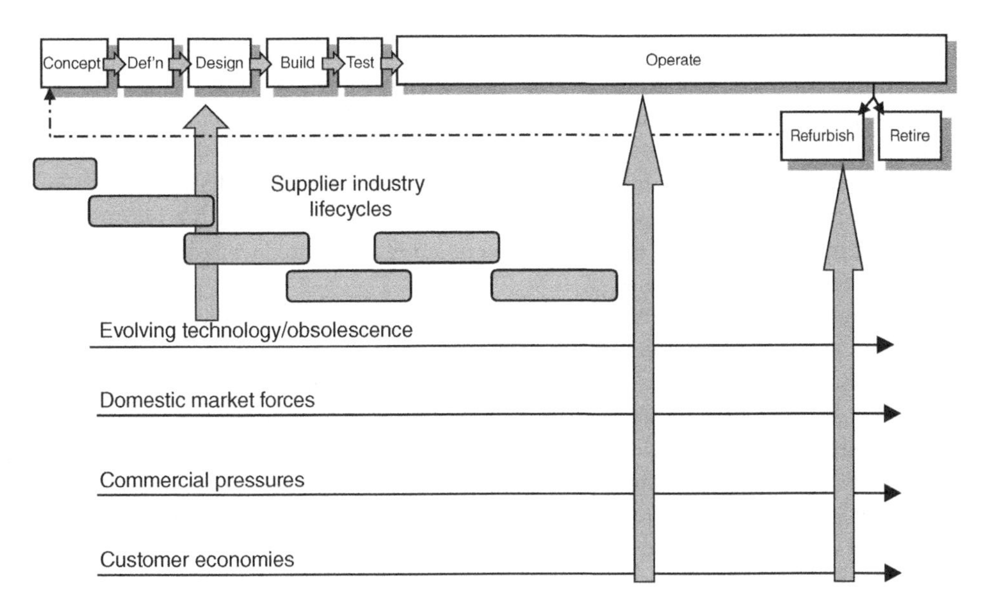

Figure 3.3 Some external influences on the lifecycle.

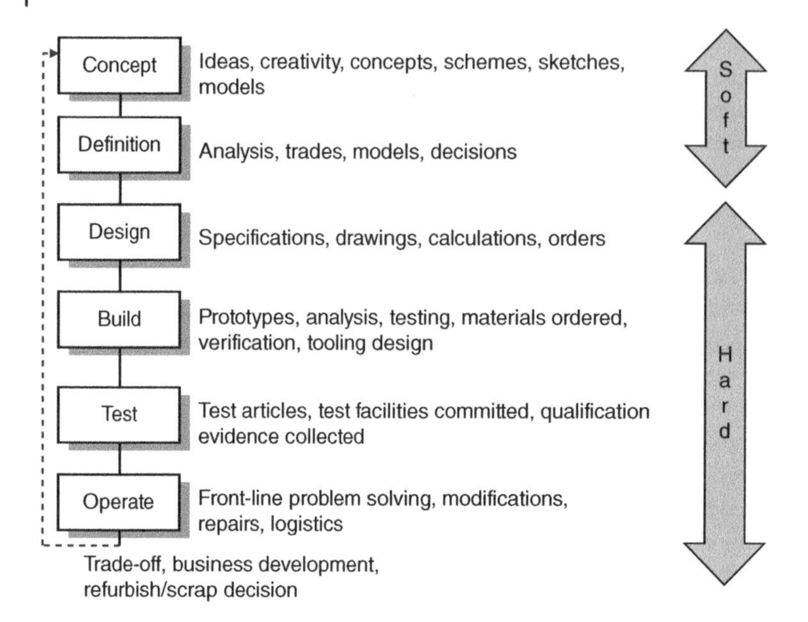

Figure 3.4 How the system develops in the lifecycle.

There are many examples of products that have failed to meet a customer's needs. Part of the challenge of working a good engineering process is ensuring that does not happen and that errors and misunderstandings are detected and eliminated at each phase of the lifecycle. This is particularly important in concurrent working, where the number of customers within the organisation is large and each will bear the cost of correcting errors. The cost of correcting errors is closely linked to the maturity of the product as it develops through the lifecycle, as shown in Figure 3.4. Early lifecycle products are 'soft' and easy to change – ideas, schemes, notes, rough calculations. As the lifecycle develops the products become 'hard, either because they become more physical, such as scale models or prototypes, or because there is an increase in the number of people or stakeholders using common or shared information. This increasing dependency on shared information means that more design work must be repeated if the information database changes. For this reason it is necessary to exercise control over the configuration of the information sources. An explanation of how to manage the configuration of a product design throughout the lifecycle is given in Chapter 9.

An illustration of the cost of correcting errors found in the lifecycle is shown in Figure 3.5. This shows that costs of correction are relatively small when the product definition is largely confined to paper or is used by a small group of people, but rise rapidly when something has been manufactured. The cost of correcting errors whilst the product is in service is magnified by the need to call back products for change and to maintain the customer's service to his own customers. There is a hidden cost in loss of goodwill and poor publicity, particularly if the customer suffers a loss of capability or revenue during the time the product is out of service.

Some studies show that the cost of correcting errors found when a product is in service may be more than 1000 times more expensive than for errors found early in the lifecycle.

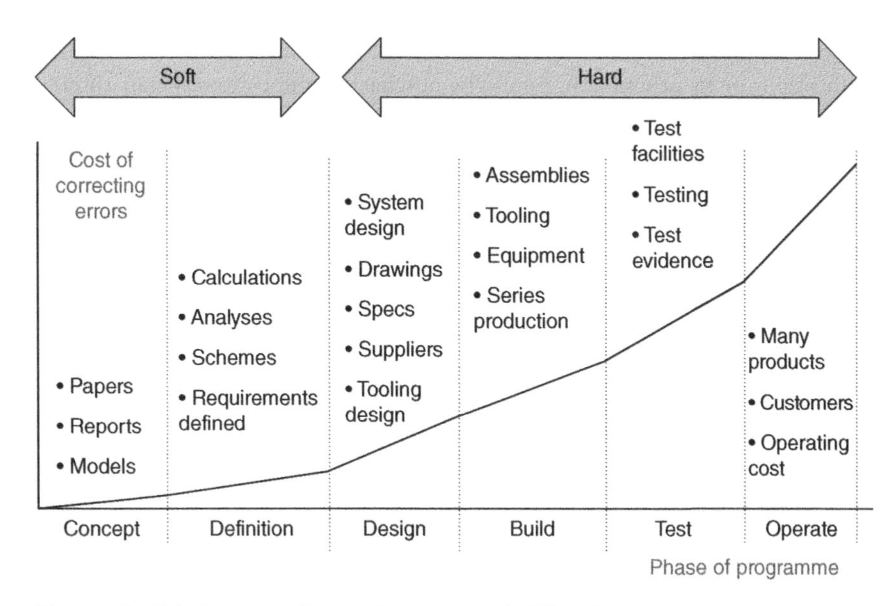

Figure 3.5 Relative costs of correcting errors in the lifecycle.

A mechanism for addressing the proliferation of errors in large and complex systems has been successfully used in contemporary programmes. This involves a combination of risk management and maturity management. One assesses the risk of applying technology and in proceeding through the programme with perceived uncertainties, and the other attempts to measure the maturity of design, asking the question: How certain is the design team that there are no uncertainties remaining at each phase? A common forum for assessing this is a team of project independent specialists and managers representing the prime contractor, the suppliers, and the customer, occasionally supported by 'grey-heads' to provide independent wisdom and to inject 'lessons learned' from previous projects. This should happen before the completion of every process phase as well as at each formal lifecycle review.

Each of the lifecycle phases requires various parts of an organisation to do work to produce a range of deliverables. These deliverables may be in the form of reports, drawings, test data, financial information or hardware, items that are required by other parts of the organisation. The work requires an understanding of the overall engineering process to be employed, the work to be carried out within each phase (a sub-process), the deliverables required, and the schedule of delivery. A mixture of skills is needed to discharge this work, and engineering teams will be made up of people with differing skills working together. The mixture of skills in the team will change throughout the process. The initial set of skills will be based on an understanding of requirements and broad conceptual solutions, and this will develop into skills covering a number of specific engineering domains to develop individual system designs. Then follows the ability to turn these designs into hardware and software solutions, and to test them singly and then progressively assemble them as a whole before committing to series manufacture and release to the customer.

As well as nurturing individual skills, significant efficiencies can be obtained by fostering an understanding of skills possessed by other team members. For example, an engineer who understands the purchasing and legal process will use this knowledge in the compilation of specifications or in negotiation. Similarly, purchasing and contracts staff should understand the engineering process in order to deal sensitively with supplier and contractual issues. In the following descriptions of the lifecycle phases an indication of the required skills will be given.

Some of the expected outputs from the lifecycle include the following, which will be described further in the description of each phase:

- concept – definition of technology
- definition – schemes, plans, preliminary hardware and software specs, supplier selection, models, mock-ups
- design – drawings, hardware and software specifications, materials
- build - build plan, tooling, facilities, materials, resources
- test – plans, schedules, test rigs, test results
- operate – flight test plans, schedules, results.

3.4 Concept Phase

Figure 3.6 illustrates the key engineering activities associated with this phase of the lifecycle. The concept phase is about understanding the customer's emerging needs and arriving at a conceptual model of a solution to address those needs. The customer continuously assesses his current assets and determines their effectiveness to meet future requirements. The need for a new military system can arise from a change in the local or world political scene, or a perceived changing threat that requires a change in defence policy. The need for

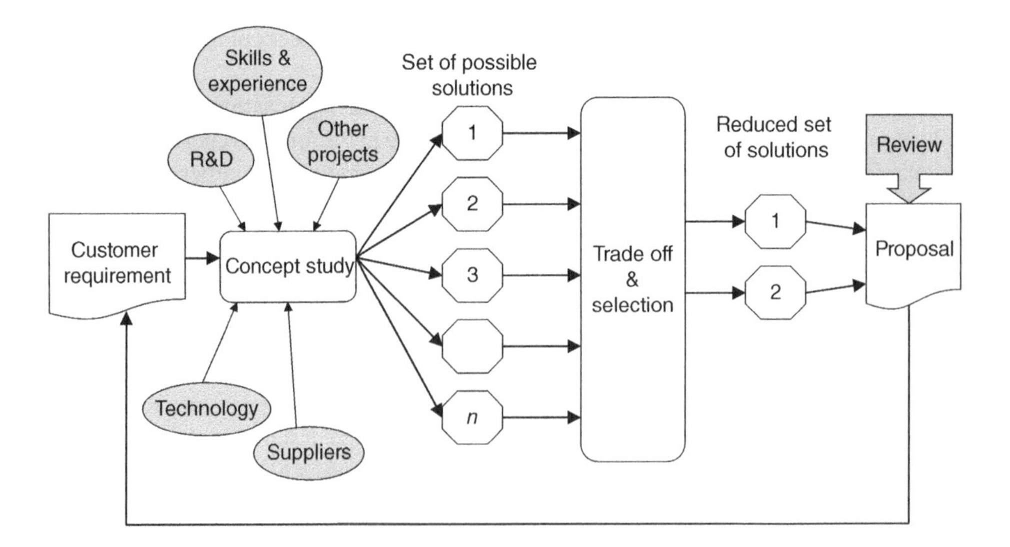

Figure 3.6 The concept phase process.

a new commercial system may be driven by changing national and global travel patterns resulting from business or leisure traveller demands.

The customer requirement may only be simply stated for some projects and the concept study uses all the resources of the company to understand it better and to generate a number of potential solutions. Some of these solutions will be discarded during the trade-off process, leaving a small set, preferably only one. This solution is reviewed and offered to the customer. This phase is focussed on establishing confidence that the requirement can be met within the bounds of acceptable commercial or technological risk. The establishment of a baseline of mature technologies may be first solicited by means of a request for information (RFI). This process allows possible vendors to establish their technical and other capabilities and represents an opportunity for the platform integrator to assess and quantify the relative strengths of competing vendors and also to capture mature technology of which he was previously unaware for the benefit of the programme.

A key function of this phase is to generate ideas using all means at the disposal of the concept team. An example idea-generating process used in lean product development is shown in Figure 3.7. It is a process like this that generates the solutions in Figure 3.6, the trade-off and down-select part of the process should produce the output to the next phase. It is important that discarded ideas are archived for future teams and as a resource should this phase of the process need to be re-visited.

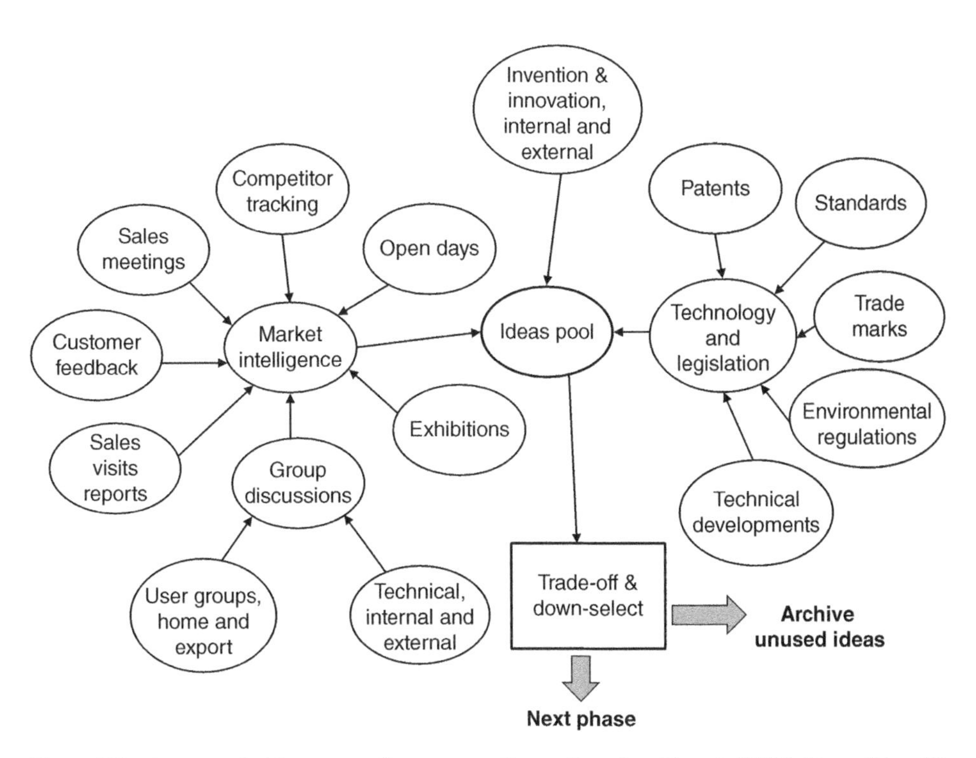

Figure 3.7 An example idea-generating process. *Source:* Based on Mynott (2011), Figure 7.1, p. 63.

3.4.1 Engineering Process

The customer's requirement will be made available to industry so that solutions can be developed specifically for that purpose or can be adapted from the current research and development (R&D) base. This is an ideal opportunity for industry to discuss and understand the requirements to the mutual benefit of the customer and his industrial suppliers, to understand the implications of providing a fully compliant solution or one which is aggressive and sympathetic to marketplace requirements. Not all R&D is driven by the customer, nor is it all customer funded. Industry will, as part of its forward-looking strategy, seek to identify and carry out speculative, self-funded research. This may be to support current projects or to reduce the risk of proposing innovative solutions. It may also be non-project related 'blue skies' research.

Typical considerations at this phase are:

- establishing and understanding the primary role and functions of the required system
- establishing and understanding desired performance and market drivers such as:
  - range
  - endurance
  - routes or missions
  - technology baseline
  - operational roles
  - number of passengers
  - mass, number, and type of weapons
  - availability and dispatch reliability
  - fleet size to perform the role or satisfy the routes
  - purchase budget available
  - operating or through-life costs
  - commonality or model range
  - market size and export potential
  - customer preference
- establishing confidence that the requirement can be met within acceptable commercial or technological risk
- developing an understanding of a solution that can be manufactured. This will lead to proposed aircraft shapes, and interior and exterior configurations together with preliminary system architectures.

3.4.2 Engineering Skills

In this phase the key skills are related to the ability to visualise options and solutions to meet the customer's requirements. Typical skill areas include the following:

- Understanding the requirement: using the customer's information and business intelligence to determine what the customer needs as a solution and how to express that as a directed business strategy that can be accomplished to meet performance, cost, and schedule constraints.

- R&D: the investigation of new concepts, processes or technologies and their insertion into current or future projects. Key skills are to determine which technologies to pursue, when to direct and apply R&D into a particular domain, and to ensure that the activity is focussed upon providing a solution that benefits the business.
- Conceptual thinking: to work from brief requirement statements and work in abstract concepts, slowly developing these towards realistic solutions.
- Proposal writing: the ability to describe the solution in a clear and succinct form, often to meet a restricted word or page count. This must include any technical solutions as well as a definition of cost and the time required to implement a solution.
- Modelling: an ability to visualise draught concepts as models or simulations to demonstrate such aspects as performance, viability, mass, cost, etc. to aid understanding and to provide a sound basis for comparison or different concepts. Models may be physical scale models of solutions, three-dimensional computer-aided design models or mathematical models on lap-top PCs or main frame computers.

The output from this phase is usually in the form of reports, drawings, mathematical models or brochures. The customer may use these to refine his initial requirement by incorporating new information or by taking into account the risks identified. As implied by the title of the phase, the output is a conceptual design and does not necessarily guarantee that the proposed system is optimal or that it could be manufactured. The output is intended to be sufficient for the customer and industry to jointly agree to move on to a more detailed definition phase. In fact the outcome may be a number of potential solutions from which a choice has to be made using cost benefit analysis, and in extreme cases building and flying prototypes in a competition. A recent example of this was the project to establish a design for the Joint Strike Fighter in the USA where two types, the Boeing X32 and the Lockheed X35, entered a 'fly-off' competition. The Lockheed X-35 was the successful outcome and is now in production as the F-35 Lightning II.

Figure 3.8 shows some typical products that emerge during this phase. The fighter aircraft shapes were produced in response to a particular threat. Four different shapes with different engine configurations were considered suitable and these configurations were 'frozen' to enable a review to take place. Following this review one of the candidates was selected for further design. The P1A model shown was tested in the wind tunnel to determine the most suitable location for the tailplane – it is shown in the position finally selected. Modelling clay was used to determine the shape of the control column handle. It was originally moulded to meet the shape designed by the human factors team, and then offered to several pilots for them to mould it to what they perceived to be an optimum shape to reach all the switches and still be comfortable. Again one model was selected for manufacture. The angle of the seat in a highly manoeuvrable aircraft can improve tolerance to high g; high angles can degrade visibility of instruments and may not provide the best ejection clearance. A number of different configurations were analysed before selecting the angle considered most suitable.

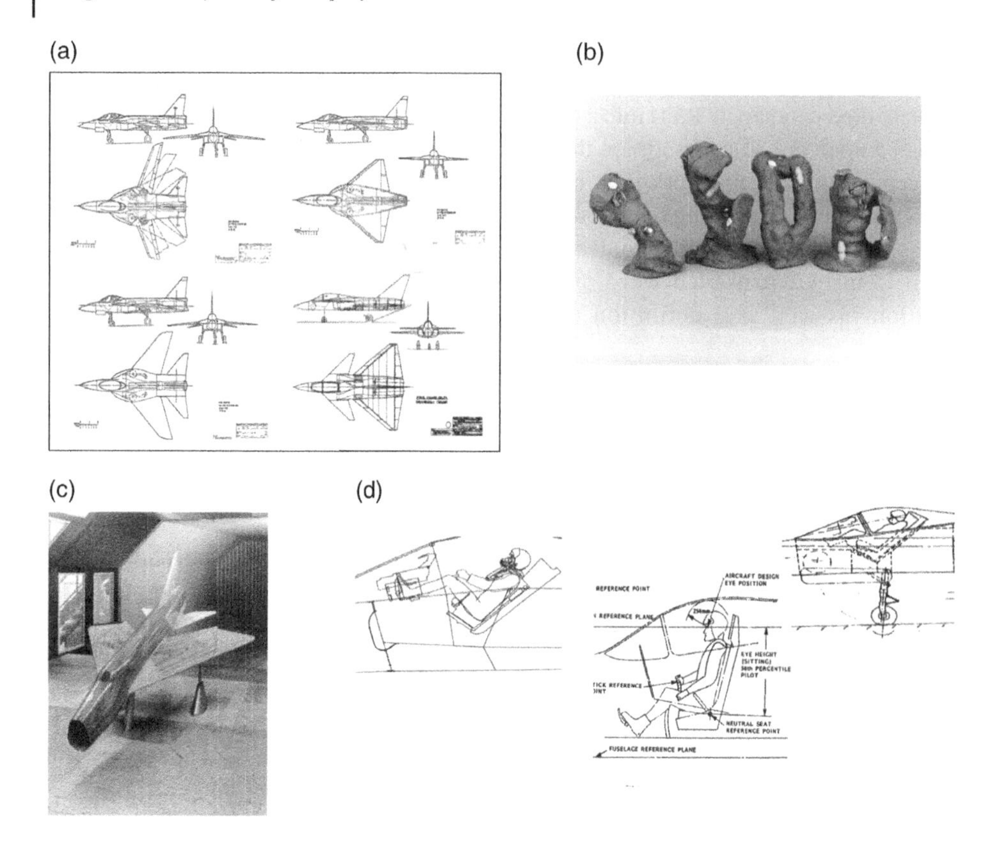

Figure 3.8 Typical products of the concept phase process: (a) candidate shapes for a fighter project, (b) concepts for a control column handle using modelling clay, (c) P1A wind tunnel model, and (d) selecting the right seat back angle (photos: BAE Systems).

3.5 Definition Phase

Figure 3.9 illustrates the key engineering activities associated with this phase of the lifecycle. This phase takes as its input baseline the reviewed concept and develops that to examine the practicality of developing a complete and definitive design. The concept is firmed up into a set of documents that define the emerging product in order to feed it into the design phase. This is the preliminary phase to detailed engineering design, a phase in which multiple concepts are firmed up into a single definition.

Figure 3.10 shows some outputs of this phase. The graphic cockpit model develops sketches and drawings from the concept phase in a pseudo three-dimensional layout which will be amended in discussion to achieve a satisfactory layout of the components parts. The landing gear model has been taken from the preliminary sketches of the retraction sequence. With animation in the whole aircraft model it is possible to observe clearances and potential fouls in the process, and to synchronise the gear retraction with the doors. The four line replaceable units (LRUs) were fabricated in wood but to scale with representative connector and handles. Filled with ballast to the correct weight, these were used to check the installation in the aircraft mock-up to check physical installation and health

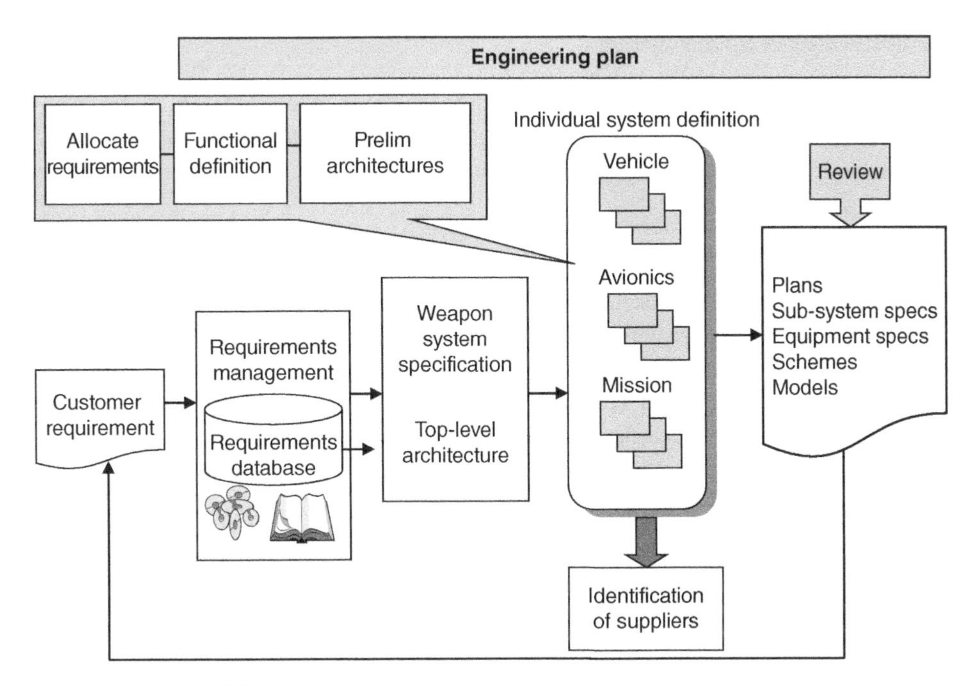

Figure 3.9 The definition phase process.

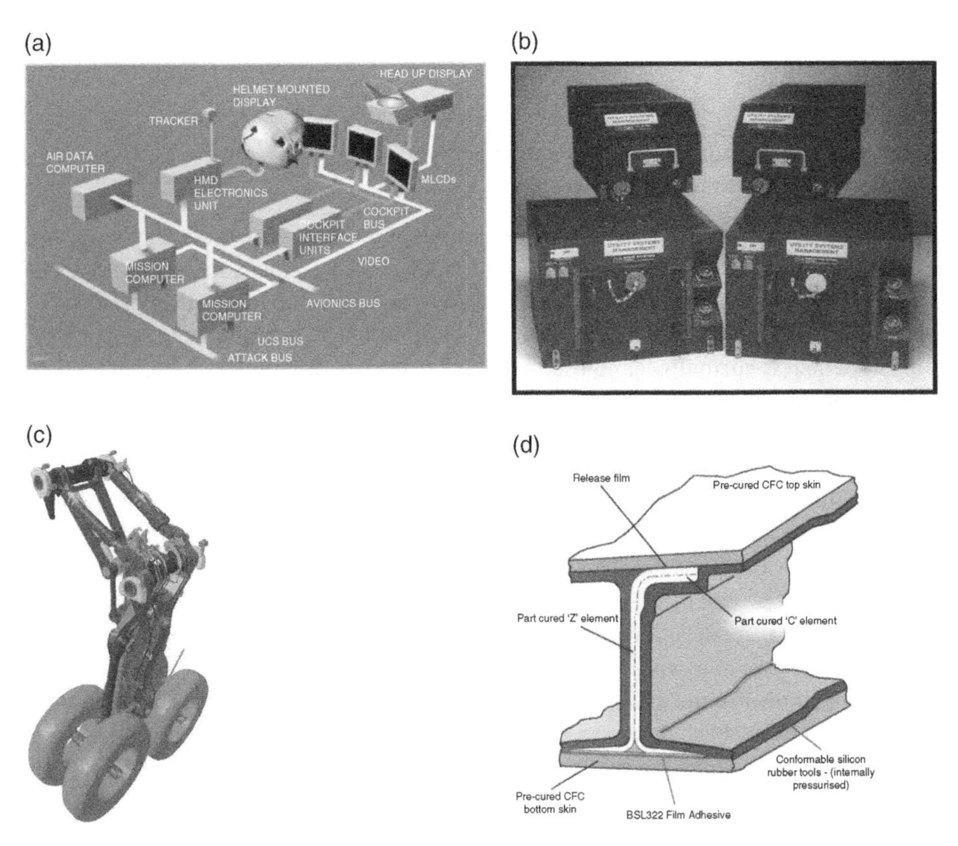

Figure 3.10 Typical products of the definition phase process: (a) cockpit graphic model, (b) full-scale wooden replica of LRUs (photo: BAE Systems), (c) landing gear main leg animated CAD model, and (d) proposal for bonding the wing structure (photo: Jim Banks).

and safety issues associated with weight and clearance on the bay doors. Preliminary schemes of detailed structural components may be produced for debate as a preliminary to a more formal computer-aided design definition.

3.5.1 Engineering Process

The customer will usually consolidate all the information gathered during the concept phase to firm up his requirement. This often results in the issue of a specification or a request for proposal (RFP). This allows industry to develop their concepts into a firm definition, to evaluate the technical, technological and commercial risks, and to examine the feasibility of completing the design and moving to a series production solution. Typical considerations at this stage are:

- developing the concept into a firm definition of a solution
- developing system architectures and system configurations
- re-evaluating the supplier base to establish what equipment, components, and materials are available or may be needed to support the emerging design
- defining physical and installation characteristics and interface requirements
- developing models of the individual systems
- quantifying key systems performance measures such as:
  - mass
  - volume
  - growth capability
  - range/endurance
- identifying risk and introducing mitigation plans
- selecting and confirming appropriate technology.

This phase of the process demands the beginning of a rigorous methodology to record the requirements and establish traceability of change. Requirement management tools exist to enable the requirement to be unambiguously stated and to record the design solutions, and one example is DOORS (Dynamic Object Oriented Requirements supplied by Rational), which is currently being used on many contemporary projects. In conjunction with this tool it is possible to start to model the design using tools such as Unified Modelling Language (UML) or Systems Modelling Language (SysML). These tools record the design and form a valuable input to the qualification phase.

The output from this phase is usually in the form of feasibility study reports, performance estimates, sets of mathematical models of individual system behaviours, and an operational performance model. This may be complemented by breadboard or experimental models, as well as with mock-ups in three-dimensional computer model form or wooden and metal physical models developed from the concept stage models. In some circumstances the customer may wish to proceed to prototypes and, if funding is available, may ask two competitors to enter a 'fly away' competition in which two prototypes maybe developed and flown to establish the best solutions. This approach is relevant to very large-scale production contracts where the risk of a single untested solution maybe untenable.

An example of this can be seen in the Joint Strike Fighter programme in the USA, where two aircraft companies were each tasked to produce a prototype to demonstrate capability and performance by flight trials. This led to the customer selection of a single solution.

3.5.2 Engineering Skills

In this phase the key skills are related to the ability to convert the conceptual solutions into a single defined product to meet the customer's requirements. Typical skill areas include the following:

- Requirements management: the capture, manipulation, and management of systems requirements, including the management of traceability between levels of design. This often involves the use of database tools to manage large amounts of data and to enable a trace to be accomplished between requirements and various phases of design and test. There is a skill to acquiring a 'top level' view of the requirement and flowing down requirements into the project teams and the suppliers to build a progressively more detailed understanding of the customer's needs. This analysis leads to a better understanding of how to construct a complete solution that fully satisfies these needs.
- Process capability: includes necessary design tool development and tailoring, development of suitable training materials, etc. Process support includes performing the role of system design responsibility (SDR), process query resolution, 'help-desk' provision, etc.
- Design process engineering: develops, deploys, and controls a recognised process for the various disciplines of engineering to use, and defines interfaces for support tools. Adherence to a controlled process leads to a consistent approach throughout the project lifecycle. This is especially important if the teams are geographically dispersed, as is often the case in multi-national partnerships.
- Systems integration: the structuring and partitioning of complex systems, usually to minimise interface complexity between sub-systems whilst maintaining a 'whole systems view' in order to ensure that the end product meets its requirements.
- System architecture design: building up a design architecture to meet the requirements, and to partition or allocate functions to elements of the architecture. This usually starts with a simple block diagram of the system with indications of function locations and data flows. Once agreed, this architecture can be developed to show increasing amounts of detail.
- Behavioural design engineering: analysis of requirements, identification of potential solutions, and selection of the most cost-effective solution (unless directed otherwise by the requirement) from a system behavioural viewpoint. The expression of the requirements or solutions may be in several forms. These include functional, state, transition, and object oriented.
- System safety engineering: those aspects of systems engineering that address the certification requirements and safety liability associated with aircraft systems. Safety engineering includes the identification of hazards, hazard risk assessment, definition of safety requirements, safety assessment of designs and implementation, production of safety cases, and the analysis and assessment of the system design process with respect to safety management. Safety engineering requires familiarity with the requirements of standards, contract, and legislation in addition to best practice in the field under consideration. Knowledge of aircraft behaviour in service and practical hazards should be built up and recorded.
- Performance analysis: analysis of system behaviour from a performance viewpoint, understanding what the total system should do and what numerical targets have been set. A key skill is judging the proportion of the subject system to be modelled, and the

means by which the required analysis result can be obtained in the most cost-effective manner. It includes performance budgeting, characterisation, statistical analysis, scheduler analysis, etc. Another key skill is in selecting and using the tools available to model individual systems and combining these models to represent the complete solution.

- Mission analysis: analyses the mission requirements, defines mission types and phases or segments, and defines the mission timeline and availability targets. Mission is defined as a specific type of operation from pre-flight briefing to post-flight de-briefing, i.e. a combat mission for a military aircraft or a routine airport-to-airport flight for a commercial aircraft.
- Human factors and cockpit/flight deck integration: identification of human factors issues of the system, the identification of potential solutions and their management and implementation to ensure that the human (operator and maintainer) and the system are successfully integrated.
- Modelling and simulation engineering: analysis of design requirements and solutions in order to determine the most critical characteristics of a system and to simulate those characteristics in the most cost-effective manner.
- Reliability engineering: analysis of the design and requirements, and application of techniques, methods, and technologies to assure and demonstrate acceptable reliability and 'fault tolerance' of the product. Reliability engineering must be flowed down through specifications and into the supplier skill set. Examples include analysis of availability targets versus 'affordable' technology capabilities, trade-off against safety design requirements, support, spares-holding, testability requirements, analysis of the product to establish acceptable levels of achieved reliability (failure modes and effects analysis [FMECA] etc.), development of fault-tolerance mechanisms (redundancy, reversionary moding, etc), defensive programming, development and assessment of 'reliable' software, etc.
- Maintainability engineering: analysis of the design and requirements, and application of techniques, methods, and technologies to ensure the cost-effective maintainability of the product. This is often provided by ex-service or airline staff with a practical knowledge of maintenance activities. Knowledge of tools, access requirements, and ground equipment is desirable.
- Testability engineering: analysis of the design and requirements, and application of techniques, methods, and technologies to ensure the satisfactory ability to perform testing and diagnostics of the product at all levels. Includes the analysis of testability requirements to provide a framework for built-in test, pre-flight test, build test, serviceability testing, and post equipment replacement testing. Examples include designing and engineering abilities (as required) to 'test' the health of the complete system (including its components and their interfaces) at the complete vehicle level.
- Estimation, measurement and metrication: using the process, work breakdown structure (WBS), and product breakdown structure (PBS) identify the product and the work required to design, test, and build it. This enables the estimation of the cost to complete the job, the cost to completion. Identification, capture, and analysis of appropriate metrics to understand the actual cost of activities and assist process improvement, sensitivity of cost to programme risk, and currency variations. This is most often seen as a project management task.

- Design to cost engineering: identifying the relationship between the system design options and cost, and the choice of design options in order to meet cost requirements, also known as cost as an independent variable (CAIV).
- Risk analysis and management: analysing the concept and the design to determine where there are any areas of major concern or uncertainty that could jeopardise the successful completion of the project. This will include aspects of technology, long-term security of suppliers, performance estimation, novelty etc. For each risk identified a mitigation plan is prepared to demonstrate how the risk is to be eliminated and what cost should be set aside to achieve the mitigation.
- Specification and procurement: The identification, specification, and technical procurement of bought-out systems, sub-systems, and equipment. This also includes the management of these products and their integration with interfacing areas of product, definition of software-related requirements/components on the hardware (processors, board architectures), etc.
- Weapons/explosives safety, surveillance, and legislative control: scrutiny of design to ensure safe handling and carriage of explosive and pyrotechnic devices to prevent hazards to vehicle, crew, and maintenance staff in accordance with health and safety and ordnance regulations.
- Signature measurement: analysis and management of the system design for audible, optical, and electro-magnetic signature optimisation. This is of particular interest to the designer of military aircraft who has a need to design a vehicle that will escape detection as far as possible by radar, visual, acoustic, radio-frequency or infra-red sensors. This reduces the risk of an aircraft being detected and targeted by anti-aircraft weapon systems.
- Security engineering: the definition and development of techniques to ensure the integrity associated with the handling and transference of secure (i.e. classified) data and information. Includes the development of encryption techniques, Tempest proofing, etc.
- Proof of design (qualification and certification): the identification of requirement/design attributes to be demonstrated and methods to be employed in order to achieve proof of design, and the management and implementation of activities to ensure that this is successfully achieved in the most cost-effective manner. Accumulation, analysis, integration, and evaluation of evidence to verify fitness for purpose and safety of use.
- Configuration management: management of the design configuration, control, and authorisation of change, management of configuration/change management boards, and processes. This task continues for the entire lifecycle.
- Quality management/capability deployment management: the production and maintenance of the quality management systems/business management system (QMS/BMS) for the local project/business area, ensuring timely availability, awareness, and smooth deployment of capability improvements into the business, internal auditing and assurance of compliance (and support to external audits), resolution of non-compliances, and identification of capability related improvement needs and concerns.
- Project/business management: planning, network/schedule preparation, definition of performance milestones, and operation of earned value management.

3.6 Design Phase

Figure 3.11 illustrates the key engineering activities associated with this phase of the lifecycle. The design phase is often divided to produce a preliminary design, which is reviewed before committing to a detailed design, which is the input to manufacturing. It is important to note that even at the preliminary design review decisions will have been made that commit up to 80% of product costs. Some products of this phase are shown in Figure 3.12 to illustrate where in the process they are produced.

3.6.1 Engineering Process

If the outcome of the definition phase is successful and a decision is made to proceed further, then industry embarks on the design phase. Design takes the definition phase architectures and schemes and refines them to a standard that can be manufactured.

Detailed design of the airframe ensures that the structure is aerodynamically sound, is of appropriate strength, and is able to carry the crew, passengers, fuel, and systems that are required to turn it into a useful product. As part of the detailed design attention must be paid to the mandated rules and regulations which apply to the design of an aircraft or to airborne equipment. Three-dimensional solid modelling tools are used to produce the design drawings in a format that can be used to drive machine tools to manufacture parts for assembly.

Systems are developed beyond the block diagram architectural drawings into detailed wiring diagrams. Suppliers of bought-in equipment and components are selected and they become an inherent part of the process, starting to design equipment that can be used in the aircraft and systems. Indeed in order to achieve a fully certifiable design of many of the

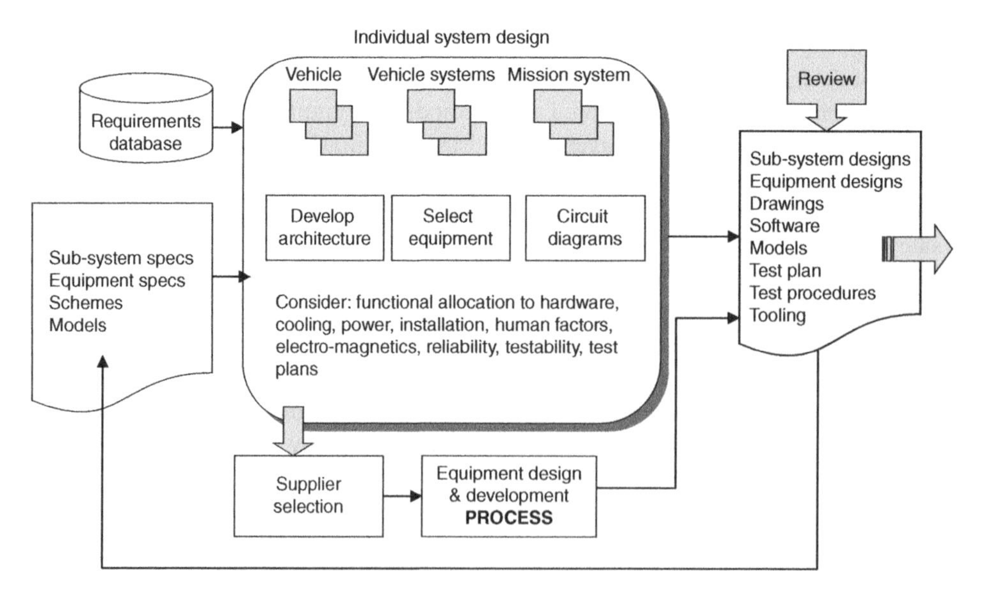

Figure 3.11 The design phase process.

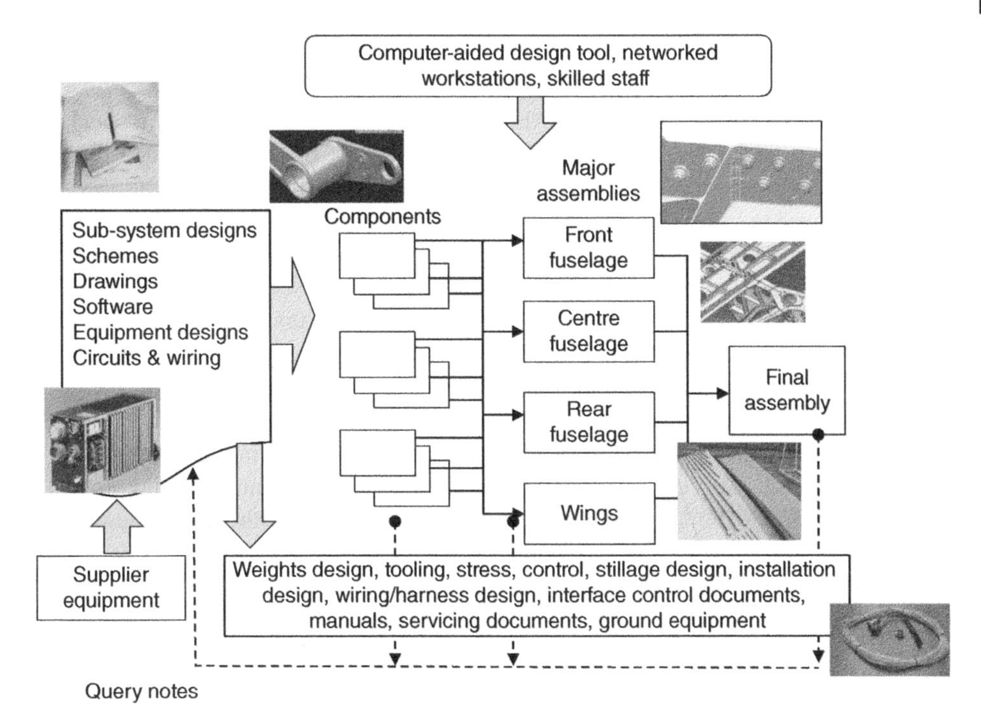

Figure 3.12 Typical products of the design phase process.

complex integrated systems found on aircraft today, an integrated design team or integrated product team (IPT) comprising platform integrators and suppliers is essential.

3.6.2 Engineering Skills

All the skills required for the definition phase are required for design phase, since this phase can be seen as an extension into a solution that can be manufactured. Additional skills include:

- Software design: Application of software design techniques such as object-oriented design (OOD) to produce preliminary and detailed designs. Note: software design also involves software-related elements of other skills such as requirements engineering, architecture design, performance analysis reliability, maintainability and testability engineering, safety, etc.
- Physical design: Analysis, definition and specification of the physical product and its physical integration dependencies, including analysis of requirements and production of specifications, production of interface control requirements and engineering design requirements (EDR), and analysis of satisfactory installation designs and environmental conditions.
- Supplier management: Selection of suppliers based on their responses to RFPs will demand that each supplier will need to be managed in the sense that they are kept informed of all project progress and decisions, and that the supplier's information is coordinated and supplied to the project teams.

3.7 Build Phase

The build phase is where the products of the design phase are used to produce a physical product. This means that all design information (drawings or computer-aided design models) have been produced, checked, approved, and stored under configuration control. They are used to order materials to plan the build line process so that all jigs, tools, and materials are in the right place at the right time. Build may take two different forms: (i) building the prototypes, if required, and (ii) the series production build which manufactures the product in quantity. Before leaving the production line each aircraft is subject to testing to ensure that it is fit for engine running and flight. At every stage of build and test any discrepancies, malfunctions, defects failures to, etc. are reported back to design in a formal query note procedure. Any rectification action that is taken is recorded in the design data (with appropriate configuration control discipline) so that the changes can be incorporated into all products from the build process. Figure 3.13 illustrates the key engineering activities associated with this phase of the lifecycle.

Some typical examples of build phase products are shown in Figure 3.14, which shows some of the many components, in this case the cockpit panels and the avionics equipment racks, to be incorporated into the equipment bays, one wing yet to be skinned, the front fuselage in its jig, and the final assembly in progress.

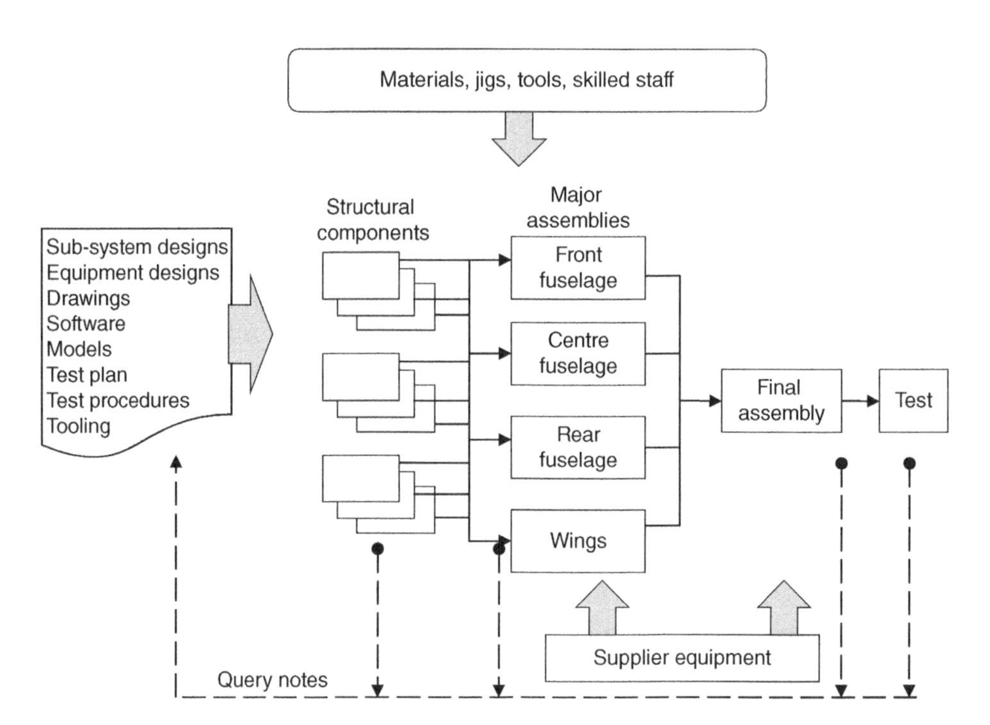

Figure 3.13 The build phase process.

Figure 3.14 Typical products of the build phase process: (a) cockpit components (photo: BAE Systems), (b) avionics racking (photo: BAE Systems), (c) front fuselage (photo: BAE Systems), (d) wing (photo: BAE Systems), (e) wiring (photo: Allan Seabridge), and (f) final assembly (photo: BAE Systems).

3.7.1 Engineering Process

The aircraft is manufactured to the drawings and data issued by design. This includes the fabrication of detailed sub-assemblies and their progressive build-up, or final assembly, into a complete airframe together with the installation of pipes, ducts, wiring harnesses, and equipment. The main systems engineering support to this phase is to provide a service to manufacturing in answering queries in instances where the solution cannot be achieved in practice or in an economical manner from a quantity production viewpoint. Prompt and effective answers at the early stages of build can reduce the probability of errors appearing in quantity production.

3.7.2 Engineering Skills

In this phase the key skills are related to providing support to the manufacturing process to ensure that problems are solved as they arise, that design errors are found and corrected, and that solutions are incorporated into the design. Typical skill areas include:

- knowledge of the design and an ability to provide answers to manufacturing problems
- knowledge of change management and configuration management
- hardware/software integration, i.e. integration and qualification of software loads in their target hardware environment/equipment
- the ability to develop methods for build testing and writing test procedures.

3.8 Test Phase

Figure 3.15 illustrates the key engineering activities associated with this phase of the lifecycle. This is an important phase in the life of an aircraft. It is the opportunity for the built design to be tested by an independent group of test engineers to ensure that its functionality is correct, and that the systems are robust and will operate in the required environmental conditions and will continue to operate for the life of the aircraft. This test regime is conducted on a number of test rigs and facilities by the aircraft company, by suppliers, and by the customer to ensure that the product is fit for purpose. It is a costly phase in both facilities and hours.

3.8.1 Engineering Process

The aircraft and its components are subject to a rigorous test programme to verify its fitness for purpose. This programme includes testing and progressive integration of equipment, components, sub-assemblies, and eventually the complete aircraft. Functional testing of systems on the ground and during flight trials verifies that the performance and operation of the equipment is as specified. Conclusion of the test programme and the associated design analysis and documentation leads to certification of the aircraft or equipment.

3.8.2 Engineering Skills

In this phase the key skills are related to the ability solve problems and to maintain the testing progress, and as before any errors in design must be solved the design data set updated. Typical skill areas include the following:

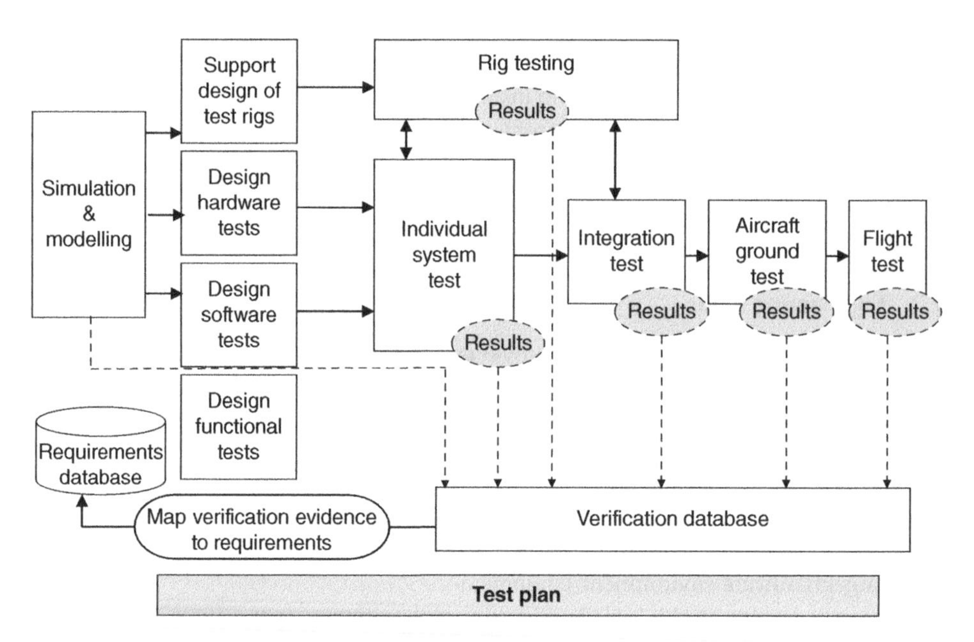

Figure 3.15 The test phase process.

Figure 3.16 Typical products of the test phase process: (a) Jaguar S2 on the uncompleted M55, (b) Tornado RF interoperability test, (c) ECS test rig, and (d) Tornado flight with canopy jettison (photos: BAE Systems).

- Test facility design: analysis of testing requirements and the scoping and scaling of the facilities required in order to complete the range of tests. This includes the design and location of individual test rigs and their location in suitable and safe buildings.
- Test preparation: definition of test specifications, methods, pass/fail criteria and procedures, test measurement techniques, and use of test equipment and instrumentation needed for hardware, software, sub-system, system, and whole aircraft testing.
- Test execution: execution of test/evaluation activities, and recording and analysis of test results for validity. Providing evidence for qualification.

Figure 3.16 shows some examples of test phase products. A Jaguar aircraft is used to demonstrate the capability of the type to land on roads, and is shown landing on the almost completed M55 near Blackpool. The last flight of Tornado P12 was used to demonstrate that the aircraft could be safely flown following a canopy jettison. An example is shown of a major test rig used for testing the environmental control system cold air unit in the laboratory. A Tornado GR4 is shown suspended in the electronic warfare chamber undergoing radio frequency (RF) interoperability testing for the collision warning system (CWS).

3.9 Operate Phase

This phase is a real test, away from the controlled environment of the test laboratories and into the random world of passengers, baggage, beverage carts, and regular cleaning. In this phase the aircraft will be subject to wear and tear on a daily basis and comments can be

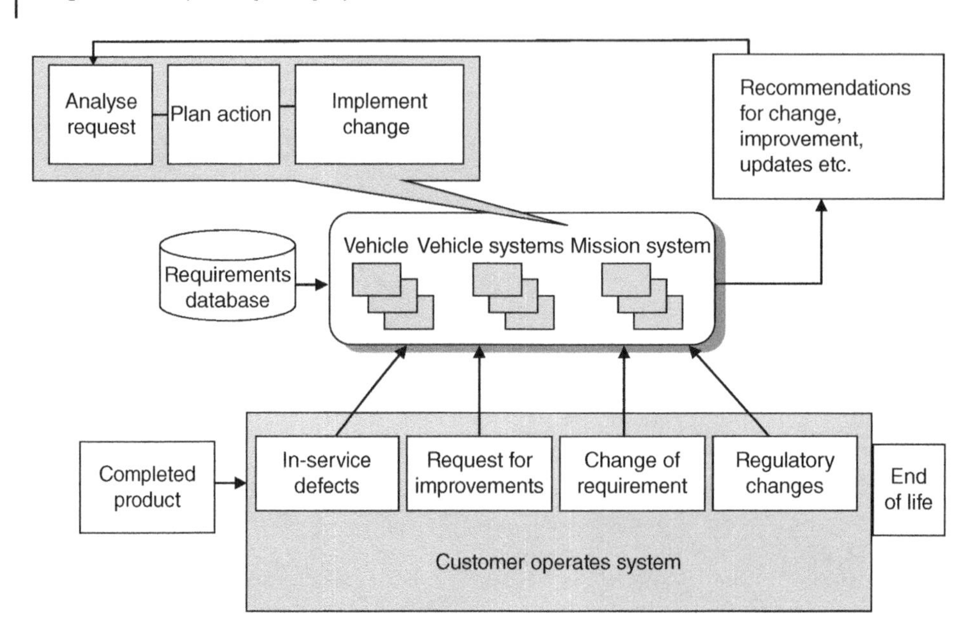

Figure 3.17 The operate phase process.

expected from passengers, cabin crew, and maintenance crews. Passenger aircraft can be expected to undergo regular upgrades to improve furnishings and services in the cabin. Military aircraft go through a similar period with upgrades to systems and weapons, and if used in theatre can be subject to harsh conditions. Figure 3.17 illustrates the key engineering activities associated with this phase of the lifecycle.

The output from this phase may be progressive improvements mainly keeping the aircraft in service. However, continuous monitoring of customer comments and demands, and of emerging technologies and regulatory issues leads to major improvements to the model. Demands for increased comfort, longer range, quieter operation, and lower emissions can lead to improvements, as demonstrated by the B737 in its journey from the 737-100 to the 737-900.

3.9.1 Engineering Process

During this phase the customer is operating the aircraft on a daily basis. Its performance will be monitored by means of a formal defect reporting process so that any defects or faults that arise are analysed by the manufacturer. It is possible to attribute causes to faults such as random component failures, operator mishandling, or design errors. The aircraft manufacturer and his suppliers are expected to participate in the attribution and rectification of problems arising during aircraft operations, as determined by the contract. Any systemic failures occurring may be the subject of design action and will require investigation to allocate liability.

3.9.2 Engineering Skills

In this phase the key skills are related to supporting the customer and his operations. An acute customer focus and the ability to solve problems rapidly in order to minimise aircraft down time is essential. Typical skill areas include the following:

- All systems engineering skills need to be available to support the operations phase on demand. The operator usually operates a query reporting system that enables in-service problems to be reported and rectification action provided.
- Engineers need the ability to write test requirements to enable the test departments to conduct regression testing for individual systems and integrated systems.

3.10 Disposal or Retirement Phase

There comes a time in the life of an aircraft when it is no longer suitable for continued operation. This may be because of the economics of operating an aircraft that is becoming increasingly unreliable or uneconomic to maintain or simply because it becomes time-expired. Retirement may come about naturally, in which case the aircraft are withdrawn from service over a period of time. Alternatively, it may be enforced as a result of a management decision or in the case of some military projects when project funding is terminated. Decisions need to be made about how to dispose of the product in a manner that meets health and safety and environmental conditions.

Figure 3.18 illustrates the key engineering activities associated with this phase of the lifecycle.

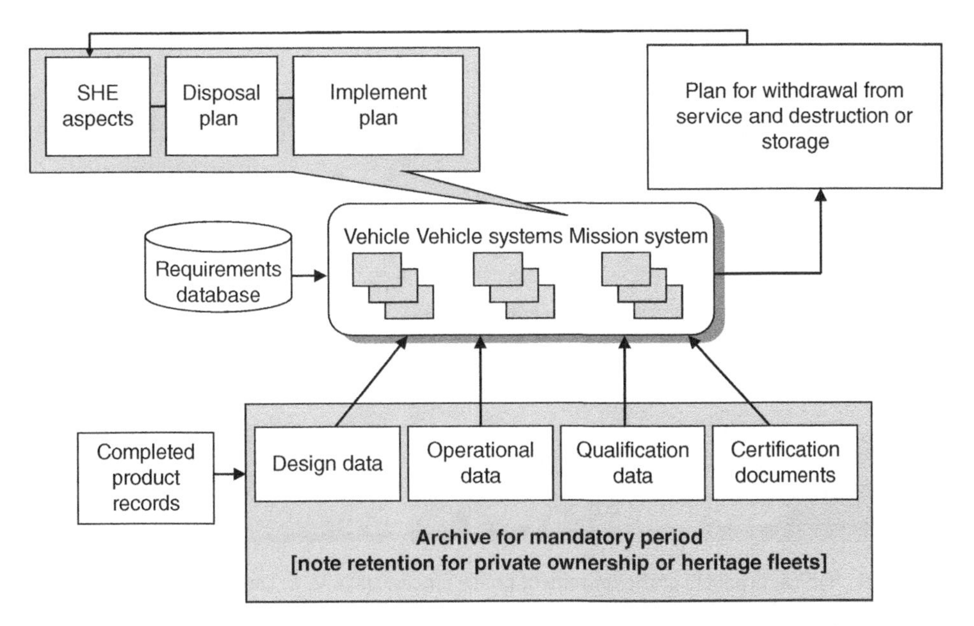

Figure 3.18 The retirement phase process. SHE, safety, health, and environment.

Figure 3.19 shows some examples of products in different situations. Experimental Aircraft Programme (EAP) was an experimental demonstrator which retired when its project objectives were met and many of its major components had reached the end of life. Because of its historical significance it was first donated to Loughborough University, where it was used as a teaching aid in aerospace courses. A decision was made eventually to move the aircraft to the RAF Museum at Cosford, where it remains today. Engineering advice was sought to enable safe dismantling of the aircraft, to transport it between sites, and to re-assemble it. Partly for historical reasons, and perhaps because of public admiration, a number of Concorde aircraft were donated to museums. Shown in Figure 3.19 is G-BOAC at the Manchester Airport at the Runway Visitors Park. Some aircraft are sold to organisations and private buyers who intend to continue to fly them. This happened to some of the Lightning aircraft returned from Saudi Arabia, but many others were scrapped, such as Buccaneer XK523. Thunder City near Cape Town has flown many retired types and offers people an experience of flight in a rare aircraft. Many other aircraft play a useful role as gate guardians at the sites of industry, museums, and the armed services. Corrosion often renders these unsafe and uneconomic to maintain, and replacements are often sought as full-scale plastic replicas. The aircraft manufacturer will be requested to provide suitable drawings of the aircraft and its markings.

Disposal is the ultimate fate of many aircraft. For nations with plenty of space there is the opportunity to store aircraft in hot dry conditions, and there are a number of desert locations in the USA that are dedicated to storage. Most aircraft are simply broken up for scrap,

Figure 3.19 Typical products of the retirement phase process: (a) EAP at Loughborough University (photo: Allan Seabridge), (b) FBW Jaguar at the RAF Museum Cosford (photo: Allan Seabridge), (c) Concorde G-BOAC at Manchester Airport Runway Visitor's Park (photo: Allan Seabridge), and (d) Lightning aircraft awaiting disposal (photo: BAE Systems).

although some organisations and private collectors bid for parts to put into museums, and many parts end up on eBay.

3.10.1 Engineering Process

At the end of the useful or predicted life of the aircraft, decisions have to be made about its future. The end of life may be determined by unacceptably high operating costs (as demonstrated by the decision to remove Concorde from service), unacceptable environmental considerations – noise, pollution, etc. – or by predicted failure of mechanical or structural components determined by the supplier's test rigs. In the military field a decision to retire an aircraft type may be driven by political expediency, for example a need to reduce defence spending, or in some cases by a recognition that a particular threat is no longer present. If it is not possible to continue to operate the aircraft, then it may be disposed of – sold for scrap or alternative use, such as purchase by a museum, an aircraft enthusiast group or used at military bases as a gate guardian. If the aircraft still has some residual and commercially viable life, then it may be refurbished.

A key process component here is developing a plan to assist the customer in retiring the aircraft and ensuring its safe removal, storage or destruction in accordance with statutory and advisory requirements.

3.10.2 Engineering Skills

- Assisting the customer in identifying components for safe storage.
- Providing information to support breakdown, reassembly and transportation.
- Understanding the requirements for disposing of potentially hazardous components and consumables such as fuels, oils, greases, refrigerants etc.
- Recording the decision in project records.
- Ensuring that all design authority records of design and qualification are archived in safe storage for a period of time defined by relevant regulations. This is essential to provide advice to purchasers of redundant aircraft.

3.11 Refurbishment Phase

Figure 3.20 illustrates the key engineering activities associated with this phase of the life-cycle. There will be a stage in the life of an aircraft when a need for refurbishment becomes apparent. This may arise because the aircraft has become obsolete in its original role or it has been sold to a customer wishing to change its role. An example is the use of commercial passenger-carrying airframes being converted to freight or in-flight refuelling roles.

The VC10 and Lockheed L1011 Tristar passenger aircraft converted to in-flight re-fuelling use for military applications. The VC10 was ordered by RAF in 1961 and converted to tanker role in 1977. The fleet was retired in September 2013. Some commercial aircraft are converted to a cargo role after their useful passenger-carrying role expires. In some instances this conversion requires an extensive re-design for military operations.

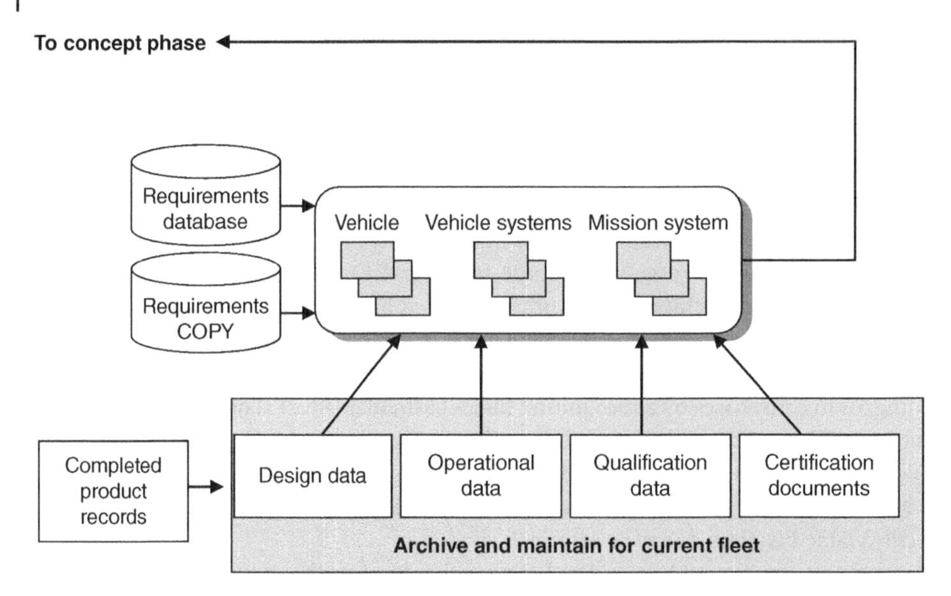

Figure 3.20 The refurbishment phase process.

3.11.1 Engineering Process

- Record the decision in project records.
- Ensure that all design authority records of design and qualification are archived in safe storage for a period of time to support the aircraft that continue in service during the refurbishment.
- Return the existing type record to the concept phase to enable the refurbishment design to commence.

3.11.2 Engineering Skills

Similar skills to the concept phase are required since the options for re-furbishment or conversion into a different role need to be considered with an open mind.

3.12 Whole Lifecycle Tasks

In addition to the specific domain engineering tasks in the processes described above, there are other engineering tasks that take place continuously throughout the lifecycle. Some of these tasks are to do with exercising control over the process, whilst others are carried out to impose a consistent approach to certain disciplines across the domains. This is an important integrating activity that ensures that all domain engineers adhere to standards and processes wherever they practice their individual skill. This integration activity was introduced in Chapter 1. In modern projects where engineering tasks may be spread across international partners, such integration is essential for project consistency. Examples of these activities include:

- *engineering management* to manage the activities of the engineering teams, to take ownership of specific domain requirements, and to ensure that the requirements are met within project constraints
- *project management* to ensure that tasks are performed in accordance with an agreed schedule, to an agreed budget, and that performance criteria are met
- *configuration management* to ensure that the configuration of the product is properly recorded and promulgated to all stakeholders, and to ensure that changes to configuration are recorded (see Chapter 9)
- *requirements management* to analyse and structure the customer's requirement in a suitable requirements management tool, to allocate requirements to specific engineering domains, and to record changes to the requirement throughout the lifecycle
- *risk management* to identify and register risks that impact on technology, completion, cost, schedule or safety, and to ensure that each risk is correctly documented, promulgated to other risk managers, and that a costed plan is in place for mitigating the risk
- *qualification and certification* to gather evidence from lifecycle activities that support the ability of the project to demonstrate that the customer's requirement will be met and that the product is fit for purpose
- *safety* to ensure that a consistent approach is made to product safety, that regulatory safety standards are applied, and that all processes related to demonstration of safety in design are adhered to (Drysdale 2010)
- *reliability* to ensure that the engineering design is analysed to ensure that the product will meet its reliability demonstration criteria and the customer's availability targets
- *maintainability* to ensure that the engineering design reflects the customer's need to service the aircraft with appropriate tools, support equipment and suitably skilled ground crew
- *testability* to ensure that factors are built into the design to allow flight crew and ground crew to perform pre-flight and post-flight checks to satisfy themselves that the aircraft is safe for flight and that defects can be rapidly isolated to support rapid rectification and repair
- *human factors* to ensure that relevant standards are applied in the ergonomic and workload-related aspects of design so that the aircraft can be operated safely by an appropriate percentile range of aircrew and maintainers without undue stress or health and safety impact
- *electro-magnetics* to ensure that the individual systems can operate without causing mutual interference, and that the systems can operate in the presence of external hazards such as high power radio frequency transmissions, high electrostatic fields or lightning (MacDiarmid 2010)
- *systems security* to ensure that military aircraft do not radiate information unintentionally, and that all aspects of loading, storing, and destruction of classified data are controlled.

3.13 Summary

The system lifecycle used is a convenient way of visualising the different phases of the development process. It is not the only way and the reader is encouraged to devise their own lifecycle, noting that it will only work if it is used across a project. It is a convenient shorthand for displaying what happens in the individual phases and also across the entire lifecycle.

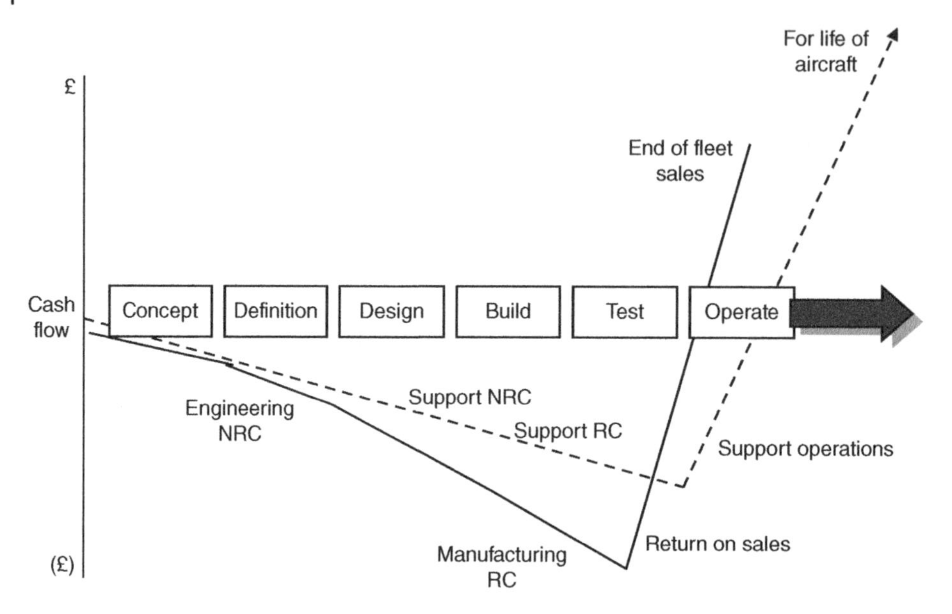

Figure 3.21 Indicative cash flow in the product lifecycle. NRC, non-recurring cost; RC, recurring cost.

One last item to note is the cash flow in the lifecycle. Figure 3.21 is indicative only of the investment required in a major project, and the fact that return on investment occurs fairly late in the lifecycle. It is often the case that it is product support that makes a profit. A good payment plan is required to try to recover costs earlier than this.

Figure 3.22 shows some examples of projects that have been through the entire lifecycle and have been in operation for varying periods of time. The Tornado at the time it was photographed had been in service for 40 years with three European air forces. The project started as an early attempt at European co-operation and three aerospace companies in the UK, Germany, and Italy worked together to a common standard to produce a multi-role combat aircraft. The resulting aircraft went through a number of updates, which included a new variant, the air defence variant, which become the F3. The aircraft featured in many campaigns in its life and still continues to serve the German, Italian, and Saudi air forces.

Finally, a lesson to be learned from past experience. There is a phenomenon known to engineers as 'requirements creep'. This is the emergence of new or 'improved' requirements that appear after a lifecycle phase has been frozen, usually by well-meaning attempts by people to improve the product. Occasionally, they will be stimulated by baser measures such as ambition, pride, jealousy, inter-department strife or commercial concerns. This 'creep' often leads to increased complexity in the product as engineers struggle to incorporate changes, often forcing changes into the system design; an event that often leads to costly rework. It has further implications because changes are incorporated after the architectural principles have been established and new functions and physical interfaces can contradict these principles, threatening robust system integration. In early 2019 a new airliner suffered fatal accidents as a result of a late change to the system that was not clearly understood by the pilots. Sound and strong project management and configuration management are needed to limit this effect.

Figure 3.22 Examples of products of the lifecycle: (a) Tornado GR4 (photo: Allan Seabridge), (b) Apache helicopter at work (photo: Leon Skorczewski), and (c) Alaskan Airlines B 737 (photo: Allan Seabridge).

Exercises

1 The lifecycle used in this book was chosen by the author as an illustration of some major identifiable periods in the development of a product that can be used to offer suitable review points. Devise your own lifecycle that is more suited to your own task. Describe the outputs from each phase. Compare and contrast this with the example given in Figure 3.2 and explain the differences.

2 Figure 3.5 shows an example of the cost of correcting errors found at different phases of the lifecycle. The cost scale has been deliberately left blank. Using your own judgement complete the scale to show the relative costs for each phase and provide your reasoning.

3 Section 3.3 included a list of performance parameter examples that could be used to establish the feasibility of a particular aircraft solution. Imagine that you were about to buy a motor vehicle or vehicles as (i) an individual owner/user, (ii) a local business transport or passenger hire fleet operator, or (iii) a multi-national vehicle hire business. Produce lists of the performance parameters for these scenarios that would influence your decision in purchasing the right kind of vehicle. Identify the differences between these lists and explain the influences on your decisions.

4 Using the same scenarios as in question 3, explain the process you would have to go through to determine how best to deal with the vehicle or fleet of vehicles at the end of its useful life.

5 What are the primary objectives of the concept phase? Identify at least six issues which can be considered in the trade-off used to narrow the number of solutions in this phase.

6 For each of the lifecycle process phases provide some examples of the outputs of each phase from a project with which you are familiar.

7 The acceptance of change differs in each phase of the lifecycle. Explain why change is encouraged in the early phases but positively discouraged in the later stages.

References

Checkland, P.B. (1972). Towards a systems-based methodology for real world problem solving. *Journal of Systems Engineering.* 3: 87–116.

Drysdale, A.T. (2010). Safety and integrity in vehicle systems, Chapter 11. In: *Encyclopedia of Aerospace Engineering*, vol. 8 (eds. R.H. Blockley and W. Shyy), 5036–5044. Wiley.

Eisner, H. (2002). *Essentials of Project and Systems Engineering Management*, 2e. Wiley.

Hall, A.D. (1962). *A Methodology for Systems Engineering*. Van Nostrand.

Hartmann, J., Meeker, C., Weller, M., Izzard, N. (2004). *Determinate Assembly of Tooling Allows Concurrent Design of Airbus Wings and Major Assembly Fixtures.* SAE Aerospace Manufacturing & Automated Fastening Conference & Exhibition.

Jenkins, G.M. (1972). The systems approach. In: *Systems Behaviour* (eds. J. Beishon and G. Peters). The Open University, Harper & Row.

MacDiarmid, I. (2010). Electromagnetic integration of aircraft systems, Chapter 412. In: *Encyclopedia of Aerospace Engineering*, vol. 8 (eds. R.H. Blockley and W. Shyy), 5045–5057. Wiley.

Mynott, C. (2011). *Lean Product Development.* Westfield Publishing.

Seabridge, A. and Skorczewski, L. (2016). *EAP: The Experimental Aircraft Programme.* BAE Systems Heritage Department.

Shisko, R. (1995). *NASA Systems Engineering Handbook.* SP-6105. Linthicum Heights MD. NASA Technical Information Program Office.

Further Reading

Assadi, M., Dobbs, S., Stewart, B. et al. (2015). Panel Assembly Line (PAL) for high production rates. *SAE Int. J. Aerosp.* 8 (1): 104–116. https://doi.org/10.4271/2015-01-2492.

Boothroyd, G., et.al. (1994) Product Design for Manufacture and Assembly.

Eisner, H. (2002). *Essentials of Project and Systems Engineering Management*, 2e. Wiley.

International Council on Systems Engineering (INCOSE), (n.d.) INCOSE. 2033 Sixth Avenue, #804, Seattle, WA 98121, USA. www.incose.org. Accessed January 2019.

Judt, D., Forster, K., Lockett, H. et al. (2016). Aircraft wing build philosophy change through system pre-equipping of major components. *SAE Int. J. Aerosp.* 9 (1): 2016.

Jukes, M. (2004). *Aircraft Display Systems.* Wiley.

Kossiakoff, A. and Sweet, W.N. (2003). *Systems Engineering – Principles and Practice.* Wiley.

Longworth, J.H. (2013, 2014, 2015). *Test Flying in Lancashire*, vol. 1, 2 and 3. BAE Systems Heritage Department.

Meakin, B. and Wilkinson, B. (2002). *The 'Learn from Experience Journey in Systems Engineering.* INCOSE 12th International Symposium, Las Vegas, July 2002.

Moir, I. and Seabridge, A. (2013). *Civil Avionic Systems*, 2e. Wiley.

Overmeyer, L. and Bentlage, A. (2014). Small-scaled modular design for aircraft wings. In: *New Production Technologies in Aerospace Industry*, Lecture notes in Production Engineering (ed. B. Dekena), 55–62. Switzerland: Springer International Publishing https://doi.org/10.1007/978-3-319-01964-2.

Saif, U., Guan, Z., Wang, B. et al. (2014). A survey on assembly lines and its types. *Frontiers of Mechanical Engineering* 9 (2): 95–105.

Schrage, D.P. (2010). Product lifecycle engineering (PLE): an application, Chapter 390. In: *Encyclopedia of Aerospace Engineering*, vol. 8 (eds. R.H. Blockley and W. Shyy), 4767–4784. Wiley.

Seabridge, A. and Skorczewski, L. (2016). *EAP: The Experimental Aircraft Programme.* BAE Systems Heritage Department.

Stevens, R., Brook, P., Jackson, K., and Arnold, S. (1998). *Systems Engineering – Coping with Complexity.* Prentice Hall.

Whitney, D.E. (2004). *Mechanical Assemblies: Their Design, Manufacture, and Role in Product Development.* Oxford: Oxford University Press.

Wise, P.R. and John, P. (2003). *Engineering Design in the Multi-Disciplinary Environment.* Professional Engineering Publishing.

4

Design Drivers

4.1 Introduction

Chapter 3 introduced the concept of design drivers, or factors that positively and strongly influence the design of a system and therefore must be taken into account. Combinations of these factors may be predominant at different phases of the lifecycle; they change and their influence varies. Not everyone involved in the design at different organisational levels will take the same view of any one factor's importance. Each will have their own personal viewpoint depending upon their particular discipline and their perception of the issue at hand – marketing, engineering, management, financial, contractual, etc. This can lead to organisational stress, differences of priority, and poor communication in the organisation as each group works, unknowingly perhaps, to their own agenda to the detriment of the whole.

A holistic systems approach (or looking at the big picture) will aim to make the design drivers openly visible to all participants, ensuring that they are all aware of the ownership and stakeholder issues relating to any factor, and to advise of altering priorities or balances and the need for the change in the 'corporate' approach. It is worth remembering that you cannot hope to understand the whole by studying a part, you need to look at the big picture.

Design drivers arise in the environment of the system as perceived by different organisational levels. The system may be considered to have a series of overlapping environments containing drivers with varying degrees of influence and crossing environment boundaries as illustrated in Figure 4.1.

To illustrate the impact of design drivers on organisational levels, the following system environments will be used to describe drivers with varying predominance:

- The business environment: the consideration of the value to the business of bidding for a contract taking into account factors within the organisation and external pressures. It is often at this stage that decisions are taken to proceed or not with winning the business. Deciding *not* to proceed may often be a sound decision, it can reduce business risk and exposure.
- The project environment: once a contract has been accepted a project team will focus in on the impact on the organisation of taking the project through its initial stages. This is

Design and Development of Aircraft Systems, Third Edition. Allan Seabridge and Ian Moir.

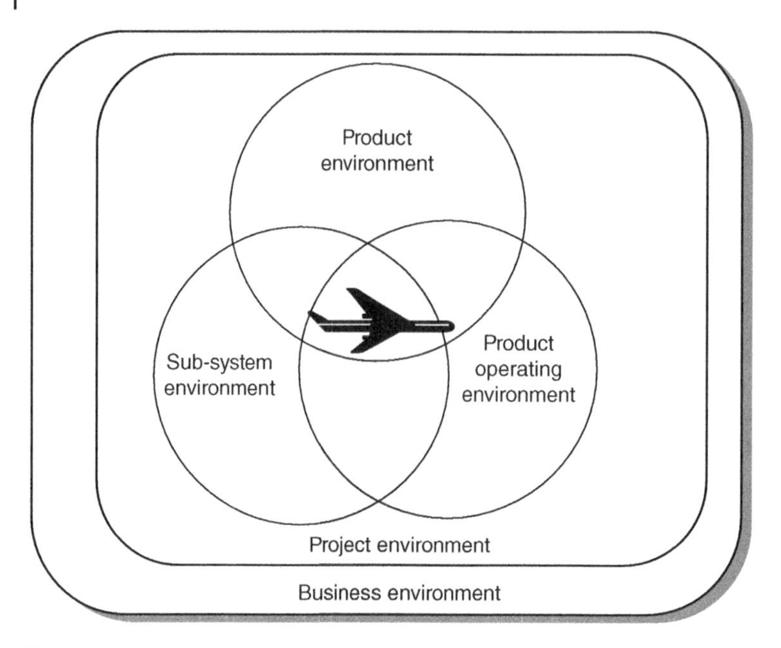

Figure 4.1 Environment considerations.

very much a risk reduction stage to ensure that the business has identified the appropriate skills, experiences, and resources to bring the project to a satisfactory conclusion.

- The product environment: the detailed design and production readiness factors that must be considered. This includes the people, facilities, and assets that exist and any future investment required to bring them to an appropriate state of readiness.
- The product operating environment: ensuring that the design incorporates all known factors likely to be encountered when the product enters service. This requires a comprehensive understanding of the customer's operating scenario.
- The sub-system environment: the detailed factors of sub-system and component design.

These levels represent phases of design from concept though to detailed design and installation of hardware and systems.

Figure 4.1 illustrates high-level drivers which business and project teams must consider in the early stages of a lifecycle, and the overlapping drivers closely associated with the product. In order to illustrate the role of design drivers it has been assumed that they will originate in a particular environment and play a dominant role in that environment. However, they will also flow down through, and make an impact on, all successive environments – in other words decisions made even at very early stages in the project will continue to have an influence. This can be beneficial, but it can also be a considerable threat if poor or undesirable decisions are not recognised and corrected. There is a need, therefore, to review the decision-making process to record all decisions made and their rationale.

As Chapter 2 made clear, the environment boundary is not impenetrable, nor are the drivers described below confined to any one environment. This is illustrated in Figure 4.2, which shows that the original business and project drivers persist throughout, whilst others only apply at later stages.

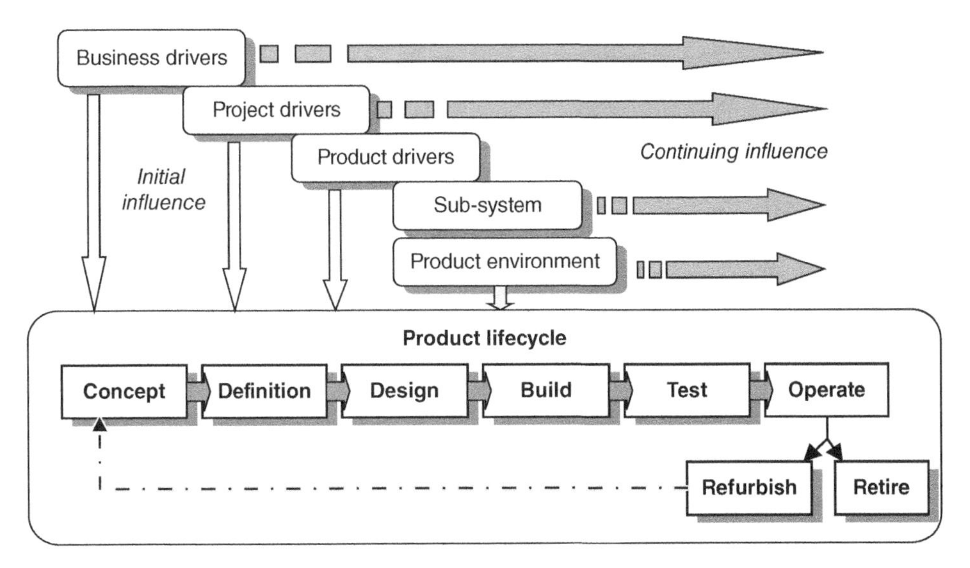

Figure 4.2 Influence of design drivers in the lifecycle.

It is a truism that the practical realisation of each element of a sub-system or system is often a compromise rather than an ideal solution. This chapter will provide an insight into the many conflicting requirements, desires, aspirations, and realities that are the daily life of the systems engineer. Example drivers will be described in this chapter to illustrate typical considerations that need to be applied by the systems engineer. Design drivers will be presented below in bullet point lists for clarity. The lists of examples are by no means exhaustive, and the wise engineering team will brainstorm its own design problem statement and specify its own design drivers, making use of the customer's requirement and their own company business strategy. In other words, is the company capable of providing what the customer wants and how will fitness for purpose and customer acceptance be measured?

4.2 Design Drivers in the Business Environment

The business environment contains drivers that concern the ability of the business to satisfy its shareholders as well as its customers, employees, and the local community. The shareholders cannot be disregarded – they are the people financing and investing in the business. At the same time customers cannot be ignored since it is they who are buying the product for which the investment was made. These drivers predominate in the conceptual phase of a project where the business is concerned with the nature of the business it can acquire, the investment that will required, and the magnitude and impact of risk. They are factors which the business will continually review before committing further funding to a project.

These drivers remain valid throughout the remainder of the lifecycle and will be flowed down to the project teams. Some typical design drivers are shown in Figure 4.3 and are described below.

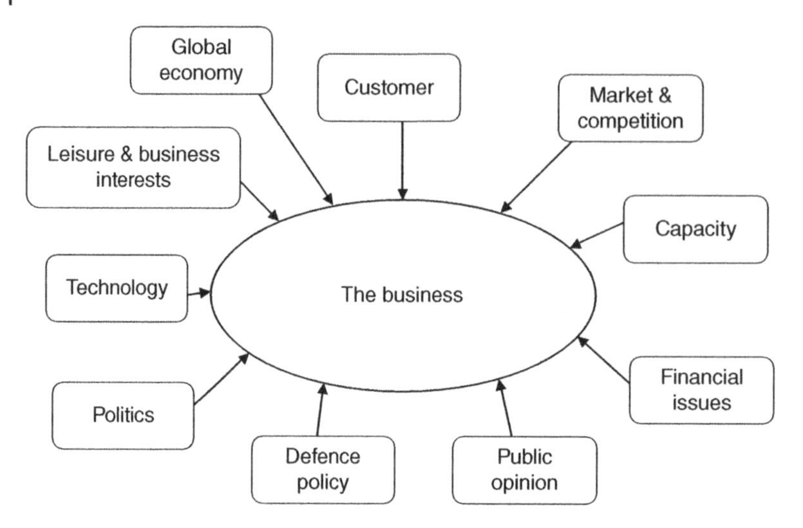

Figure 4.3 Design drivers in the business environment.

4.2.1 Customer

The customer is all-important and it is vital that their needs are understood, continuously monitored or tracked, and satisfied. In our everyday life we are all consumers who purchase products and govern our future purchases by whether we like the product or service, how expensive or otherwise it is, how reliable it has been in use, etc. It is important to note that the customer, in this context, is not simply the agent who purchases the product. There are internal customers and suppliers within the engineering team, as well as formal customer–supplier relationships with external suppliers. Typical considerations include the following:

- The customer's requirement should be tracked and monitored at all stages of the requirement definition process.
- The customer's requirement must be clearly understood and the supplier's interpretation checked frequently with the customer to confirm mutual understanding.
- A sound relationship must be established with the customer(s) and this must be maintained throughout the lifecycle.
- It is often useful to acquire early knowledge of a customer requirement, even by helping the customer to develop his requirement.
- The customer's budget must be understood: How big is it? When is it available?
- If the customer has experience of the product, is it good or bad? How can it be improved?
- Internal customer relationships within the company are important to ensure mutual understanding of information flows within the project.

4.2.2 Market and Competition

It must be recognised that the market for products is finite. For a product to be successful it must satisfy a demand, it must be fit for purpose (i.e. do the job for which it was intended), it must be priced correctly, and it must be seen by potential customers as representing value for money. Typical considerations include the following:

- Is there a market for the product?
- How widespread is the market, how many customers?
- Is there room in the market for another product?
- Is the market stable? Are there any indicators that the market will change, increase or decline in the life of the project?
- What is the likely market share?
- Can the market be developed to include other consumers or to develop variants of the product?
- How many competitors are there, is there room for another player?
- How good are the competitors (gather market intelligence and customer preferences)?
- Is the product pricing understood – investment, quantities, time to break even?
- What is the probability of winning and continuing to win?
- Can the product be developed after initial market penetration?

4.2.3 Capacity

Before accepting a contract it is vital to determine that there is sufficient capacity in the business to bring the project to a satisfactory conclusion. This requires an understanding of the current project status, and the status and certainty of contemporaneous bids to ensure that there is not 'too much' business. Typical considerations include the following:

- What is the status of current skilled resources deployment and spare capacity?
- What training will be required?
- Is there flexibility and reserve in the current facilities?
- Has the supplier base a suitable capacity?
- Is outsourcing required?
- Are there opportunities for work sharing with strategic partners?
- Is it possible for a new project to be performed concurrently with existing projects?

4.2.4 Financial Issues

Although the potential return can look good there are many aspects to be considered before committing to a bid or tender, and most of these will be the subject of a business review. Typical issues include:

- investment in new technology
- investment in infrastructure, facilities, resources, and capacity
- balancing investment against funding for existing programmes
- availability of financing and interest rates
- return on investment and break-even point
- security and proprietary rights restrictions that may limit the exchange of information, e.g. International Traffic in Arms Regulations (ITAR) and the UK Export Control Act.

4.2.5 Defence Policy

Government defence policy has an impact on the sale and continued use of military airborne assets. Defence policies can be shaped by global strategic political situations as well as by local tactical scenarios. For example, the Cold War led to products suited to strategic nuclear strike and defence of Northern Europe. More recent conflicts have increasingly made use of rapid reaction forces, peace keeping, and swift resolution of conflicts. In many of these conflicts the determined use of tactical air power has been a decisive factor. A more recent change to 'asymmetric warfare' will influence defence policy and there may be a move away from deployment of capital armaments towards those more suitable to urban terrorist suppression, network security, internet protection, and intelligence gathering. Typical considerations include:

- defence policy changes in response to changing world-wide political conditions and the impact of current and future orders
- continued global pressure on defence budgets resulting in fewer and smaller orders for traditional combat aircraft, for which there is already increased competition from established manufacturers
- monitoring of global strategic defence reviews and studies to observe and react to trends
- aircraft product lifecycles are so long that the ability to adapt products to meet changing requirements is desirable
- encouraging the customer to take a long term, wider business view of products, support, infrastructure, training, facilities, etc., which places responsibility on the government in that it has to provide stability so that major suppliers do not invest heavily in projects only to find that the original requirement has been diluted or has disappeared.

4.2.6 Leisure and Business Interests

The commercial aircraft market is driven by business and leisure traveller needs and can be subject to economic trends and other factors, such as the 2003 severe acute respiratory syndrome epidemic. Customer and airline loyalty are important for long-term business security. Typical considerations include the following:

- The commercial aircraft market must respond to consumer demands and must monitor trends in business travel and in holiday destination expansion.
- Both business and leisure travel are sensitive to political threats such as terrorism or trade embargoes.
- Fare structure is an important factor in attracting business and leisure customers.
- The impact of environmental legislation of fare structures (taxes) and the use of acceptable materials and consumables.
- The impact of environmental awareness on long haul aspirations of certain passenger groups.
- The commercial aircraft field will begin to dominate aerospace activity as a result of new airliners designed to meet challenging environmental legislation.

4.2.7 Politics

Local and international politics play an important role in both military and commercial aircraft sales and operations. Apart from the political acceptability of the product, the nature of technology and its dissemination between countries may also be affected. Typical considerations include the following:

- The political situation in the country of origin and the customer's country could lead to a situation in which trade embargoes could affect export potential.
- Political situations and trade embargoes can also affect the transfer of technology or materials between countries.
- Changes in the world economic situation may significantly affect defence budgets for military products.
- Changes in the world economic situation may have a severe impact on both business and leisure travel, causing fluctuating demands which affect aircraft range and size.
- Environmental legislation will demand 'greener' solutions to aircraft design and economic operations.
- Environmental taxes on air travel may reduce demand, as may environmental awareness as people change their leisure travel habits to reflect their concern for the environment.

4.2.8 Technology

Technology is a key driver in aircraft projects in which technology insertion has long been a key to sustaining the aerospace industry. There is a delicate balance to be struck between technology that is appropriate to improving performance whilst minimising obsolescence, and that which poses too high a development and maturity risk. Modern electronic technology is advancing at a rate that is dictated by the relatively unstructured and fast moving information technology (IT)/personal computer (PC) industry. This means that many generations of technological advance are seen in any one single aircraft project lifecycle. Furthermore, rather than being the principal driver of electronic component technology, as it was 30 years ago, the aerospace industry now follows the trends established in the IT, telecommunications and leisure/games development industries.

It is important to note that technological advance does not merely provide a solution to technical issues. It also demands more experts and specialists, and generates a complementary demand for administrative, regulatory, and approval processes.

The adoption of particular technologies needs to take into account the following:

- Technology must be available and affordable within the known project time-scales – it must be realistic.
- If technology needs to be developed specifically for a project, then what are the risks of cost or failure to demonstrate its fitness for purpose?
- Investment in industrial research and development (R&D) and focussing of key technologies is required.
- Plans must be made for obsolescence to understand its impact and respond in a timely manner to maintain product currency and continuing support throughout the lifecycle.
- It must be recognised that projects making use of electronic products and technologies with very rapid product cycles, often driven by commercial markets, will be subject to component availability and price variation in the market place.

4.2.9 Global Economy

The global economy has an influence on all of the above as it impacts on local and overseas economies with some specific effects:

- investor confidence
- cost of loans
- travel and vacation markets and airline stability
- defence budget restrictions with potential impact on projects
- employment and availability of skills
- training budget restrictions.

4.3 Design Drivers in the Project Environment

These are drivers closely aligned to the early definition phases in assessing the requirements to be met, the standards to be applied, and the resources required to complete the project within cost, time, and performance limitations. These are important aspects in the planning of the project. Some typical drivers are shown in Figure 4.4 and are described below.

4.3.1 Standards and Regulations

The design of an aircraft and its systems is subject to many rigours and must be performed in accordance with standards and regulations. These standards and regulations have been established by the aerospace industry over many years to impose a measure of consistency and visibility in the design process. However, the exact standards that apply to specific projects will vary from customer to customer, many being determined by national requirements. For example there are clear US, UK, French, and Scandinavian specific standards,

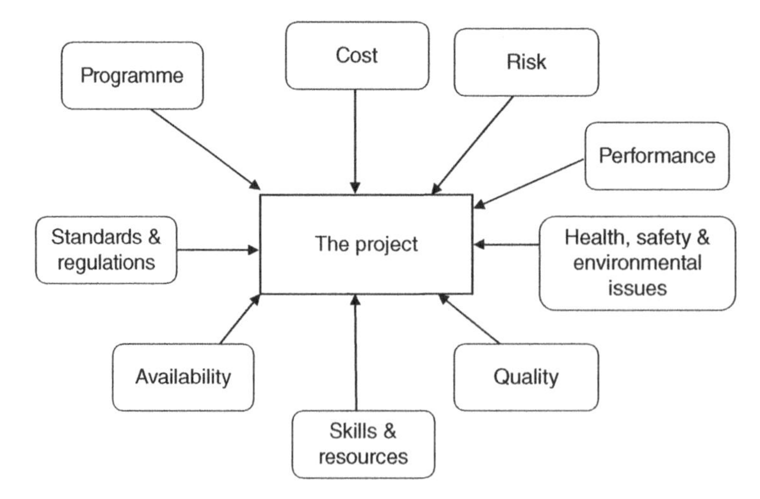

Figure 4.4 Design drivers in the project environment.

although many military aircraft designed for the North Atlantic Treaty Organisation (NATO) tend to rely heavily on US-derived standards, and commercial aircraft around Federal Aviation Agency (FAA) and other standards.

Standards tend to be sponsored, developed, issued, and maintained by a number of widely recognised agencies, for example:

- Society of Automotive Engineers (SAE)
- FAA
- Joint Airworthiness Authority (JAA), now the European Aviation Safety Administration (EASA)
- Air Transport Association (ATA), now A4A
- Radio Technical Committee Association (RTCA).

These agencies will provide information in the form of regulations, advisory information, and design guidelines whereby aircraft and system designers may satisfy mandatory requirements. Typical considerations to be applied when using standards include the following:

- The customer will usually specify the standards they wish to be applied.
- Some standards exist to be applied to the letter whilst others offer guidelines and advice.
- Such guidelines can be used to generate a project-specific specification or plan.
- The issue of a standard used for a project needs to be recorded in order to track changes to that standard during the project lifecycle and any potential impact on the design.

4.3.2 Availability

Customers expect a reasonable number of aircraft in their fleet to be available at all times to enable their operation with minimum disruption to military missions or airline schedules. This operational availability depends on a number of factors, including reliability, aircraft in routine maintenance, aircraft in for repair, and aircraft already in service. It can be expressed in numerical terms and is used to determine the size of the fleet required and the type of aircraft required to perform a role, as well as to establish targets for reliability and system integrity. Typical considerations include:

- the cost of providing accommodation or alternative travel for delayed customers
- the impact of customer dissatisfaction on future business
- the cost of unscheduled maintenance remote from the main operating base
- failure to complete military missions and consequent impact on mission success rates.

4.3.3 Cost

The amount of money invested in a project depends on the expectation of a return on that investment. The costs accrued in market assessment, bidding, and design are non-recurring costs (NRC); the costs of manufacture or series production and support are recurring costs. The business must decide how it apportions or amortises the cost of marketing and research and development, and how it recovers these costs from the price that it charges customers.

Stringent cost controls must be applied within the business to ensure that financial targets are met. Typical considerations include the following:

- All work must be described in a statement of work (SOW) and must be formally estimated against a schedule to ensure that time and costs are correctly apportioned.
- Costs must be constantly monitored against designated achievement points or milestones.
- Change, re-work, and errors lead to unnecessary expenditure, increase cost, and reduce profit margins, especially at later stages in the lifecycle.
- Thorough analysis, checking, and testing of the design in the very early stages of the lifecycle is a valuable investment.
- The adoption of a formal review process that includes scrutiny of costs and schedule is vital.

4.3.4 Programme

The project will be defined to be started and completed within specific time constraints and major project targets will be set to measure achievement. This represents a top-level programme (or schedule, in US parlance) from which lower-level programmes can be constructed. Typical considerations include:

- the number of activities in the programme to be separately identified
- major milestones – payment, review, phase completion
- dependencies between stakeholders
- critical paths.
- risk.

4.3.5 Performance

The customer will define the programme performance parameters that are expected in meeting the stated requirements. These requirements must be converted into a specification of performance parameters and tolerances for the design teams to flow down into their designs and for the test teams to assess during trials. Performance points will become part of the contract and their demonstrated achievement will largely determine the success or failure of a project. Typical considerations include the following:

- Are the customer's expectations achievable with available technology?
- How is each performance point to be demonstrated – by inspection, analysis, modelling, test or field demonstration?
- To what extent can system performance, not simply function, be modelled to reduce costly testing?

4.3.6 Skills and Resources

The availability of people with the right types of skills, training, and experience can have a major impact on what type of work can be done within a project. It is not uncommon for projects to fail because the appropriate resources were not available. The technical content of the programme must be balanced against the demands of the time-scale to ensure that the contract can be fulfilled. Typical considerations include the following:

- What skills are available?
- Are there any scarce skills?
- Are there any obsolescent skills?
- What training can be put in place to ensure that appropriate skills are available?
- Does work need to be subcontracted?
- Can the right types and numbers of people be made available at the right times in the programme?

4.3.7 Health, Safety, and Environmental Issues

Health and safety must take into account the needs of all those people associated with the project – the staff designing, building, and managing the project, as well as the staff and people involved in its operational use, such as crew, maintainers, and passengers. There are statutory regulations directing organisations to provide a duty of care to all employees and users to which the organisation must abide, quite apart from its moral responsibilities.

Some of the environmental aspects of aircraft have attracted sufficient media and public attention that companies are morally obliged to adhere to policies that help to reduce environmental impact. They are also legally and contractually obliged to adhere to regulations and standards enforcing certain restricting criteria. Typical considerations include:

- consultation on and adherence to health, safety, and environmental regulations – the customer will expect this
- providing a safe condition of offices, plant, and premises
- provision of safe working processes
- consideration and notification of new and potentially hazardous materials, treatments, and finishes
- discharge of pollutants into the local environment
- disposal of waste materials
- recycling policies
- public awareness of environmental issues, leading to increasing regulation of noise, emissions, pollutants, and use of materials
- consideration of international protocols as well as national agreements, particularly for products aimed at a world-wide market
- growing concern about the impact of emissions on the environment, for example the impact of fluoro-carbons on the ozone layer (Montreal Protocol 1987).
- design of the product to minimise noise, energy consumption, visible and invisible emissions, and disturbance to the environment
- reducing environmental impact during normal operation and whilst being serviced
- the constant drive to achieve fuel-economic operations for cost savings as well as on environmental grounds.

The environment has become a major factor in the aircraft industry and the debate on this issue has grown steadily over the years, but it is clear that there is no universal agreement.

Does the aerospace industry in general pollute the environment to a detrimental extent and if so by what mechanisms? More importantly can the effect be measured, the impact predicted with certainty, and what can be done about it? This debate has been conducted in the press, in textbooks, in the political arena, and in journals. In so doing a set of polarised views has emerged, and it is difficult to separate scientific and popular viewpoints, objective and vested opinions, serious and provocative statements.

The range of topics is wide, covering innovative aircraft and engine designs, aerodynamics and propulsion, aircraft operations and air-traffic management, alternative fuels, lightweight high-strength materials, onboard auxiliary power units, noise reduction technologies, and non-conventional aircraft (electric, solar, hydrogen).

4.3.8 Risk

A business must continuously assess the risk to the successful completion of a project. Risk assessments will start very early in the lifecycle to establish a complete database or catalogue of risks and their impact on project performance. The probability of risks arising in varying combinations can be assessed using risk analysis tools based on a statistical analysis or modelling techniques, e.g. Monte Carlo analysis. Many projects use a living risk log or register in which risks are prioritised on the basis of probability of occurrence and the severity of impact on the programme should the risk arise. This log will be carefully controlled and reviewed on a regular basis. Typical considerations include:

- technology risk in terms of availability in the right time frame
- the probability of failure of technology
- a mitigation plan for each risk identified
- quantification of each risk in terms of the impact on performance, schedule, and cost.

4.4 Design Drivers in the Product Environment

These design drivers are closely aligned to the design of the product and its sub-systems and components. They are especially relevant to the design phase of the lifecycle. Some typical design drivers are shown in Figure 4.5 and are described below.

4.4.1 Functional Performance

To meet the customer's requirement a large number of functions will have to be performed. Some of these will be performed by the crew, but the majority will be performed by the systems, either in response to crew demands or completely automatically. In the case of unmanned aircraft the crew will be remote from the vehicle with commands received by data link.

The systems engineer must determine what functions need to be performed and how they should be allocated to individual sub-systems and items of equipment. The function must be defined in terms that enable its performance to be measured. These terms include:

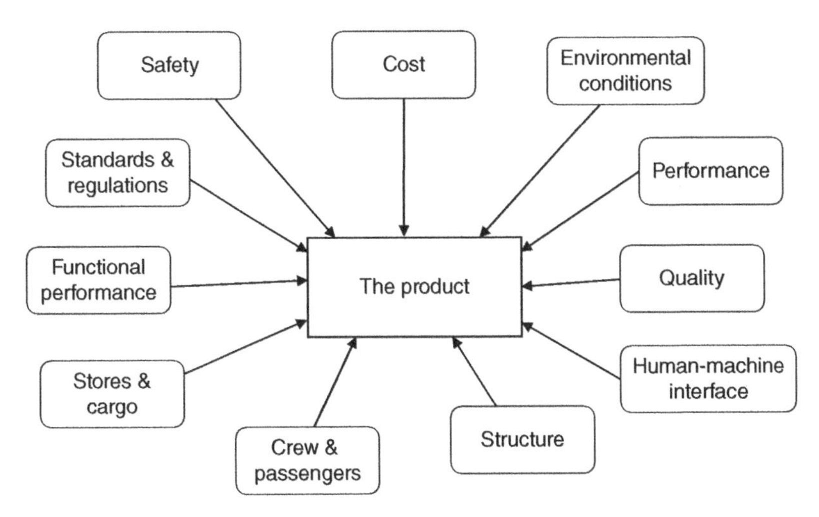

Figure 4.5 Design drivers in the product environment.

- requirements description in an unambiguous language statement using appropriate requirements techniques and tools
- time or duration of events
- repetition rate or data update rate
- data requirements from sensors and other systems
- data requirements to effectors and other systems
- data accuracy, range, and scaling.

4.4.2 Human–Machine Interface

The modern aircraft is a complex machine and it is vital that the interface between the machine and the operator is designed for maximum effectiveness so that the aircraft can be operated safely at all times. The range of operations covers normal stress-free flying, high workload combat flying, and high workload, high-stress flight under emergency conditions. In all these cases the crew must work in a clear, well-designed, and intuitive environment. Typical considerations include:

- definition of the interface between the operator and the aircraft
- interfaces between aircrew and the cockpit controls and displays:
  - reach, feel, forces, damping, tactile recognition, range of crew sizes
  - colour, audio, rates, display sizes, fonts, character recognition, and other presentational issues
- interfaces between the maintainer and the aircraft:
  - equipment mass, handling, health, and safety
  - ease of access for removal, replacement or adjustment
  - ease of access for connection of test equipment
  - ingress and egress requirements
- use of flying suits, immersion suits, chemical/biological protection.

4.4.3 Crew and Passengers

The crew and passengers must be housed safely and comfortably. The flight crew need to be comfortable to endure long duration flights without loss of vigilance, and the passengers are paying for the ride and may choose to travel by another carrier if they dislike the aircraft. This has led to airlines providing higher levels of comfort and service in leisure travel flights. Typical considerations include:

- seating and restraints
- cabin conditioning and air quality
- general and personal lighting
- baggage stowage space
- in-flight entertainment/business systems provision
- food preparation areas such as galleys
- toilet and washroom areas
- exits and safety labelling/lighting
- emergency equipment – smoke hoods, escape chutes, rafts, life vests
- emergency oxygen
- ejection seats.

In contrast to some of these considerations, a number of 'budget' or low-cost airline operators have dispensed with in-flight catering and sacrificed leg room to reduce ticket prices.

4.4.4 Stores and Cargo

Many military aircraft carry external stores – a term that covers items such as weapons, fuel tanks, reconnaissance pods, and target drogues. These stores have an impact on performance because of their mass and drag, but the physical size of many aircraft prevents internal carriage. The commercial passenger aircraft carries stores internally in the form of baggage, mail, or commercial freight items, often housed in standardised containers or pallets. Cargo aircraft, both military and commercial, carry vehicles or containers. Typical considerations are:

- attachments or launcher hard points for external stores
- jettison capability
- impact on performance (mass and drag)
- ordnance safety
- cargo container standards and interfaces
- restraints
- baggage-handling systems
- door/ramp access
- ground handling equipment
- attention to safe passage of live animals such as pets, horses, and legitimate animal trade
- associated ground facilities such as veterinary care and quarantine.

4.4.5 Structure

The installation of systems, stores, and sensors has an impact on the aircraft structure. Anything that makes a hole in the structure compromises its integrity in some way and it is important that systems engineers and structure designers understand what needs to be installed and what design constraints exist in installing some items of equipment. Typical considerations include:

- attachments for external stores
- internal equipment mounting
- hatches for access
- gaskets for pressure hull sealing for antennas, windows, external pipe connections, etc.
- holes in internal structure for harnesses, connectors, pipes, ducts, etc.
- bonding and earthing.

4.4.6 Safety

Safety is of paramount importance – that of passengers, aircrew, ground crew, and the overflown population. Systems are designed according to processes which make safety endemic in the design. Independent hazard and safety analyses are conducted at system, equipment hardware, and software levels, and the overall system design is scrutinised to eliminate errors or failure modes that affect flight safety. Typical factors include:

- elimination of single events leading to a catastrophic failure
- elimination of common mode failures
- incorporation of a robust software design process to ensure that there are no events that may cause the system to perform in an unsafe manner.

4.4.7 Quality

A single and powerful method of ensuring high and consistent quality in the project is to ensure that all design teams are aware of the common standards and processes that apply and are aware of the need to abide by them. A process of rigorous checking of documents and regular review of engineering documents at each phase of the lifecycle ensures that independent observers have the opportunity to review and improve the design process in a constructive manner. A quality management system should be in place defining the organisation, responsibilities, processes, and procedures used as well as a regular review policy.

4.4.8 Environmental Conditions

Customers will define the areas of the world in which they expect to operate the aircraft, and this will largely determine the climatic conditions to which the aircraft will be exposed. However, to design specifically for that operating environment may restrict sales to other areas of the world and hence it may be cost effective to design for world-wide operations. The conditions that aircraft and systems must withstand are well understood and there are standards of testing to verify designs under a wide variety of environmental conditions, many of them extremely severe.

The conditions of use of the aircraft will determine the local environment that will affect structure, systems, and inhabitants, introducing such aspects as vibration, shock, temperature, etc.

Combinations of these environmental aspects will be used by the systems engineer to scope his design and test requirements. A handbook or database of such conditions will be of use to systems engineering teams to ensure a consistency of approach within any one project. Typical considerations include:

- the areas of the world in which the aircraft will be used
- the impact of designing for world-wide operations to increase the market
- the impact the conditions of use will have on internal equipment and inhabitants, and translation of this into engineering parameters
- the various environmental conditions that exist for different zones or compartments in the aircraft
- definition of all engineering requirements in a handbook or database.

4.5 Design Drivers in the Product Operating Environment

These are factors that influence the design of the product to ensure that it is able to operate in a defined environment for life. The operating environment is determined by the conditions of use to which the product is put, and the areas of the world (or beyond) in which it has to operate. Some design drivers are shown in Figure 4.6 and are described below.

4.5.1 Heat

Heat is a waste commodity generated by inefficiencies of power sources, by equipment using power, by solar radiation, by crew and passengers, and by friction of air over the

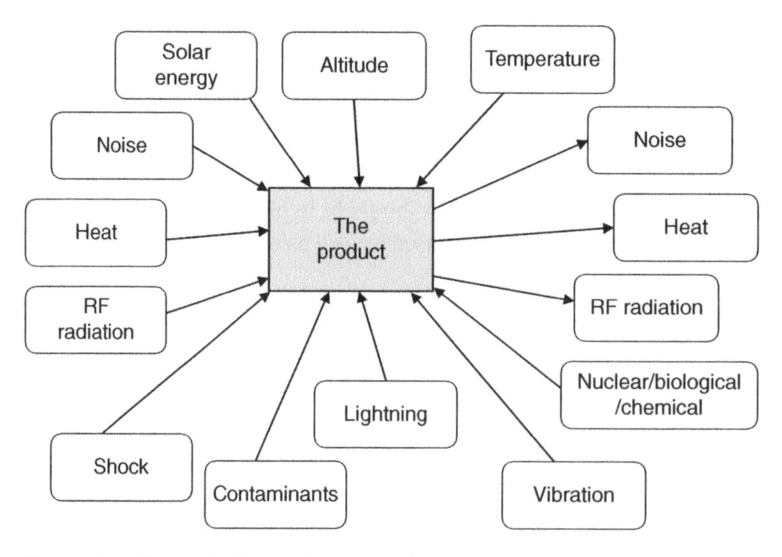

Figure 4.6 Drivers in the product operating environment.

aircraft surface, especially during high-speed flight. Thus, all human and physical occupants of the aircraft are subject to the effects of heat and they themselves radiate heat. These effects range from those affecting the comfort of human occupants to those that cause irreparable damage to components of equipment. Typical considerations include the following:

- If the system or systems component is likely to be affected by heat, then it should not be installed near to a major heat source or it should be provided with cooling.
- The aircraft environmental control system (ECS) can cool equipment using air or a liquid coolant.
- Some systems produce heat in performing their function and must be isolated or insulated from other systems, e.g. engines and high power transmitters.
- Some systems produce heat by performing work in energy transformation, e.g. hydraulics fluid for flight control systems, and this heat needs to be removed.

4.5.2 Noise

Noise is ever present in an aircraft environment. It is produced by the engines or auxiliary power units, by motor-driven units such as fans and motors, and by air flow over the fuselage. It can cause discomfort to passengers and crew, whilst high noise levels external to the aircraft can cause damage. Typical considerations include the following:

- High sound pressures or acoustic noise levels can damage equipment. Installation in areas subject to high noise levels should be avoided. Typical areas are engine bays, external areas subject to engine exhaust or bays likely to be opened in high-speed flight, e.g. bomb bays.
- Equipment can produce noise that is likely to be a nuisance to aircrew, contributing to fatigue and loss of concentration. Examples are fans and pumps or motors installed in the cockpit. Measures must be taken to install equipment so that excessive noise can be avoided and crew efficiency maintained.
- Some noise can be reduced by the use of noise-cancelling headsets for the crew, ear plugs for passengers, and ear defenders for ground crew.
- Health and safety regulations must be met.

4.5.3 RF Radiation

Radio frequencies (RFs) are radiated from equipment and from the aircraft, either deliberately or accidentally. As far as aircraft systems are concerned, RF emissions generally occur in the electromagnetic spectrum from 10 MHz to tens of GHz. Accidental radiation occurs when equipment or wiring is badly installed, or inadequately or incorrectly screened. Deliberate radiation occurs during radio transmissions, navigation equipment transmissions, and operation of radars and other communication equipment. RF radiation can cause interruption or corruption of a system function by affecting system component operation or by corrupting data. Typical considerations include the following:

- Equipment should be protected from the effects of RF radiation by the application of an electromagnetic health (EMH) strategy. This involves the use of signal wire segregation,

screening, bonding, separation of wiring and equipment, and RF sealing of equipment. This will obviate the effects of some of the key electromagnetic effects:
  - electromagnetic interference (EMI) resulting from the effects of local equipment on board the aircraft
  - lightning strike on the structure or in the vicinity of the aircraft
  - high intensity radio frequency (HIRF) from local high-power transmitters such as airfield primary surveillance radar or domestic radio transmitters.
- Radiated transmissions can disclose the presence of an aircraft to enemy forces, which can be used as intelligence or as a means of identifying a target for attack.
- In the military field, analysis of signals by an electronic support measures (ESM) team can provide valuable intelligence about deployment of military assets.
- It is generally acknowledged that signals intelligence is one of the most prolific sources of intelligence during peacetime, periods of tension or conflict. Its contribution to diplomatic or military success can have an effect far outweighing the relatively small investment required to gather and analyse information (Schleher 1999).
- Incorrectly screened secure communications can cause 'leakage' of classified information from the aircraft that can be detected by enemy forces.
- The EMH strategy is also intended to reduce the risk of equipment producing RF emissions from local on-board equipment and suppliers must be fully aware of the need to demonstrate compliance.
- Each project will have an EMH plan defining the strategy to be adopted for that project.
- There is a risk of mutual interference between transmitters and receivers. Care must be taken in the design of RF systems to prevent this.

4.5.4 Solar Energy

Sunlight will impinge on the surface of the aircraft and will enter through windows and canopies, thereby exposing some parts of the interior. Prolonged exposure at high altitudes to unfiltered ultra-violet (UV) and infra-red (IR) radiation is likely to damage some materials. UV exposure is also experienced when parked for long periods on the tarmac. Typical considerations include the following:

- The UV and IR content of solar radiation can cause damage to plastic materials, such as discolouration, cracking, and brittleness. This can affect interior furnishings such as display bezels and switch/knob handles.
- The items most affected are those situated on the aircraft outer skin, e.g. antennas, where high-altitude, long-duration exposure is experienced.
- Cockpit items are also vulnerable if they are in direct sunlight in flight or whilst the aircraft is parked – cockpit temperatures have been known to reach over 100 °C in some parts of the world. All such items must be designed to withstand such effects and must be tested.
- Glare and reflection will affect crew visual performance and may adversely affect display visibility.

4.5.5 Altitude

Many aircraft operate between sea level and 40 000 ft, Concorde routinely operated up to 50 000 ft, and some military aircraft routinely operate well above this altitude. Missiles and spacecraft will enter the stratosphere or operate in a vacuum. The interior of conventional aircraft is maintained at a pressure that is tolerable to crew and passengers so that there is a pressure differential across the aircraft skin. Any disturbance of this pressure differential may cause a rapid or explosive decompression, which is a potential cause of damage. Typical considerations include the following:

- The aircraft can routinely operate at altitudes up to 50 000 ft and sometimes beyond. This will expose inhabitants and equipment to ionising radiation from cosmic radiation, which can affect human health and can temporarily change the state of dense flash memory cells. This may become more of an issue as memories grow denser. Crew hours must be monitored and recorded to avoid exceeding regulatory radiation dose rates.
- Equipment and aircrew must be capable of operating at pressures representative of altitudes from sea level to 50 000 ft.
- Differential pressures can affect the performance of sealed components.
- Although the cockpit, cabin, and equipment bays are normally pressurised, rapid or sudden decompression (rapid rates of change of pressure) can lead to component failures. This can occur as a result of damage, failed seals, canopy loss or battle damage.

4.5.6 Temperature

All aircraft are expected to operate in a wide range of temperatures from arctic to desert conditions. The effect of ambient temperature on internal temperatures can affect equipment, especially when powered. The effect of external ambient temperatures is a key design consideration. Typical considerations include the following:

- The aircraft will be expected to operate in extremes of temperature ranging from −55 to +90 °C. The range depends on the part of the world in which the product is expected to be deployed. In some cases the environment may be even more severe after a hot or cold soak. In some parts of the world −70 °C is not uncommon.
- An aircraft is expected to operate in world-wide conditions and to experience temperature extremes, and in some cases experience gross deviations during normal operating regimes between different climatic zones, e.g. northern Canada, Iceland, Norway, Saudi Arabia, Arizona.
- It may be economical to design and develop the system for world-wide operation to increase market potential, avoiding re-design or re-testing.
- The aircraft may be parked for extended periods of time in hot or cold conditions (hot or cold soak) or subject to direct sunlight. Key equipment maybe expected to operate immediately in such conditions, but not necessarily entire systems.
- Hot and cold soak have an impact on refuelling, especially on the capacity of wing tanks, with potential for fuel spillage.
- In very cold regions the type of fuel will be selected by local providers and may affect fuel calculations.

4.5.7 Contaminants, and Destructive and Hazardous Substances

The aircraft exterior and interior surfaces, and the installed equipment can be contaminated by substances that in a normal environment can cause corrosion damage or malfunction. Contamination may occur by direct means such as spillage, leakage or spray, or indirectly by being handled with contaminated hands. Equipment and furnishings must be specified and designed to minimise the effects of contamination. Typical contaminants to be taken into account are:

- fuel
- oils and greases
- de-icing fluid
- windscreen wash fluid
- hydraulic fluid
- beverages – coffee, tea, soft drinks (some are extremely corrosive)
- ice
- rain and moisture ingress
- sand and dust
- fungus.

4.5.8 Lightning

Most aircraft are expected to operate in all weather conditions, and it may not be possible to schedule flights or routes to avoid lightning conditions. Measures must be taken to limit the impact of lightning strike and associated structural damage and induced electrical effects. Typical considerations include the following:

- Lightning can be expected at all and any times of the year.
- Lightning strike can damage structure locally and induce very high transient voltages in aircraft cables.
- Lightning-induced effects can destroy entire systems.
- All equipment must be bonded and on carbon composite surfaces special foil inlay is used to provide a conductive path.
- Aircraft constructed from multiple materials (aluminium, composite, titanium, plastic) require care to ensure a consistent bond.
- Equipment and complete aircraft are lightning strike tested.

4.5.9 Nuclear, Biological, and Chemical Contamination

Military aircraft in particular may enter a theatre of combat in which deliberate contamination by chemical agents is a real possibility. The aircraft and its equipment must survive such contamination and the decontamination process. Typical considerations include the following:

- *Biological agents*: living micro-organisms or toxins delivered by bomb, missile, or spray device. Contamination of the aircraft and its equipment can harm air and ground crew.

- *Chemical agents*: compounds which, when suitably disseminated, produce incapacitating, damaging or lethal effects delivered by bomb, missile or spray device. Contamination of the aircraft and its equipment can harm air and ground crew.
- *Nuclear effects*: blast, radiation, and electro-magnetic pulse, which can damage aircraft, equipment, communications, and personnel.
- Ionising radiation will damage electronic components.

4.5.10 Vibration

All equipment is subject to vibration coupled into the mountings from the airframe. This vibration can, in turn, be coupled into circuit cards and components, leading to fractures of wiring, connector pins, and circuit boards. The effects are more severe if resonant modes occur. Typical considerations are:

- vibration encountered in normal operation:
  - 3-Axis vibration that can be randomly or continuously applied
  - sinusoidal vibration at fixed frequencies and directions
  - specific vibration regimes as determined by the aircraft zone in which equipment is installed
- gunfire vibration in fighter aircraft and attack helicopters
- anti-vibration mountings, which can be used in certain installations
- flexible equipment racks.

4.5.11 Shock

Violent or sharp shock can cause equipment and components to become detached from their mountings. They may then become a loose article hazard capable of causing secondary damage to other items of equipment or to occupants. Shock may also cause internal components of equipment to become detached, leading to malfunction. Typical causes of shock are:

- violent aircraft manoeuvres
- heavy landings
- aborted take-off
- crash conditions
- accidental drop during manual handling.

4.6 Interfaces with the Sub-system Environment

Drivers in the sub-system environment affect equipment and components of the sub-system directly. These are drivers that impact on interfaces: equipment to equipment, equipment to structure, and equipment to crew. These interfaces give rise to many of the derived requirements described earlier. Some typical design drivers are shown in Figure 4.7 and are described below.

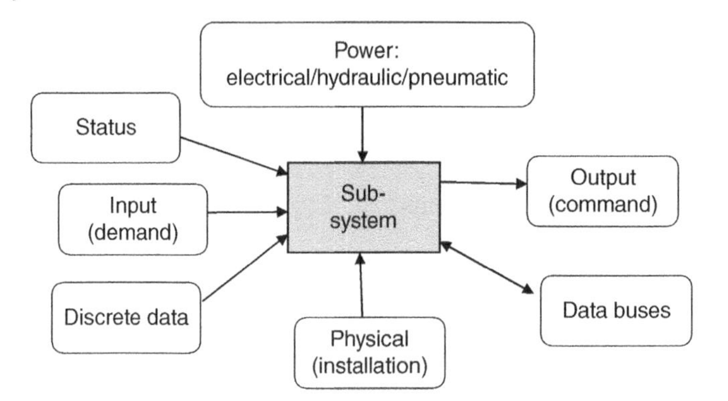

Figure 4.7 Design drivers in the sub-system environment.

4.6.1 Physical Interfaces

These are interfaces that affect the installation of equipment in the aircraft. They are important to the designers and directly affect manufacturing and assembly. Any errors at this stage may creep into quantity manufacture and may be repeated many times. Typical considerations include:

- mass and centre of gravity (cg) or centre of mass
- dimensions and aspect ratio of equipment with respect to the space envelope available within the aircraft zones or mounting trays
- hold-down mechanism/attachment
- connector types, numbers, and style
- cable/pipe connections
- need for access for adjustment/repair/removal of equipment and components
- orientation of equipment or components
- excrescences/ducts/orifices.

4.6.2 Power Interfaces

A system component will be connected to a power source to provide energy for its operation and for conversion of the energy from one form to another. Typical considerations include:

- electrical power: AC or DC power with the appropriate current rating and protection
- selection of a suitable power protection device
- bonding and earthing: this is an important consideration for good electrical and signal screening and to ensure that all parts of the structure, metallic and non-metallic, are bonded
- protection and segregation of hardware
- correct wire size and insulation material
- wire/bundle sizes and connection
- hydraulic power: pressure, fluid, pipe connection type, properties of fluids, temperature of operation

- pressure/pressure loss/flow
- pneumatic power: pressure, temperature, pipe connection type.

4.6.3 Data Communication Interfaces

Data is often sent between systems by means of a data bus or data link in serial or parallel electrical formats. The interface with the data bus will usually be made by means of a control unit to which a number of components are connected. The data bus is efficient in reducing the number of wires required to transfer data. Some important points to be noted include:

- *MIL-STD-1553B (Def Stan 00-18, Stanag 3838)*: two wire transformer coupled bus command/response, military applications, type specific data formats, message formats using 1 Mbit/s transmission rate.
- *ARINC 429*: well-established commercial avionic bus standard with defined protocols, message formats, and data names/tags. Usual use is 110 kbits/s.
- *ARINC 629*: civil standard using 2 Mbit/s data rates. Similar technology to MIL-STD-1553.
- Modern data bus types including IEEE-1498, CANBus, ARINC 664, etc.
- Data bus transfer must be such that delays are not introduced into control loops (data latency) sufficient to degrade system performance.
- The integrity of the data bus architecture must be appropriate for the purpose for which it is provided.

4.6.4 Input/Output Interfaces

The system functions are often performed in control units that require signals from system components to measure characteristics such as flow, movement, pressure or temperature. The control unit will issue demand signals to cause transfer of energy into motion on some components or to provide signals to the crew.

- Discrete input: switch (on/off) type input varying from one fixed state to another, e.g. 0 to 28 V to signal landing gear UP to DOWN or to signal landing lights ON or OFF.
- Analogue input: continuously variable analogue signal to demand a variation in a condition, e.g. throttle demand for increased or decreased thrust, demand for a change in aircraft altitude.
- Electrical output to drive an actuator or motor, e.g. fuel valve, pump.
- Hydraulic output to move a surface or door, e.g. elevator, bomb bay door.

4.6.5 Status/Discrete Data

- Status data indicates to a system component the status of other system components, e.g. operating or failed, and is usually generated as a result of a test. The test may be automatically initiated by the system, a built-in test (BIT), or by the crew.
- Warning information generated to inform the crew of a specific condition or failure, e.g. LOW OIL PRESSURE, FIRE.
- Status condition or a discrete signal type is usually in the form of a two-state signal: ON or OFF.

4.7 Obsolescence

4.7.1 Introduction

Obsolescence has long been an issue and the pace of technology is such that it manifests itself throughout the lifecycle. Techniques for management of obsolescence throughout the lifecycle have been described (Aerostrategy Commentary 2010).

In a complex aerospace product with long development timescales and long durations of in-service use there are many opportunities for obsolescence to appear. Apart from the obvious examples of obsolete technology and component parts, obsolescence is often a major factor in the initial operational requirement or, indeed, the design and manufacturing processes, the design toolset, organisational skills, support, and operations as well as the information processing system that supports this.

A failure to recognise where obsolescence can exert an influence can lead to major costs in the project lifecycle as a result of a failure to meet the original specification or a failure to respond to competition in the marketplace.

Obsolescence represents a major and increasing risk to the aerospace and defence business in terms of capability development, sustainment, equipment availability, and through-life support services. It is strongly related to technology, especially the way in which modern technology is maturing rapidly. Ten or twenty years is roughly the time-scale associated with projects that have recently entered service for the same reason: an early commitment to a technology in order to meet the delivery date. It is largely apparent in the selection of electronic components that are used in avionic systems. Lack of continuing supply leads to the prospect of replacing equipment with new types or continuing in service with out-of-date equipment and 'soldiering on' until replacement is inevitable. This is a situation that leads to costly maintenance and the potential for reducing aircraft availability to unacceptable levels.

It does not just affect the aircraft though, major infrastructure projects are similarly affected. This means that facilities for commercial aircraft are under stress, and military aircraft basing and support may also be inadequate for long-term sustainment planning.

Modern aircraft design and manufacture represents the integration of complex structures, systems, and equipment. Many aircraft systems are defined as safety critical and sustainment dominated. These systems are designed and manufactured to stringent aerospace requirements and subjected to rigorous qualification and certification programmes. They have long development and manufacturing lifecycles and are expected to remain operational, and be maintained and supported for periods of 40–50 years.

Obsolescence is defined as the discontinuation or sudden unavailability of a component part from the original source of supply due to the cessation of production, non-availability of raw materials, impacts of legislation or withdrawal of product support services.

Obsolescence can arise during all stages of the product lifecycle, from concept through to disposal, and all forms of systems, equipment (including ground equipment, hardware, and software), and resources (people, tooling, process, material, knowledge, environment, and facilities) are placed at risk from the effects of unmanaged obsolescence.

Component obsolescence is the most obvious problem, especially in the electronics or avionics sectors, due to the speed at which technology becomes outdated in terms of functionality, performance, and availability.

Components that are presently being selected for new products have a high probability of being declared obsolete when, or before, the system enters service and being unavailable when subsequently required.

4.7.2 The Threat of Obsolescence in the Product Lifecycle

Obsolescence has long been recognised in technical areas of expertise as posing a significant threat. Charles Babbage in 1842 tried to convince Prime Minister Robert Peel to provide government funding for an analytical engine to replace the difference engine that he never completed: *The general fact of machinery being superseded in several of our great branches of manufacture after a few years is perfectly well known* (Essinger 2012).

Much later, in the beginnings of the space age, Dave Scott, Commander of Apollo 15, was speaking at the Natural History Museum, Oxford. Answering a question on the (President) Bush plans for Mars missions, he said: *The problem is where do you freeze the technology? ... By the time you're ready to go, the technology you're using is ten or twenty years old. It's a difficult problem and I don't know how you deal with it* (Martin 2005). The same is largely true of commercial airliners.

Modern authoritative documents suffer from similar issues of currency and freeze point. *For the Ordnance Survey this meant a continual stream of revisions and new maps on a variety of scales. The job was never done, and would never be done. No sooner had an OS surveyor put up their feet to toast the completion of a new series, than the work would already be out of date* (Garfield 2012). Similar issues are relevant to requirements, specifications, and standards.

Torenbeek identified the same problem in aerospace documents: *...design handbooks are essentially based on existing or even obsolete technology and may produce unrealistic results when applied to future advanced aircraft design projects* (Torenbeek 2013). From these excerpts it is clear to see that obsolescence is going to be an issue, no matter how careful the design path, and that decisions on the point at which technology and design are 'frozen' need to be thoroughly understood and risk evaluated from that point.

Obsolescence is also a major factor in the design, development, manufacture, and qualification of the aircraft and its systems. Key factors here are skilled resources, design tools, and host machines, and processes that become obsolescent during the long development lifecycle and long in-service life, especially in the rapid turnover of modern electronic systems.

Figure 3.3 in Chapter 3 showed an example of a product lifecycle illustrating some of the time-scales involved. Both military and commercial aircraft types have long in-service lives with their original buyers, and this is often extended by second purchasers, by leasing or by modification for alternative roles. Fifty years is not uncommon for some types for the interval between original design decisions and retirement. This is longer than the working life of many people, and not surprisingly obsolescence of people skills is common as new generations of engineers join the project and people move on.

Figure 3.3 also shows some factors that influence the onset of obsolescence in the main product. A major factor is the change of emphasis from bespoke (and costly) aircraft components and equipment towards components developed for the commercial and domestic markets, where large turnover drives prices down. To satisfy the domestic demand these markets have evolved very short development time-scales and rapidly changing technology lifecycles that result in the acceptance of an almost fashionable obsolescence for short-lived products that does not fit easily with the extending lifetimes of aircraft products.

Figure 4.8 illustrates where opportunities exist in the aircraft system lifecycle for obsolescence to strike. It is commonly assumed that materials, components, and technology are the prime causal factors for obsolescence and ageing. However, there are many factors that exert influences on any complex aerospace project that lead to obsolescence and a holistic view of the system is useful in understanding the total risk.

With reference to Figure 4.8 these influences include the following items, which are discussed further below:

A) requirements specification
B) people
C) regulations
D) design, development, and manufacture
E) the supply chain.

4.7.2.1 Requirements Specification

The customer forms a requirement from their emerging future needs and usually expresses this in the form of a competitive tender to the aerospace industry. In the military field the requirement will be based on an understanding of known or suspected threats in the light of the existing inventory of weapon systems. It is important to strike a balance between the

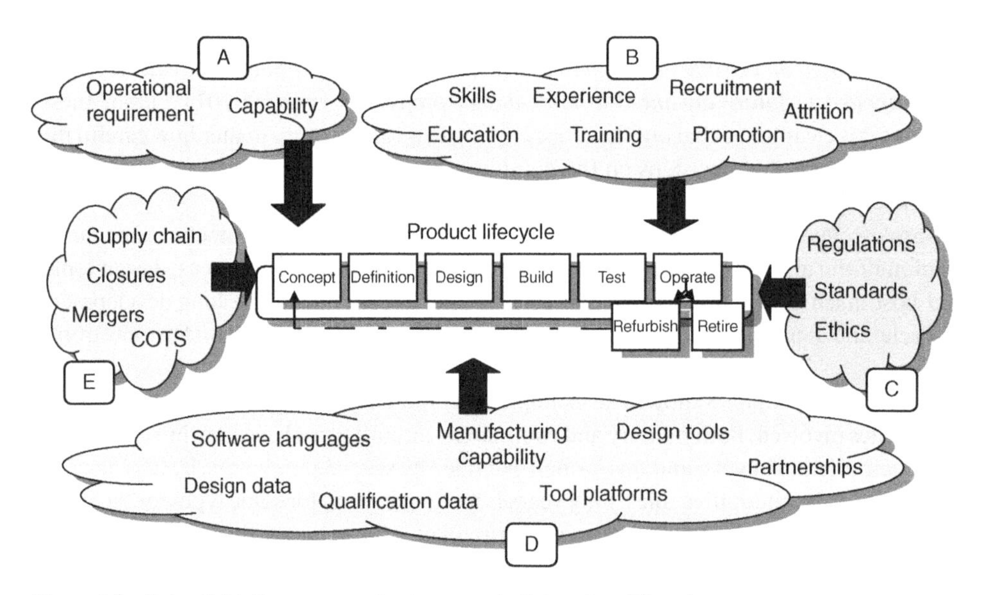

Figure 4.8 Potential influences on obsolescence in the system lifecycle.

introduction of a new product and the withdrawal of existing assets to maintain the correct balance of capabilities. It is also important to ensure that there is a low-risk introduction of new technology. This sounds simple, but it has in the past led to lengthy programmes to develop a requirement, to launch it, to develop the product, and to introduce it into service – in some extreme cases this can take 20 years or more.

Hence it should be no surprise that the customer's requirement itself can become obsolete. The most recent example is the withdrawal of the threat from the former Soviet Union which has led to the entry into service of weapon systems designed for a mechanised war in Europe featuring air superiority aircraft, aircraft carriers, deep strike weapons, nuclear armed submarines, and strategic missile defence systems. Not many of these are suited to the so-called 'asymmetric warfare' confronting many nations today, where war is waged in urban situations or in close proximity to civilian populations.

As a result nations are dealing with an obsolete capability forced on them by an obsolete threat, and an arms inventory which costs a lot of money to maintain in a state of operational readiness.

Obsolescence in the military world also arises as the result of developments in the inventory of nations posing a military threat. Radar systems can become obsolete because of improvements in stealth technology; some weapons may become obsolete because of improvements in countermeasures technology or changes to enemy tactics. It is always wise to include the enemy in stakeholder analysis of military systems because that is where major threats to defence capability emerge.

In commercial aviation the airline requirement is driven by business issues such as trends in leisure and business travel. The current trend is towards large, trans-continental aircraft, which have a major impact on the design of airports and passenger-handling facilities. Fashion has seen a decline in regional and supersonic transport systems, which, in the UK at least, has seen a decline in the commercial aircraft industry. Budget airlines are tempting customers with low prices and 'no frills' travel, a move which creates new routes but also makes other routes obsolete.

Public and political pressures are leading to demands for more economic and less polluting aircraft in a bid to reduce global environmental damage. This will eventually lead to cleaner aircraft, but the capital cost of existing types will mean that obsolete and polluting aircraft will continue to remain in service in some form. As a result of these pressures obsolescence is likely to affect certain routes, aircraft types, fuels and lubricants, and airports.

4.7.2.2 People

The main influence that people exert is in the capability of the workforce to complete the project to meet the original requirement. Given the long development times for modern projects there are always going to be issues with maintaining appropriate skills and experience throughout the lifecycle in which the application of skills varies according to the tasks to be performed. People with the skills and experience to provide the original concept definition – innovation, flair, original thinking, a grasp of concepts, etc. – are needed for a relatively short time, and they do not necessarily possess the appropriate skills to turn that concept into reality – detailed design, product definition, understanding of standards, etc. It is also true that people with detailed design knowledge and product definition skills are not necessarily good at concept design.

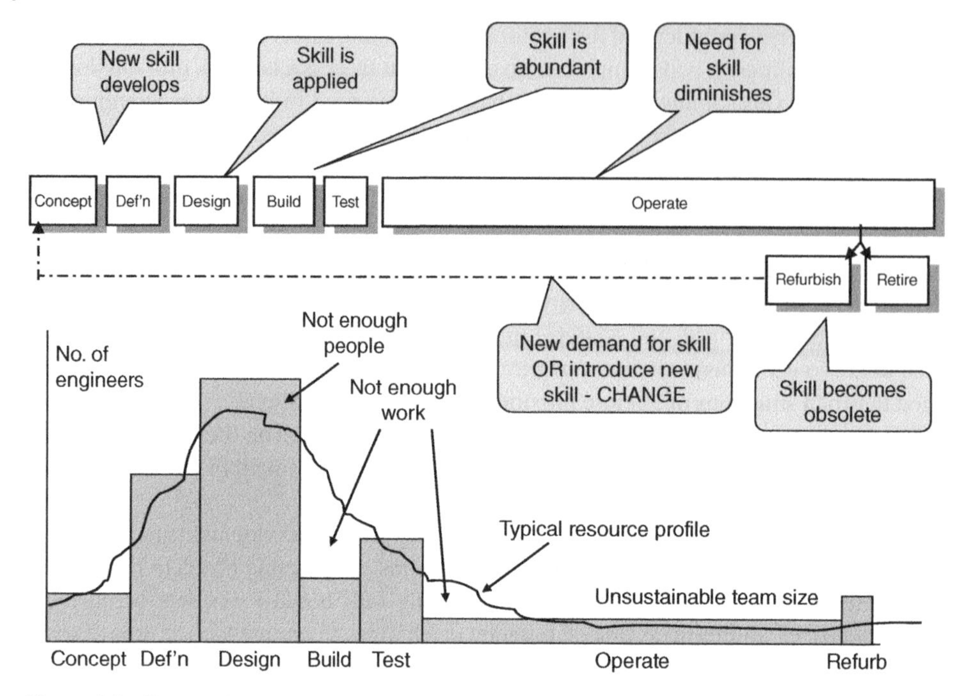

Figure 4.9 The varying demand for skills in the lifecycle.

Management of skills and education throughout the project programme is essential to maintain the correct workforce balance and to control costs of labour and training, and importantly to form and run a good team environment. An illustration of the variation in demand for skills in the lifecycle is shown in Figure 4.9.

This figure shows that a new project in its early stages often demands new skills which the workforce acquires by suitable training or by the importing of new staff with the appropriate skills. As the project develops, the need for increased numbers of staff with these skills increases and remains steady throughout the design and development phases, declining as the product goes into manufacture and into operation. The skills may need to be exercised in order to deal with initial entry into service queries, but will soon cease to be practiced.

It is during this descent into obsolescence that staff movements can arise – from lack of motivation, failure to see a way forward, and a desire to retain their skill. This can lead to staff transfers to other projects, staff leaving the company, and promotions that almost inevitably take people away from engineering and into management.

Particular skills also decline because of obsolescence in that skill generally. An example of this is the use of Ada as the preferred software language for military projects. In the commercial world languages such as C++ began to gain popularity, universities and colleges taught the new language in preference to Ada and the supplier of Ada scaled down its support of compilers. Thus Ada became obsolescent and new military avionics equipment began to be programmed with C++. Users of military systems must continue to support aircraft in service with Ada and somehow retain these skills and maintain compilers. Eventually this will lead to a limit to further expansion of systems capability because of the limitations of Ada and Ada-compatible hardware and software.

4.7.2.3 Regulations

Regulations change in response to technology and to pressure in the technical, commercial, and political environments. This can lead to changes in standards and their applicability, which in turn lead to changes to technical and manufacturing processes and procedures. This can affect the choice of materials for construction of aircraft assemblies (e.g. magnesium alloys) and electronic components (e.g. beryllium), the choice of treatments to protect materials (e.g. cadmium), and the choice of consumables such as fuel (e.g. benzene) or refrigerants (e.g. chloro-fluoro-carbon compounds [CFCs]). Some of these changes are enacted for environmental reasons and some for health and safety reasons. In either case manufacturers must abide by the regulations and demonstrate compliance. Thus many materials become obsolescent during the life of an aircraft and are generally allowed to remain in service whilst being forbidden for use in new projects.

A number of substances and materials are subject to regulatory examination for reasons to do with environmental or health and safety concerns. Over a period of time many substances have been prohibited or restricted in their application to new projects, and now can be used on heritage projects only after declaration of their intended use and the granting of a concession. Such substances include:

- lead solder
- cadmium plating
- cleaning agents such as trichloroethylene
- volatile oil compound (VOC) based paints
- halons in fire suppressant systems
- coolanol in liquid cooling systems
- chloro-fluorocarbon based fluids (CFC) in cooling systems
- yttria powder
- older carbon fibre composite resins
- aluminium magnesium alloys
- copper beryllium alloys
- micarta (dust in machining)
- aluminium lithium.

The engineer must carefully consider any materials before selection to ensure that they are not subject to existing or emerging legislation. These prohibitions and continuing regulatory changes pose a major issue for manufacturers and users alike, who will have to maintain a record of materials use and monitor the health of employees and their environmental obligations. Further consideration must be given to safe disposal at end of life and to containing contamination in the event of an accident or crash.

4.7.2.4 Design, Development, and Manufacture

A contemporary project that has been in service for 50 years is likely to have its design record established in the following media:

- paper drawings – drawing board and stores
- linen – drawing board and stores
- mylar film – drawing board and stores
- microfiche – microfiche/film reader

- computer aided design (CAD) database – workstation
- floppy discs – desk-top computer and operating system dependency
- diskettes – desk-top computer and operating system dependency
- CD ROM – desk-top computer and operating system dependency
- remote memory vaults and cloud storage.

Figure 4.10 shows the periods over which the main materials, tools, and media were used, and approximately when they became obsolescent. This is not to say that some of them are no longer in small-scale use. They will be found in organisations dealing with legacy aircraft, in heritage groups, and in some cases because they are still useful for recording concept studies.

All of these types of media need to be stored until the product is withdrawn from service and for a specified period after that date to support any formal enquiries or crash investigations from post-service owners. This means that the machines which are needed to read the media also need to be stored and maintained. This needs real estate for storage and maintenance costs that must be provided by the designer. With modern systems this means keeping examples of the computing devices used throughout the design and development period as well as different versions of operating system. There may still be a need for refreshing data since there are no long-term guarantees for the stability of magnetic media and recovery of complete data. An alternative is to progressively update the record onto new technology storage systems with the associated risk of cost and transcription errors.

Obsolescence of small-scale memory devices and the transition to memory vaults will bring added risk. At the moment the owner of data can retain the physical memory storage devices in fire-proof and secure storage. Placing the data in a remote vault means that

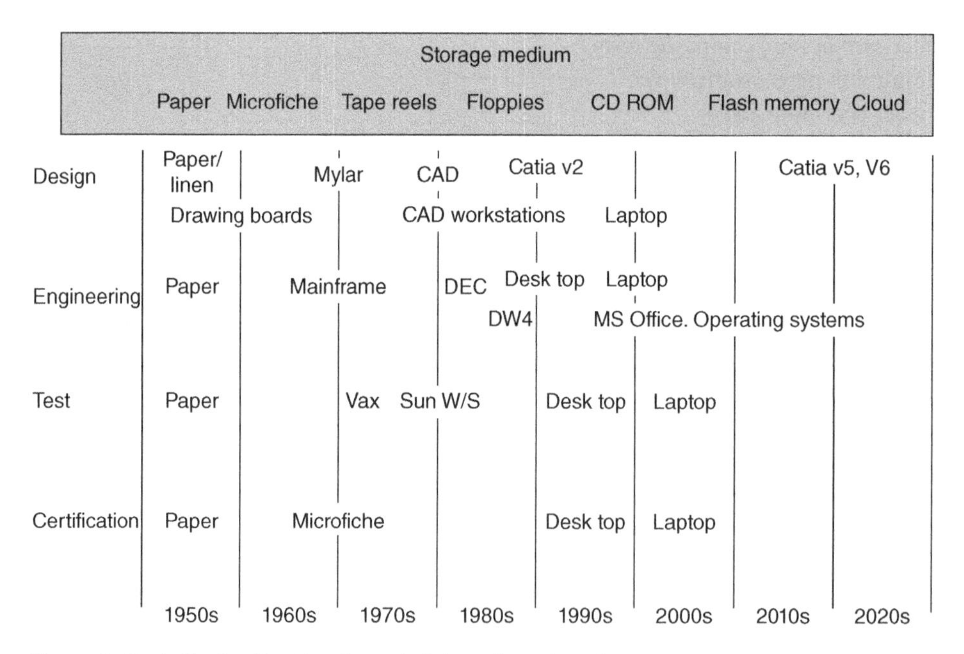

Figure 4.10 Indicative timeline for materials, tools, and media.

access is less controllable and there may be a risk of information being tampered with, stolen or misappropriated.

4.7.2.5 The Supply Chain

Inevitably all these issues are passed down the supply chain through the original requirements specifications. The supply chain will also have its own obsolescence issues to deal with as a result of specifying materials, processes, and components that are rendered obsolete by age, developing technologies (especially electronics), commercial market pressures, regulations, and changes to standards.

Planned obsolescence is the term used to describe the techniques used by manufacturers to shorten the useful life and/or limit the durability of consumer products and goods in order to stimulate the purchase of replacement goods. This is a growing problem within the aerospace and defence industry due to the increased use of COTS products, hardware and software.

An FAA report stated that commercial aircraft systems comprised of COTS components will *be in a continual state of enhancement because of commercial market pressures levied on vendors to improve product functionality and performance* (FAA 1996).

It is essential that the prime contractor involves the supply chain in obsolescence planning from the earliest stages of the project. Obsolescence of the supplier's facilities and assets is a serious issue that can arise as a result of mergers and changes of ownership. Such a change to the manufacturing process, facilities or materials can lead to the need to re-qualify equipment or components.

4.7.3 Managing Obsolescence

Managing obsolescence starts with understanding the issue – a survey of some industrial sources revealed that aircraft become obsolete:

- when no longer economical to provide a service
- when no longer economical to operate
- when too costly to repair
- when it is no longer possible to obtain parts
- when route changes and passenger trends mean you have the wrong type of aircraft, for example operating on a short regional route such as Manchester to Heathrow with an A380 is not feasible
- when military capability has been superseded by the enemy's capability, for example radar detection capability vs enemy countermeasures development, which can result in constant escalation.

Having understood the issue it is important to recognise the sources of obsolescence, drawing on previous experience, market predictions, confidence in the applied technology and its maturity, general knowledge, and judgement. From this is will be possible to implement an obsolescent management strategy:

- Produce a management plan for the product.
- Agree the plan with the customer.

- Design to resist obsolescence.
- Identify and evaluate risk.
- Contract with suppliers to spread the risk.
- Regular reviews of obsolescence status and risk.
- Training and continuous development for engineers to maintain the skill and experience base.
- Monitor the project and markets, and adjust the plan accordingly.

A strategy for managing obsolescence issues actively:

- Analyse the system.
- Identify all obsolescence 'hazards'.
- Identify the issues.
- Describe them.
- Identify all stakeholders.
- Determine the qualification and certification route.
- Is there going to be a Statement of Design?
- Who owns it?
- Assign responsibilities.
- Regularly review the process.

Figure 4.11 shows a rudimentary scheme for managing obsolescence. It is vital to consider obsolescence even at the concept stage, and to start to develop plans to manage obsolescence throughput the lifecycle. It is also vital that plans consider all aspects of the topic, not just bought-out components but skills and organisational infrastructure issues.

Boeing have responded to the issue of electronic component obsolescence and have reported on the subject of obsolescence management (Boeing Aero 2000).

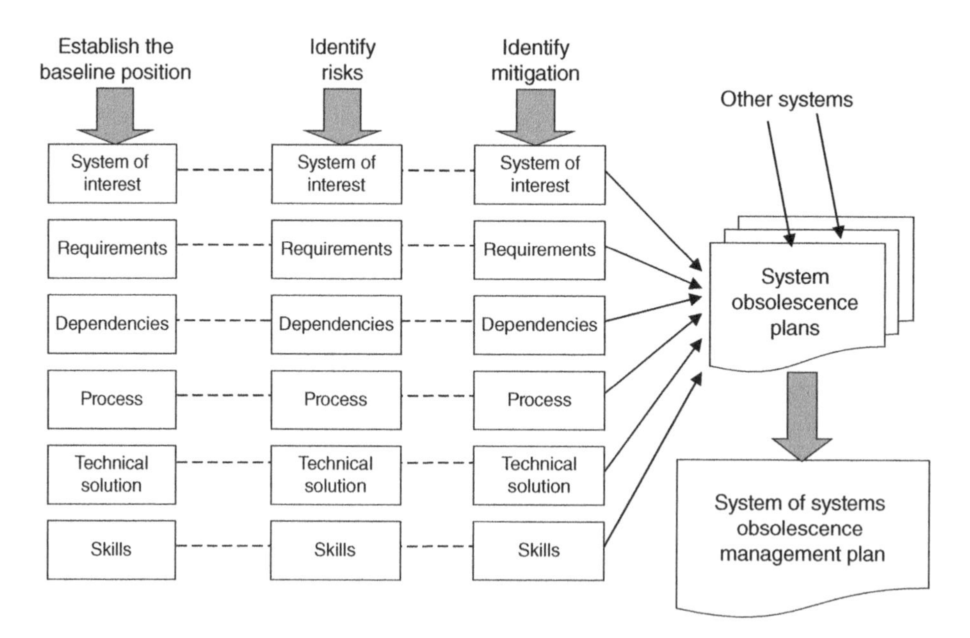

Figure 4.11 Simple scheme for managing obsolescence.

A simple tool can be constructed using a spreadsheet in order to discipline the management process and to form a record of decisions made and the reasoning. Two simple examples are shown in Tables 4.1 and 4.2.

Why is all this necessary? Obsolescence is an inevitable and insidious process; it will slowly become an issue that not only makes continued operation of a product expensive, it can also reflect badly on the quality of the product and its provider. Figure 4.12 shows some of the cost impacts in the product operating lifecycle.

Table 4.1 An example table for monitoring avionics items.

Item	Cause	When	Decision	Risk	Owner
Processor/memory	Commercial pressures, domestic markets				
Data bus	Commercial pressures				
Components	Commercial pressures				
Semiconductors	Materials advances				
Regulations	Regulatory changes				

Table 4.2 An example table for monitoring airframe items.

Item	Cause	When	Decision	Risk	Owner
Materials	Shortages of resins and fibres				
Fasteners	New standards				
Finishes/plating	Health and safety				
Solvents	Environmental issues				
Regulations	Regulatory changes				

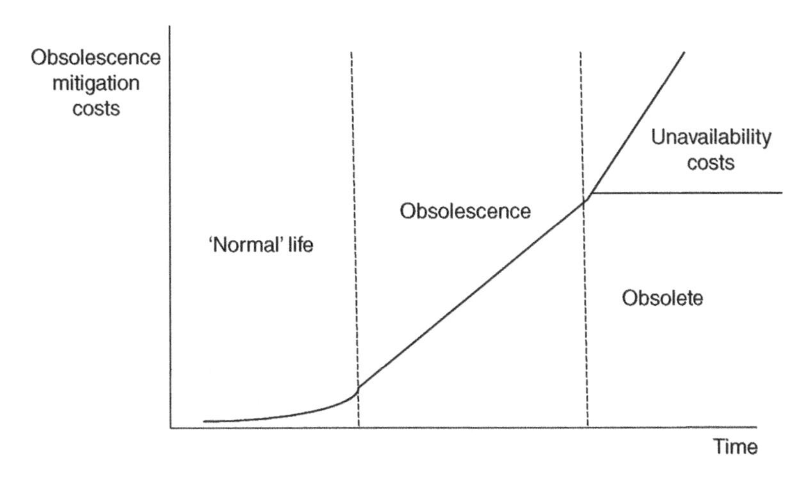

Figure 4.12 The impact of obsolescence on cost.

During 'normal' life the costs of mitigating obsolescence tend to increase as a result of implementing strategies that may not have been thoroughly understood. Much initial activity tends to be directed towards buying stocks of items to offset the initial impact of diminishing unavailability or looking for alternative sources of supply. For commonly used items there will a period of 'competitiveness' as all users rush to buy up stocks. As obsolescence sets in other issues begin to arise:

- difficulty in obtaining a continuous supply of items
- scarcity leads to opportunistic increased costs
- timing of deliveries is difficult to guarantee
- counterfeiting becomes an issue
- time must be set against identifying replacement items.

When an item is declared obsolete then different issues arise:

- replacement items will be need to be sourced
- they will need to be suitable qualified
- during any period of unavailability product performance may degrade
- company reputation may be affected
- support costs will escalate.

4.8 Ageing Aircraft

4.8.1 Introduction

Ageing aircraft have been included in this chapter because it is quite likely that engineers will encounter ageing aircraft in their careers at different stages in the lifecycle. Typical areas in which people may be employed include:

- design of modifications to in-service existing aircraft
- mid-life update programmes
- research/trials platforms
- conversions to alternative roles after an active life – freight, military, tanking
- crash investigations
- de-commissioning and scrap
- certification for private or museum use.

An aircraft or fleet that has been in service for 15–20 years is normally regarded as an ageing aircraft type. This typically applies to large commercial aircraft and military aircraft types still carrying the majority of their original airframe or equipment, although they may have had their systems updated from the original airframe. Aircraft can reach this situation because they may have a low utilisation rate, or they may cease to be attractive to new customers or their particular application to a route or a mission has been superseded. In such cases the aircraft may still have a viable life, but it may well be in another role.

4.8.2 Some Examples

Commercial aircraft may be used for long periods because of the need to recover capital costs versus operating costs. Modifications and improvements or upgrades may well increase the potential life. Changing travel patterns may cause the aircraft to be moved to different routes that are more popular. At one time second-tier airlines would see older aircraft as an attractive purchase proposition, although most low-cost airlines now lease their own new models. Conversion to alternative roles such as freight, mail, and parcel delivery services is still a viable option and a last resort is to sell the aircraft for conversion to military roles such as troop transport or tanking. There are types that have had a long and honourable life in such roles.

Military aircraft are subject to mid-life update programmes to increase their capability, often being fitted with new weapons and new sensors in response to changing threats. Sometimes this is because of the cost of developing replacements in times of tight defence budgets, or because the change invokes more expensive training and logistics costs where it makes good sense to continue with the same basic platform.

Aircraft at the end of their operational life may be sold off to private users and museums, where they may be used as museum pieces, gate guardians or for private flying as heritage pieces or for private hire.

Thus an aircraft in the latter stages of its career may have features originating from its original design, such as structural materials and fabrication techniques, together with new sub-assemblies and materials. The aircraft may also be equipped with new systems. Some examples include the following:

- The Boeing 737-200, first produced in 1970 and used by Alaska Airlines until 2007 and by Canadian North until the present. Many versions of the 737 have been produced over the years with extensive modification from the original type.
- The Lockheed L-1011 Tristar began service in 1970 and retired from RAF in-flight refuelling tanker service in April 2014.
- The KC-135 has been in US Air Force (USAF) service since 1957, and still refuels USAF aircraft today with use planned for several more years.
- The Panavia Tornado began service in 1980 as the Royal Air Force (RAF) primary strike and air defence aircraft, but was retired from the RAF in 2019. Other users will continue to operate the type.
- The Sikorsky Sea King (S-61) has been operating since 1961 and is still used by several countries for search and rescue as well as anti-submarine warfare operations.
- A recent example in the UK is the development of the Nimrod maritime patrol aircraft as an interesting case study of an aircraft with an intended life of nearly 100 years. This started as the de Havilland Comet airliner, which was originally designed in 1948, and became the subject of the HS801 requirement for the Maritime Comet in February 1965 using the Comet IV as the baseline for conversion. The first Nimrod MR1 prototype flew in May 1967 and deliveries of 38 aircraft took place from October 1969 to August 1972. A conversion of these MR1s to Nimrod MR2s started in April 1975, and these aircraft remained in service until the early 2010s. Following an industrial tender work began to convert the MR2s into MRA4s in 2010, with an intended end-of-life date of 2040. The resulting design was a mixture of new and original fuselage structure, new engines, new and original systems, and an enhanced capability to perform maritime operations. The contract was cancelled in 2014.

4.8.3 Systems Issues

Ageing aircraft research is dominated by structures and there is a lot of well-documented research and data available. Some aircraft fly beyond their original structural design life to enable operators to reduce cost by continuing to fly rather than replace at the end of design life, and this has an impact on the on-board systems. Flying beyond the fatigue model clearance is a significant risk. Typical structural and materials issues include corrosion, multi-site damage, and widespread fatigue damage.

The systems in the aircraft may suffer from a number of different problems depending on the care taken of the airframe as it approaches old age. This may not simply be neglect but because the aircraft may have been laid up or stored for a period of time before being re-used.

Pneumatic systems may suffer from corrosion of air ducts, especially at the high-pressure end, and deterioration of seals. There is potential for heat damage in leading edge piccolo tubes and internal surfaces, and hence an increased risk of corrosion and weakness.

Hydraulic systems may also suffer from deterioration of seals and leaking of pipe joints. Breakdown of fluid and contamination with moisture and debris is not uncommon. Breakdown of the fluid may occur as the result of over-pressure or over-temperature and from long-term use. The fluid may have become obsolete, and older types of fluid are likely to be flammable and may not conform to modern standards based on phosphate ester fluids, which do not support combustion. Over a period of time environmental and health and safety issues may have arisen with disposal of fluid leading to the need for greater care being taken during system maintenance.

Electrical systems may suffer from contact wear, although this issue is less likely with smart contactors or solid-state switches. Battery performance is always going to be an issue, but more serious is battery bay corrosion, which can lead to structural failure or fire. Wiring can suffer from a number of problems and these problems will be difficult to locate in congested fuselage spaces and large harness bundles. Although carefully selected at the beginning of a project, insulation material can deteriorate in service to reveal bare wires. This can occur from contamination, breakdown of the material to a brittle compound, or simple vibration and heat effects. Alternatively, it has been known in some tropical climates for animal and insect pest damage to occur. Damage at clipping points is another cause of failure.

Fatigue of conductor material from bending or vibration may be seen as a result of inadequate clipping of harnesses or simply from a severe vibration environment. Heat may also cause brittleness of the conductor, particularly where harnesses have been routed close to heat sources in the airframe or suffer from self-heating because of poor design of wire bundles. Damage during repair, servicing, and maintenance is not uncommon and conductors may be stretched after they have been used as handholds or footholds by inexperienced maintenance crew. Wear or fatigue at crimp connections can lead to high resistance or intermittent connections. Finally, dirt and dust deposition can be an issue.

Fuel systems suffer from wiring problems, which may lead to bare wires in the circuit or in tanks. This, and deterioration of bonding, can lead to a reduction in intrinsic safety with the potential for fire or explosion. Leaks can occur from deterioration of tank sealing compounds or tank material. Fuel tanks contain surfaces and inclusions which can be invaded by colonies of anaerobic bacteria.

ECSs suffer from deterioration of ducts and seals, and severe deterioration of high-efficiency particulate (HEPA) filters. Microbial contamination of ducts and cabin vents may pose a health and safety problem.

4.8.4 Certification Issues

The initial specification usually assumes a maximum life – flying hours from which total operational hours can be deduced. The selection of materials and equipment passes that assumption to suppliers as a firm requirement. All items of equipment are rigorously tested to gather test evidence based on that assumption for environmental testing and endurance testing defined in hours/cycles related to flying hours. This is reflected in statements of design, declarations of design and performance, and design certificates that together form the design data set. To extend this clearance means extrapolation by calculation, repeat testing or replacement with consequent impact on the life of test rigs.

The following issues at least will be encountered when working with ageing aircraft, and solutions will need to found:

- difficulty of access to original design data and the type record
- difficulty of access to usage data
- demonstration of compliance with modern standards
- obsolete materials, obsolete fluids, lubricants, and protective treatments
- obsolete build techniques
- exposure to hazardous materials, e.g. Halon, Freon, cadmium plating, lead solder, beryllium, etc.
- contravention of modern health and safety standards.

Exercises

1. For your own project select some aspects where technology is going to have a major impact. Put together a plan to show how you are going to prepare for the application of this technology.

2. Examine the impact of obsolescence in the domestic world where sound systems have moved from individual radio and record-playing systems to integrated radio, tape, and record-playing systems, through to iPods and streaming. Chart the complete history of this and then in parallel chart the history of the medium used from vinyl discs to cloud storage. How is this influenced by obsolescence?

3. Using this gradual obsolescence of storage medium and reproducing equipment describe how this is mirrored in the storage of the aircraft design data.

4. Look at the ways in which industry can determine what skills it requires over the lifetime of a project. How can industry work with the education sector to ensure that those skills are continuously available? Are there any internal training programmes that can be used? Are there any risks associated with this?

5 Explain the phenomenon of self-heating in large wire bundles. How does this occur and how can it be reduced in severity? Prepare an instruction to guide designers towards a safe design for wire bundles.

References

Aerostrategy Commentary September (2010). *From Tooth-to-Tail and Back Again: Military Sustainment's Difficult but Possible New Mission.*

Boeing AERO No 10 – March 2000. http://www.boeing.com/commercial/aeromagazine/aero_10

Essinger, J. (2012). *A Female Genius – How Ada Lovelace Started the Computer Age.* Melville House.

Federal Aviation Administration, (1996). *Report of the Challenge 2000 Subcommittee of the FAA Research, Engineering, and Development Advisory Committee, March 6th 1996, Use of COTS/NDI in Safety Critical Systems.*

Garfield, S. (2012). *On the Map.* Profile Books.

Martin, A. (2005). *Moon Dust.* Bloomsbury.

Montreal Protocol (1987) *The Vienna Convention for the Protection of the Ozone Layer and the Montreal Protocol on substances that Deplete the Ozone Layer.* United Nations Environment Programme. www.unep.or/ozone.

Schleher, C. (1999). *Electronic Warfare in the Information Age.* Artech House.

Torenbeek, E. (2013). *Advanced Aircraft Design – Conceptual Design, Analysis and Optimization of Subsonic Civil Airplanes.* Wiley.

Further Reading

Agarwal, R., Collier, F., Schäfer, A., and Seabridge, A. (eds.) (2016). *Green Aviation.* Wiley.

Bamford, J. (2001). *Body of Secrets.* Century.

Davidson, P.E. (1996), *Coral Reach: USAF KC-135 Aging Aircraft Program, Aging Combat Aircraft Fleets – Long Term Applications*, AGARD Lecture Series 206.

DeFazio, M.S. (1996), *F-16 System/Structrual Upgrades, Aging Combat Aircraft Fleets – Long Term Applications*, AGARD Lecture Series 206.

Enloe, S.M. (1999), *Achieving Total System Aging Aircraft Solutions,Second Joint NASA/FAA/DoD Conference on Aging Aircraft*, January 1999. NASA/CP-1999-208 982.

Flight Safety Digest, (1994). 'Landing Gear Topped List of Aircraft Systems Involved in Accidents during 35-year Period'.

Lombardo, D.A. (1998). *Aircraft Systems – Understanding your Airplane.* PA: Tab Books Inc., Blue Ridge Summit.

Marlow-Spalding, M.J. (Capt RAF). (1996). *Ageing Aircraft – Managing the Tornado Fleet, Aging Combat Aircraft Fleets – Long Term Applications.* AGARD Lecture Series 206.

Moir, I. and Seabridge, A. (2013). *Civil Avionic Systems*, 2e. Wiley.

Pappas, J. (Capt USAF) and Ward, R. (1999). *The C/KC-135 Functional Systems Integrity Program, Second Joint NASA/FAA/DoD Conference on Aging Aircraft*, January 1999. NASA/CP-1999-208 982.

Rudd, J.L. (1996). *USAF Aging Aircraft Program, Aging Combat Aircraft Fleets – Long Term Applications*. AGARD Lecture Series 206.

Sampath, S.G. (1996). *Introduction to Lecture Series, Aging Combat Aircraft Fleets – Long Term Applications*. AGARD Lecture Series 206.

Sandborn, P. (2004). Beyond reactive thinking – we should be developing pro-active approaches to obsolescence management too! *DMSMS Center of Excellence Newsletter* 2 (4).

Sandborn, P., Mauro, F., and Knox, R. (2007). A data mining based approach to electronic part obsolescence forecasting. *IEEE Transaction on Components and Packaging Technologies* 30 (3): 397–401.

Sandborn, P. (2007). *Designing for Technology Obsolescence Management. Proceedings of the 2007 Industrial Engineering Research Conference*

Singh, P., Sandborn, P., Lorenson, D., and Geiser, T. (2007). *Determining Optimum Redesign Plans for Avionics Based on Electronic Part Obsolescence Forecasts*. SAE.

Sandborn, P. and Singh, P. *Forecasting Technology Insertion Concurrent with Design Refresh Planning for COTS-Based Electronic Systems*. CALCE, Department of Mechanical Engineering, University of Maryland.

Singh, P. and Sandborn, P. *Obsolescence Driven Design Refresh Planning for Sustainment Dominated Systems*. CALCE, Department of Mechanical Engineering, University of Maryland.

Solomon, R., Sandborn, P., and Pecht, M. (2000). Electronic part life cycle concepts and obsolescence forecasting. *IEEE Transactions on Components and Packaging Technologies* 1: 707–717.

Young, D. *Lifecycle management and the impact of obsolescence on military systems*. http://vmcritical.com/columns/military

5

System Architectures

5.1 Introduction

The system architecture is an important tool in the design and development engineering process. It can form a part of the early visualisation of the concept stage by enabling requirements to be mapped at the top level onto elementary building blocks. Block diagrams are frequently used as 'scribbling pads' to play around with ideas of functions and data flows, as well as functional dependencies. A firm architecture can then be developed to add more detail, to incorporate functional to physical mapping, and to agree on functional allocations. This is a suitable stage to make decisions on which functions are to be put out to tender to suppliers.

The architecture is also a useful tool for 'fixing' external constraints. For example, a decision to use a particular commercial avionic standard, such as ARINC 429, will automatically determine some architectural principles, and the same is true of other data bus types. Other design drivers may include a decision to use of commercial off-the-shelf (COTS) components or customer inventory items that will similarly constrain the design. These restrictions can be recorded on the architecture diagrams and notes.

The system architecture is a representation of the conceptual shape and form of a system which can be visualised quite independently of any physical implementation. It is an invaluable device for making a simple and easy to understand representation of a system using a block diagram format as a convenient shorthand notation. This simple visualisation allows a concept to be represented clearly and acts as a mechanism for promoting discussion between various engineering disciplines to reach agreement on interfaces, functional allocations, and standards. From such simple basic representations it is possible to develop the architectures further without the need to move to excessive detail of wiring interconnections or detailed components. This is true for physical and functional representations in terms of software and hardware building blocks. It is especially useful for setting and agreeing boundaries and interfaces.

Apart from allowing design decisions to be made, system architectures are an ideal tool to assist in identifying candidates for early trade-offs and simple models using spreadsheets to perform cost, benefit, and performance comparisons between different architecture designs.

Design and Development of Aircraft Systems, Third Edition. Allan Seabridge and Ian Moir.

5.2 Definitions

The terms 'architecture' or 'system architecture' are much used in systems engineering. These terms owe much to their origin in civil engineering and building design. When systems engineers speak of the architecture of a system they do so in pretty much the same way as an architect speaks of the concept of a building. In civil engineering terms architecture is defined in the *Oxford English Dictionary* as:

- *The art and science of designing and supervising the construction of buildings.*
- *A style of building or structure.*
- *Buildings or structures collectively.*
- *The structure or design of anything.*

The *Oxford English Dictionary* defines architecture as *The special method or style in accordance with which the details of the structure and ornamentation of a building are arranged.* An architect is defined as *One who so plans, devises, contrives or constructs so as to achieve a desired result.* The architect envisages the form and structure of a design, often from a blank sheet of paper. He then flows down the guiding principles and standards that apply to the basic structure into the individual components. This ensures that the integrity of the design is preserved throughout the development of the product. This integrity includes aesthetic qualities such as style or fashion, as much as functional aspects such as habitation, heating, services etc. and adherence to regulatory aspects such as building regulations, health and safety, and all legal considerations.

This appreciation of pattern, form, and structure has uses beyond civil architecture. The decipherment of an ancient language (Linear B) was achieved by an architect, not a linguist, which led to an observation on the characteristics or competence of an architect:

> *The architect's eye sees in a building not the mere façade, a jumble of ornamental and structural features; it looks beneath the appearances and distinguishes the significant parts of the pattern, the structural elements and framework of the building.* (Chadwick 1987)

It was this ability to observe pattern rather than detail that led the architect to conclude that the language represented by Linear B was Greek. Linguists had embroiled themselves in detail and in philosophical debate, and had missed the key point.

The above definitions and observations lead to the conclusion that form, structure, and order are essential characteristics of an architecture, rather than detail. An architect must therefore possess skills to deal with these characteristics and to develop a form and the standards that apply so that the constructors that follow his design can produce a sound structure. In aircraft systems engineering the fundamental architectural principles flow down through layers of system design into the very items of equipment that make up the hardware and firmware solution. This is illustrated in Figure 5.1.

For the purposes of this book the term 'architecture' has been used to identify a part of the design process that can be used to provide a preliminary pictorial description of a system. It is of most use in the early stages of the design lifecycle, especially the concept

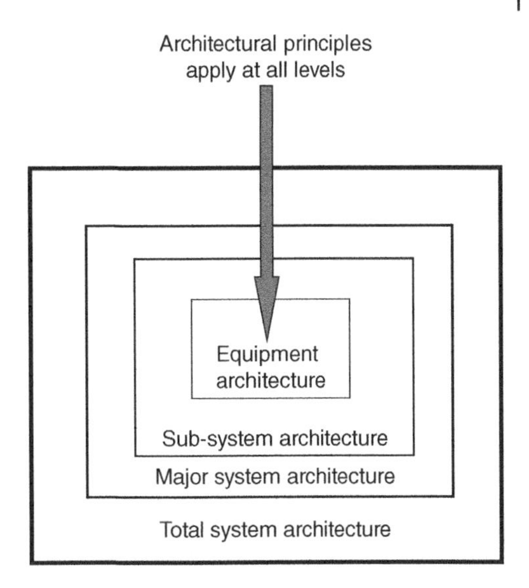

Figure 5.1 Flow down of architectural principles.

phase and the definition phase, but it remains a sound mechanism for recording the progress of design in the remaining stages of the process. It is especially useful for illustrating marketing material and in presentation as a clear and illustrative shorthand to support system descriptions.

In concept and design phases the architecture is used to introduce and develop initial concept design and establish interfaces and relationships with other systems. The architecture has a simplicity and clarity that stimulates discussion with other parties or stakeholders in the project. The architecture develops in stages, with agreement at each stage, until the design phase, where it is necessary to start to define the product to a stage where manufacturing can begin. This is where simplicity and clarity are subsumed in the level of detail required for specification and build.

5.3 System Architectures

In systems engineering at the early stages of design it is most convenient to think of form or structure rather than detailed engineering solutions. It is at this level of abstraction where decisions can be made about the major functional building blocks that are required and the means of communication between these blocks. For example, in computing systems the computer architecture is defined as the design and structure of the hardware components of computer systems. The term embraces general considerations, such as whether a system is based on serial, parallel, or distributed computing, in which several computers are linked together. It also covers more detailed aspects, such as a description of the internal structure of a central processing unit (CPU). A micro-computer is often described as having a 16-bit, 32-bit or 64-bit architecture according to the length of data word that can be processed by the CPU and the width of the data bus.

Similarly, when designing systems, engineers often speak of functions, processing standards, interface standards, software languages, and standards of data bus to connect the

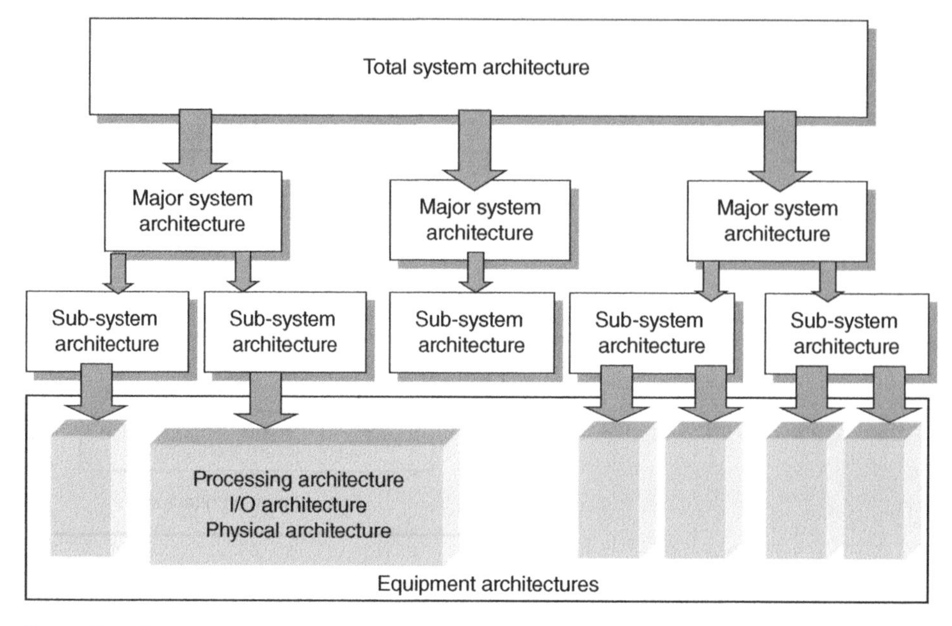

Figure 5.2 Example generic system architecture.

functions. Once agreed at this level the principles are then applied at each succeeding level of detailed design by all parties involved.

A system architecture often starts life as a simple block diagrammatic representation of a system, a block diagram. This allows one to visualise the main functions to be performed, the mechanism for data interchange, and interdependencies between functions. When basic rules have been established for naming the major high-level functions and standards for data interchange, then blocks of the architecture can be developed in more detail. This is somewhat analogous to defining the ground plan and outline of a building, and agreeing standards to be observed for construction and provision of services. Figure 5.2 illustrates this process, showing the flow down of architectural principles in a generic system architecture.

The engineer who undertakes to control or set the standards for the architecture is known as a system architect. An architect is someone who practices the skills required to design a building and can be said to look for simple representations rather than detailed design. The intention of such simple representations is to create a medium that expresses high-level views of the design in a form that is simple and clear. This can be used as a means of stimulating debate, reaching agreement, and recording the stages of a design that all parties in a project can use as a sound basis for their work.

An example of a starting point for a total system architecture is shown in Figure 5.3. In this diagram the aircraft systems have been allocated to specific groups with a common means of intercommunication:

- vehicle systems
- avionic systems
- mission systems
- cabin systems
- data bus.

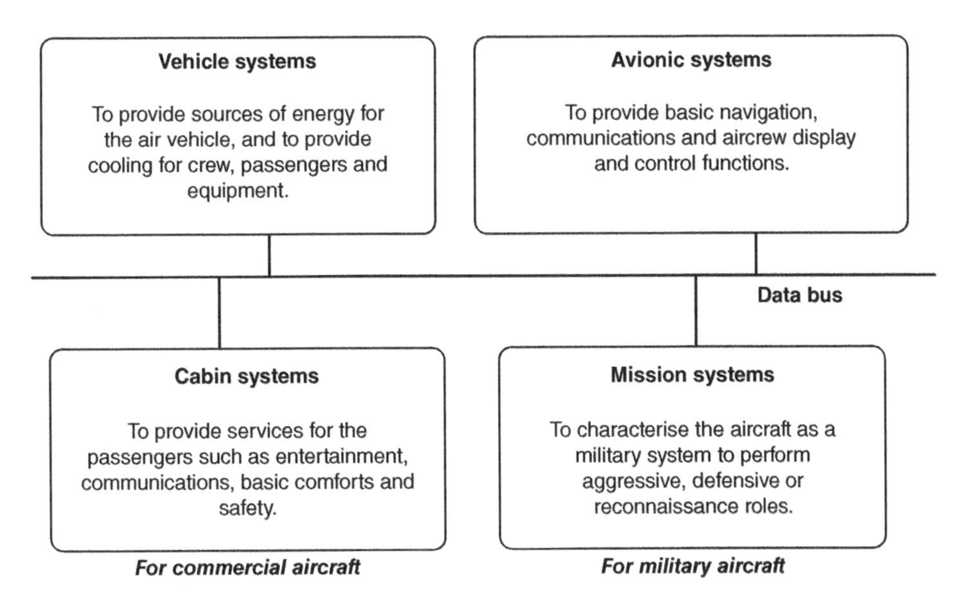

Figure 5.3 Example of a top-level system architecture.

A brief description of the functions is provided in each group. The adoption of a common data bus standard means that interfaces and data formats can be defined. The grouping is not arbitrary – the requirements for integrity for each group differ greatly. Vehicle systems are usually safety critical and must be designed so that failure is remote since it may endanger the aircraft and crew. Avionic systems are safety involved: their loss may hazard the aircraft. Mission systems failure will result in reduced performance or a failure to achieve the desired mission success rate. Loss of cabin and entertainment systems can be tolerated from a safety point of view, although customer satisfaction may diminish.

Although this is a very simple starting point it is an enabler for various teams to identify their responsibilities and their approach to design, and to further define the architectures in more detail. It is at this stage that standards for communication, safety, integrity, availability, and design for manufacture are established that will be applicable to all subsequent levels.

Figure 5.4 shows the architecture as it may look after each group has been developed further.

5.3.1 Vehicle Systems

The vehicle systems group has been developed to show the individual systems that will be required. The majority of such systems have a major mechanical content, such as pumps, tanks, levers etc., but it is quite valid to show the need for the systems in this architecture. Since vehicle systems need an element of control and monitoring this has also been included. To communicate with the rest of the aircraft a data bus connection is needed, since this is the agreed form of intercommunication. However, vehicle systems are vital to the continued safe operation of the aircraft, and to ensure that this robustness is maintained

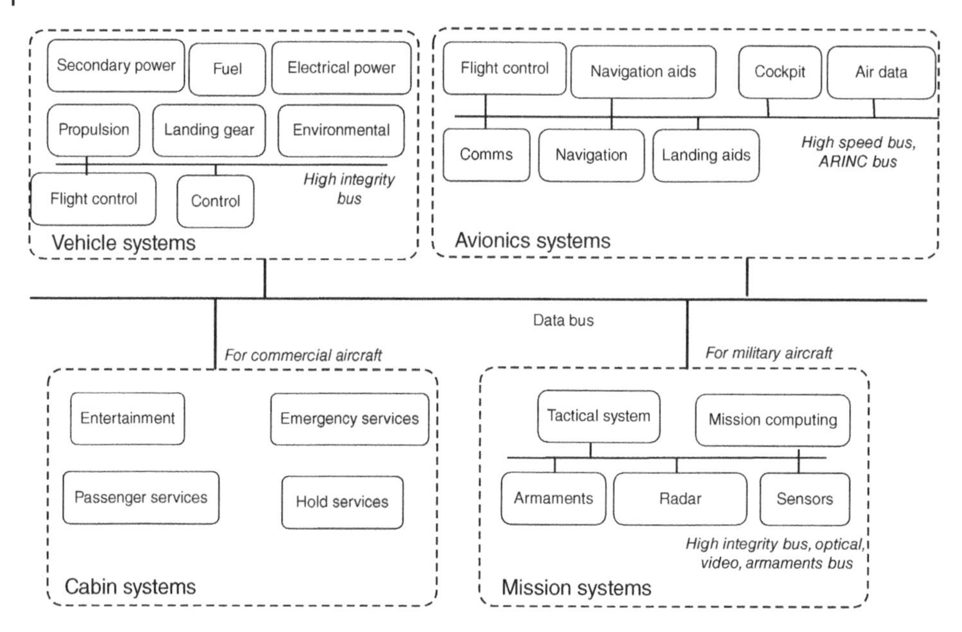

Figure 5.4 Aircraft systems.

a need has been established for a high-integrity bus. This may be the same physical implementation as the aircraft bus, but may include a different level of redundancy or a different message scheduling protocol. An example of the development of a vehicle management system (VMS) architecture can be seen in Section 5.7, Example 4.

5.3.2 Avionic Systems

The avionic systems group has been similarly developed to show a requirement for a high-speed bus for intercommunication between the cockpit display systems, and the use of a commercial standard ARINC bus for off-the-shelf avionic sub-systems, e.g. ARINC 429 or ARINC 664. Military avionics systems may need to conform to standard military bus standards such as MIL-STD-1553.

5.3.3 Mission Systems

The mission systems group is relevant for military aircraft and may make use of high-integrity weapons buses as well as optical and video links. The avionics, vehicle management, and mission systems will be designed to the same set of military standards in the majority of military projects. However, in aircraft that are based on a commercial platform, such as surveillance aircraft, it may be advantageous to retain the platform commercial avionics and make a connection to the mission system. An example of this can be found in Section 5.7, Example 1. This is especially the case where weapons are carried to ensure that the weapons circuits are isolated from all other systems.

5.3.4 Cabin Systems

The cabin systems group may have special requirements for the in-flight entertainment system such as high-quality video and audio communication. This may demand video or optical data connections. In many instances it is permissible to lose some of the cabin systems, for example shedding the load of the galleys. Although this may be annoying for the passengers, their safety is not affected.

5.3.5 Data Bus

In each case the decisions with respect to the selection of the main data bus will be preserved and its message protocols observed.

This is once more a simple representation of the systems, but the definition has advanced and agreements have been made. Figure 5.5 shows the vehicle systems group developed to show more detail of the control system and the connections of the general systems. This development demonstrates the following features:

- The flight controls system is quadruplex and directly connected to the data bus. It has deliberately not been incorporated into vehicle systems control as a project decision because its integrity is higher than that of vehicle systems.
- Control of the vehicle systems is performed by a four-computer sub-system connected to the data bus.
- Connection of the vehicle systems components (actuators, sensors, etc.) is made by discrete wiring feeding.
- The propulsion system is duo-duplex and is directly connected to the data bus but is also *not* part of vehicle systems control.

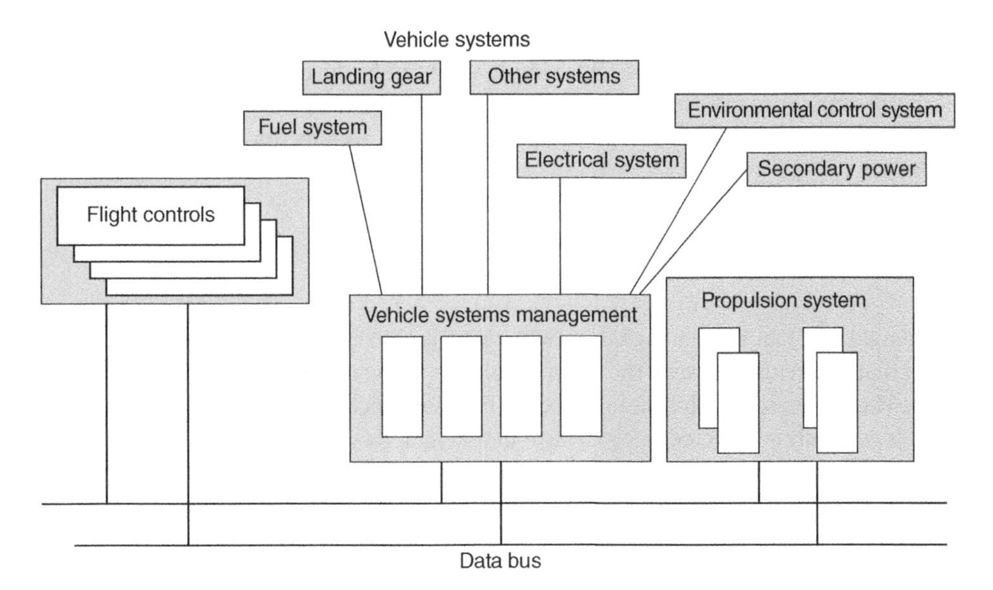

Figure 5.5 Examples of vehicle system architecture.

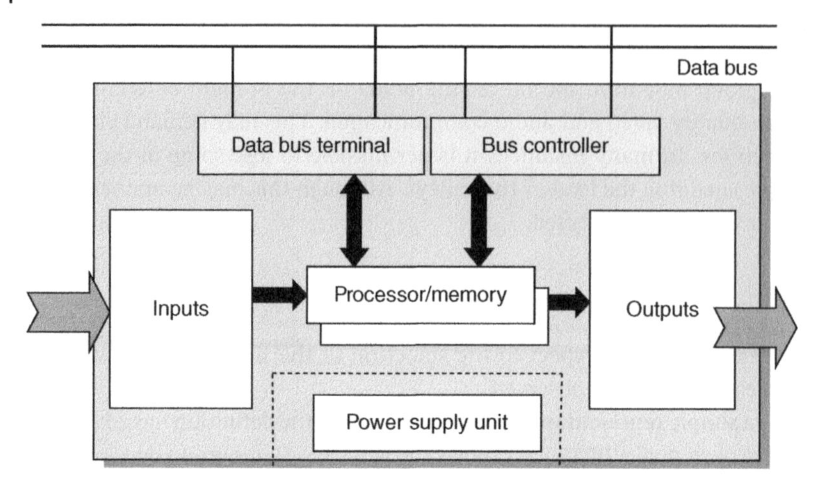

Figure 5.6 Architecture of a vehicle system management processor.

- Connection to other avionic systems, including the cockpit display and controls, is by means of the data bus.
- The data bus is duplicated to preserve a dual path for all communications.

This represents another step forward in defining the system architecture to the stage where it is now possible to allocate functions to the components of the general system architectural blocks.

Architectural block diagram representations are not only limited to visualisations of large systems. The techniques can also be used by the designers of equipment, such as the systems management processors shown above. The internal configuration or architecture of a processor is shown in Figure 5.6. This shows the processor partitioned into its major elements such as input and output interfacing, data bus connection and control, processor and memory, and power supplies. This simple view can be developed to ensure that the external systems principles of redundancy, segregation, and integrity have been preserved in the equipment design. Further examples of simple architectures are described in Section 5.7.

5.4 Architecture Modelling and Trade-off

An example of a fuel system is given to show how architectures can be used to define a system that can be used to seek the optimal solution using modelling and trade-off techniques. The fuel system performs a number of functions specific to the carriage and use of fuel. It is generally considered safety critical in that total loss of fuel to the aircraft engines may be catastrophic. Fuel is used by the auxiliary power unit (APU) and the main engines, and it has some subsidiary roles. It is used by the flight control system to control the aircraft centre of gravity within certain limits and it is used in the aircraft cooling system as a cooling medium, for example fuel-cooled oil coolers and heat exchangers. The system has a number of interfaces in the aircraft and many stakeholders are involved, including:

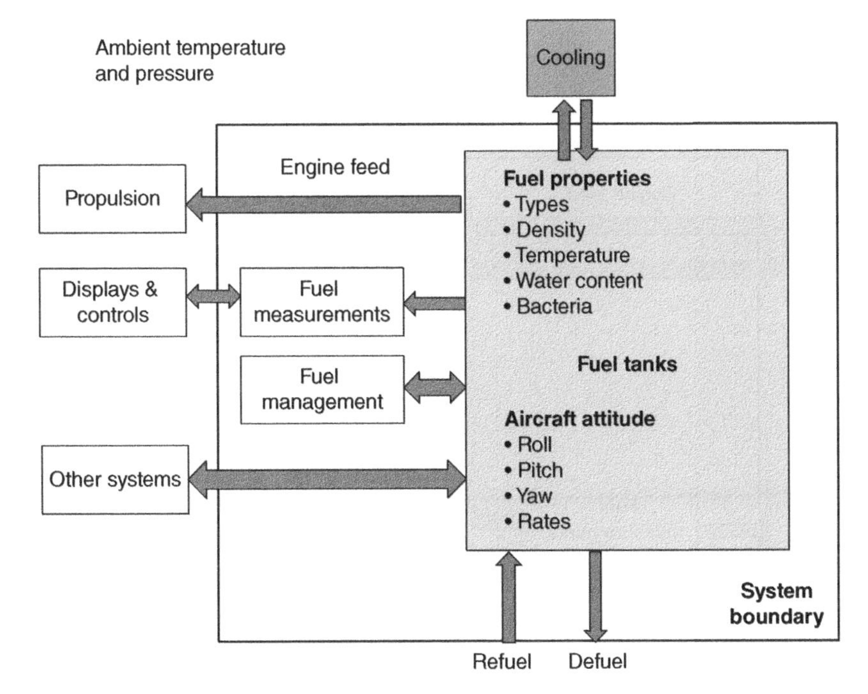

Figure 5.7 A top-level combat aircraft fuel system architecture.

- structures departments – to provide locations for tanks and pipe layout
- users of fuel – propulsion and APU
- air crew for human machine interface
- ground crew to refuel and defuel the aircraft
- cockpit designers
- systems that dissipate heat that need cooling
- safety for intrinsic safety aspects
- VMS for interfacing with system components and for hosting the system functions (note this function may incorporated into the overall aircraft computing system architecture).

This example starts with a number of proposal concept studies in which a generic aircraft is being proposed. For each study a fuel system generic architecture will be required. Figure 5.7 is a suitable architecture at this stage since changes will be made as the proposal is accepted and serious design decisions are made. Although this looks trivial, it does allow certain interfaces with other systems to be visualised.

The next step starts when a decision is made on the nature of the project configuration – in this case it is to be a supersonic fast jet fighter with two engines, with supersonic performance, a need for in-flight refuelling, and capable of world-wide operation. The concept stage architecture is shown in Figure 5.8.

From a fuel system point of view, this means that two engine-feed circuits are required, an in-flight refuelling probe and associated circuits is required, and that a rudimentary

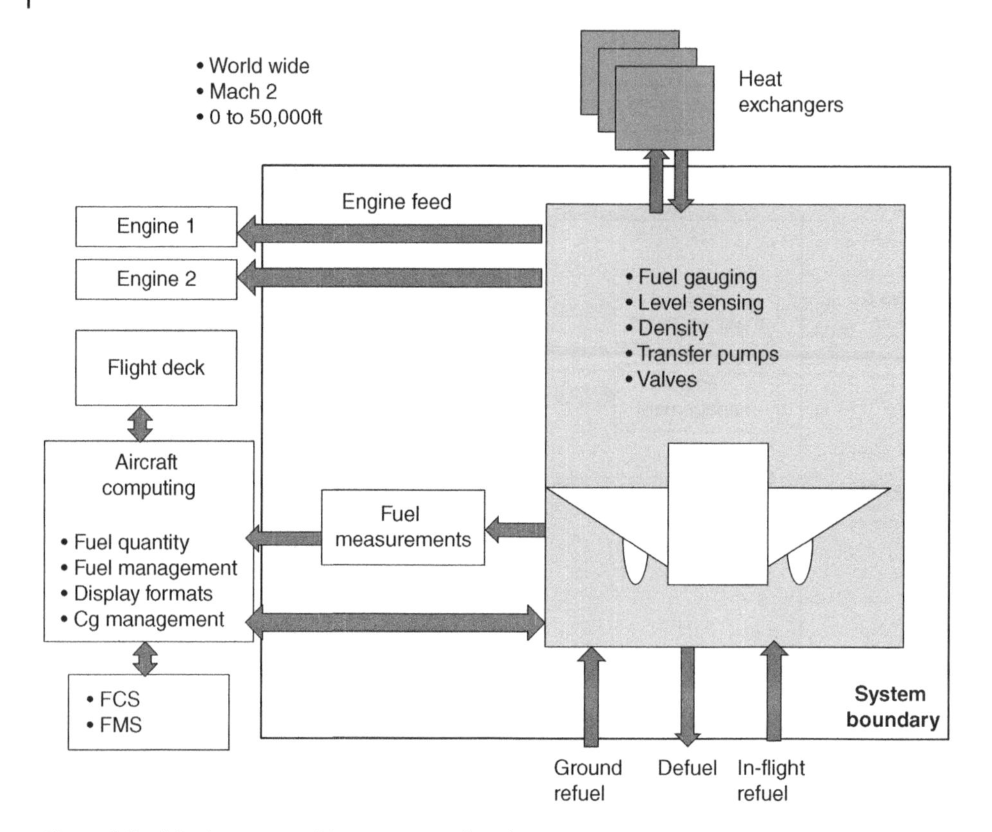

Figure 5.8 A fuel system architecture recording the concept stage agreements.

tank layout can be envisaged. The avionics groups have decided that there will be a computing system that will require data from the fuel system to perform the functions of fuel displays, interfaces with flight controls, and air data systems. From this it can be seen that the fuel system now has a firm interface with the avionic system architecture. This means that all fuel system measurements can be passed to the avionics systems and that agreement will need to be made on how the fuel systems functions are to be defined to the software teams.

Some architectural decisions can be made which can be agreed with all other teams involved at this stage and will influence other system architectures. Agreement and acceptance at this stage means that all teams can now work within their system boundaries unless a situation arises at which interface definitions cannot be achieved. Some example decisions from the fuel system perspective are:

- agreement on the location and shape of tanks in the fuselage
- agreement on the need for external tanks and a means of jettison
- agreement on the number of fuel sensors and their locations in tanks to achieve the desired fuel quantity accuracy
- agreement on the number of transfer pumps
- agreement on the size of fuel pipes, bend radius, and location

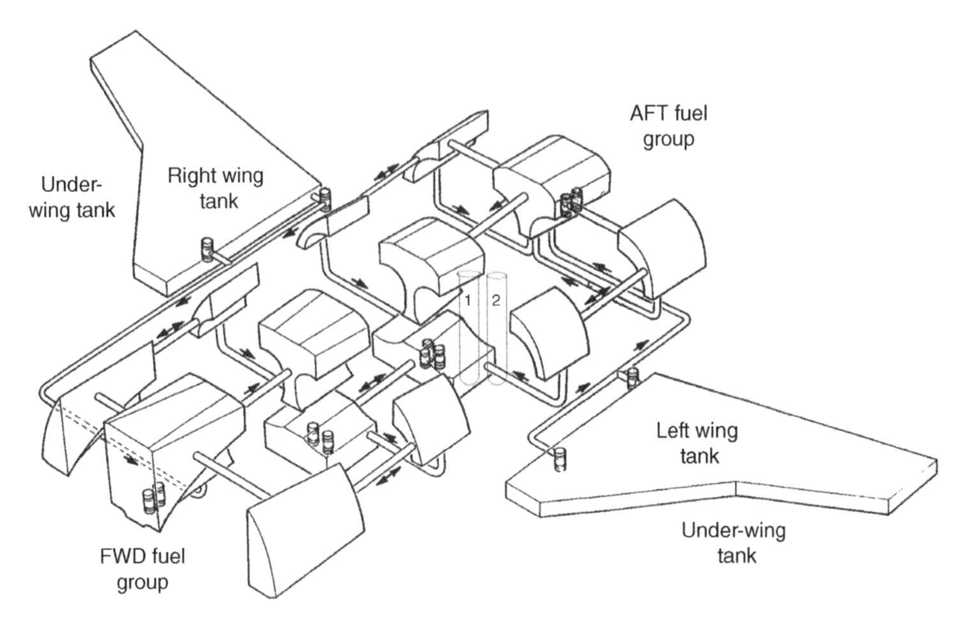

Figure 5.9 Preliminary tank layout.

- agreement on the information to be displayed in the cockpit and its form
- agreement on the interface with refuelling equipment for both the ground system and the airborne refuelling elements.

The structures teams will have defined a shape for the fuselage and wings, and will need information on which to define the location and shape of fuel tanks. With knowledge of the aircraft range, endurance, and combat performance the fuel team will know how much fuel is needed and how to design a system of pipes, pumps, valves, and measurements sensors. An initial agreement with the structures teams based on their knowledge of the volume available in the airframe for fuel storage will have led to a tank layout similar to that shown in Figure 5.9.

This level of definition is sufficient to continue the design of the quantity measuring systems to determine how many gauge probes are required and their location in the tanks. This enables preliminary modelling to take place to determine how the fuel will be managed by inter-tank transfers, and allow the airframe teams to draw the tanks so that they can be designed as an integral part of the structure. This modelling may include computer modelling of the tank shapes to perform the calculation of fuel quantity and may include models to take into account changes in aircraft attitude and calculation of the centre of mass of the fuel. This leads to the design of a preliminary circuit diagram, as illustrated in Figure 5.10.

5.5 Example of a Developing Architecture

Figure 5.3 showed a top-level system architecture in which the requirement for aircrew control and display functions are shown. This is developed further in Figure 5.4 to show other system interfaces. Figure 5.11 shows a simple architecture of a display system that

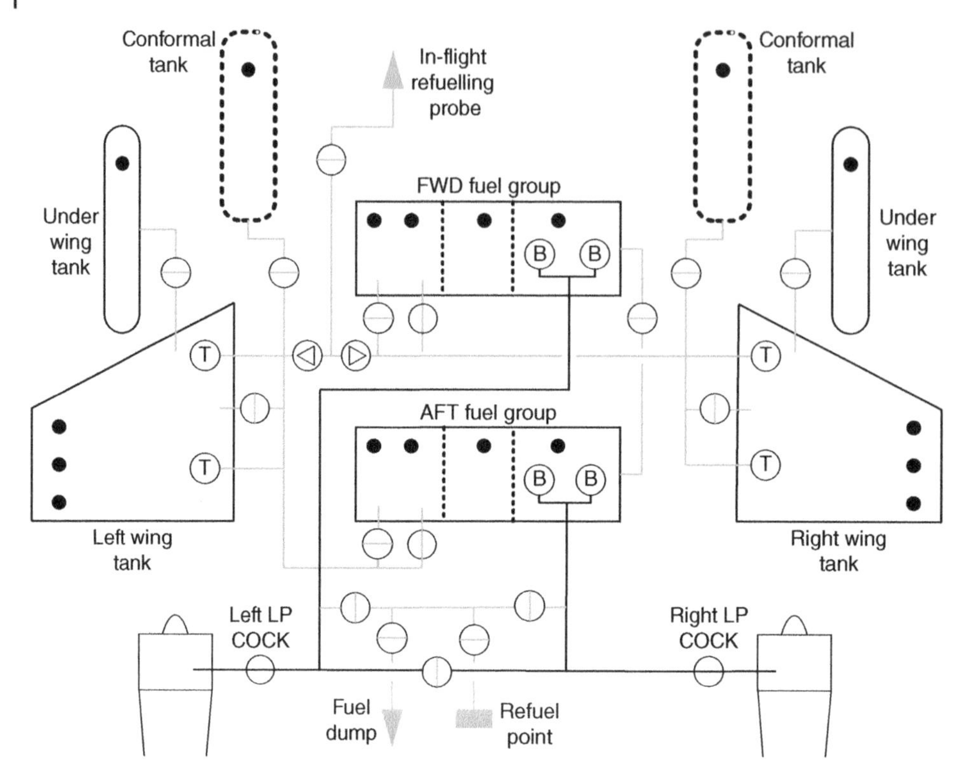

Figure 5.10 Preliminary circuit diagram.

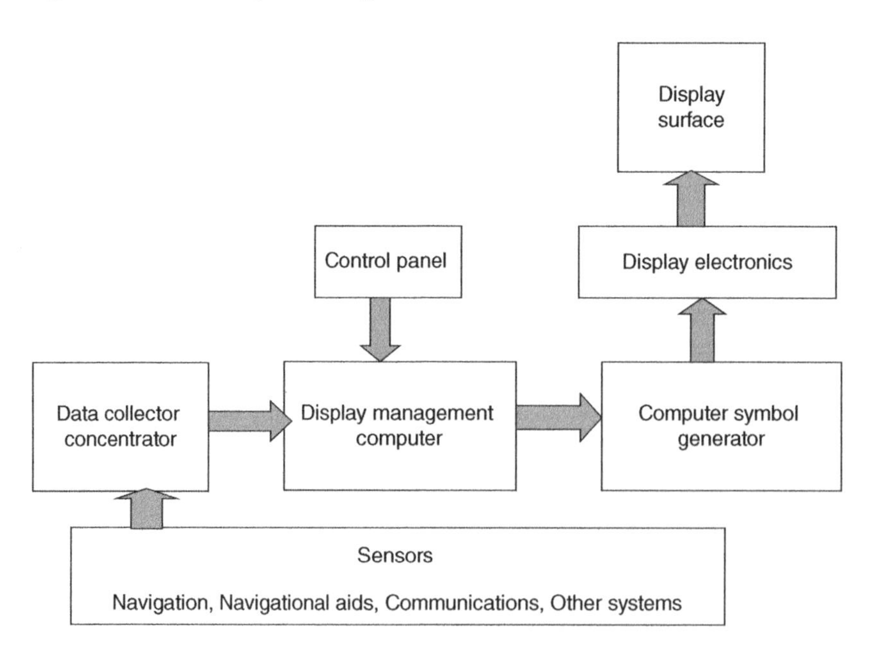

Figure 5.11 General display system architecture.

will provide control and display functions for a two-crew cockpit. The major elements of this general display system are:

- a data collection/concentrator that acquires the data to be displayed from other systems and selects the most appropriate data sources and performs data integrity checks
- a display management computer (DMC) that determines the display modes and the elements to be displayed
- symbol/graphics generator that constructs the symbology and graphics as text and symbols
- a display unit consisting of the display surface and the display device electronics.

This architecture is developed further in Figure 5.12, which shows the system exhibiting a redundancy structure of three DMCs, six displays, and a switching mechanism. This allows the crew to select the information most appropriate to the phase of flight and their roles of captain and first officer. The architecture also shows that the functions of data collection/concentration and symbol generation have been absorbed into the DMCs. Connections to the aircraft data bus structure are shown to enable connections to other systems.

Further development of this architecture is possible with ever-increasing detail until an aircraft wiring diagram emerges. The resulting implementations can be seen in the two-crew flight decks of modern commercial aircraft such as the A350, A380, and B-787.

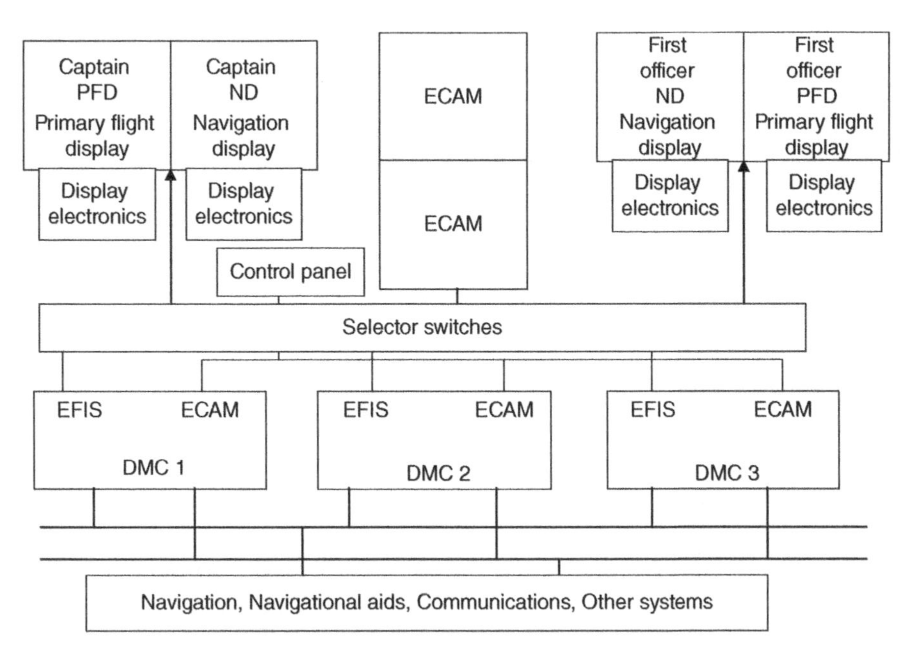

Figure 5.12 Developed display system architecture. DMC, display management computer; ECAM, electronic centralised aircraft monitor; EFIS, electronic fight instrument system; Nav Aids, navigation aids; ND, navigation display; PFD, primary flight display.

5.6 Evolution of Avionics Architectures

The application of avionics technology has occurred rapidly as aircraft performance has increased. The availability of reliable turbojet engines gave a huge performance boost to both military and civil operators alike. To utilise these improvements the aircraft avionics system rapidly grew in terms of capability and complexity, as illustrated in Figure 5.13.

This figure portrays how avionics architectures have evolved from the 1960s to the present date. The key architectural steps during this time have been:

- distributed analogue architecture
- distributed digital architecture
- federated digital architecture
- integrated modular architecture, also digital.

The evolution of these architectures has been shaped in the main by the aircraft-level design drivers that were described in Chapter 4. Their capabilities and performance have been both enabled and constrained by the avionics technology building blocks available at the time. There have also been changes in many characteristics throughout the period:

- *increases* in performance and capability, computing power, complexity, reliability, cost
- *decreases* in volume, weight, power consumption, wiring.

Key advances were enabled by the advent of digital computing technology in the 1960s, which first found application in the architectures reaching fruition during the 1970s. The availability of digital computers that could be adopted for the rugged and demanding

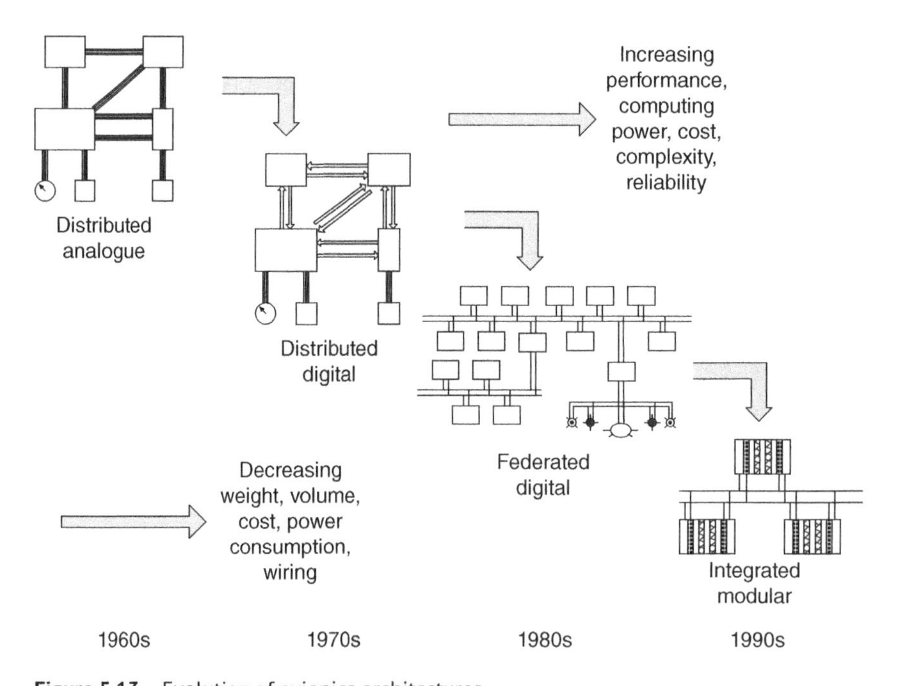

Figure 5.13 Evolution of avionics architectures.

environment of aerospace applications brought computing power and accuracies that had not been available during the analogue era. The development of serial digital data buses greatly eased the interconnection and transfer of data between the major units of the systems, units. In the early days this was achieved by means of fairly slow half-duplex (uni-directional), point-to-point digital links such as ARINC 429 and the Tornado serial data link.

The arrival of micro-electronics technology and the first integrated circuits (ICs) enabled digital computing techniques to be applied to many more systems around the aircraft. At the same time more powerful data buses, such as MIL-STD-1553B, provided a full-duplex (bi-directional), multi-drop capability at higher data rates. This enabled the federated architectures that evolved during the 1980s where multiple data bus architectures were developed to cater for increased data flow and system segregation requirements. At this stage the aerospace electronic components were mainly bespoke, being dedicated solutions with few if any applications outside aerospace.

The final advance occurred when electronic components and techniques developed in industries outside aerospace in the fields of information technology and personal computing yielded a far higher capability than that which aerospace could sustain. The use of COTS technology became more prevalent and integrated modular avionics architectures began to follow and adapt the technology developed elsewhere. The key attributes of each of the architectures is described below.

5.6.1 Distributed Analogue Architecture

The distributed analogue architecture is shown in Figure 5.14. In this type of system the major units are interconnected by hardwiring; no data buses are employed. This results in

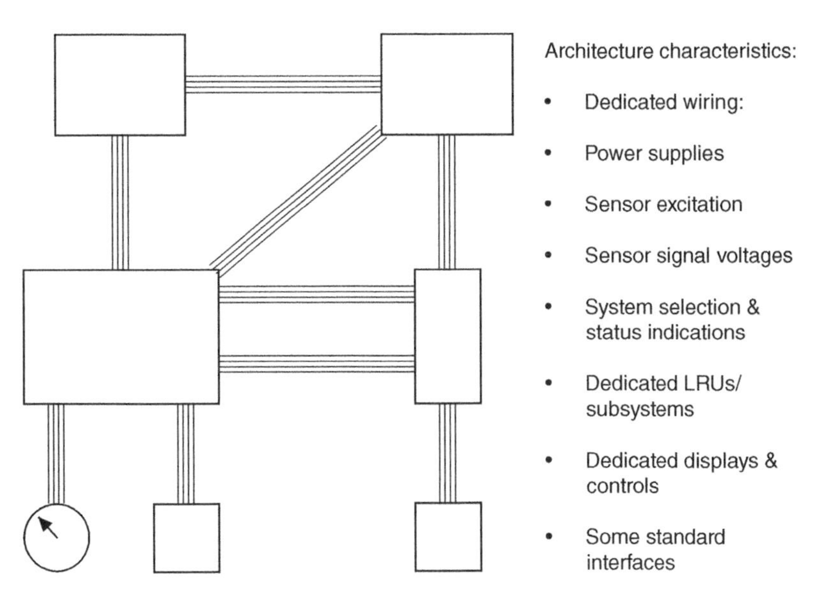

Figure 5.14 Distributed analogue architecture.

a huge amount of aircraft wiring and the system is extremely difficult to modify if change is necessary. This wiring is associated with power supplies, sensor excitation, sensor signal voltage, and system discrete mode selection and status signals.

This system has dedicated sub-systems, controls, and displays. The displays are electromechanical and often extremely intricate in their operation, requiring instrument maker skills for assembly and repair.

The use of analogue computing techniques does not provide the accuracy and stability offered by the latter systems. Analogue systems are prone to bias or drift and these characteristics are often more pronounced when the aircraft and equipment are subject to a hot or cold soak over a prolonged operating period. The only means of signalling rotary position in an analogue system is by means of synchro angular transmission systems. The older analogue aircraft – termed classic in the industry – therefore contain a huge quantity of synchros and other systems to transmit heading, attitude, and other rotary parameters. Pallet (1987) is an excellent source of information on many of the older analogue techniques, and Chapter 5 in particular encompasses a detailed description of the characteristics of synchronous data-transmission systems (synchros).

The older equipment is very bulky, heavy, and tends to be unreliable as there are many moving parts. This is not a criticism; the designers of the time did their best with the technology available and many very elegant engineering solutions can be found in this type of equipment. Another problem is that the skills required to repair and maintain some of the intricate instruments and sensors are gradually becoming scarcer and consequently the cost of repair continues to rise.

As has already been mentioned, these systems are very difficult to modify and this leads to significant problems when new equipment such as a flight management system has to be retro-fitted to a classic aircraft. This may be required to ensure that such aircraft comply with modern air traffic control (ATC) procedures, which are now far more complex than ever envisaged when the aircraft originally entered service over 30 years ago.

Typical aircraft in this category are the Boeing 707, VC10, BAC 1-11, DC-9, and early Boeing 737s. Many of these types are still flying and some, such as the VC-10 and the KC-135 (a Boeing 707 derivative), fulfil military roles. They will continue to do so for a while but gradually their numbers are dwindling as aircraft structural problems are manifested and the increasing cost of maintaining the older systems takes a toll.

5.6.2 Distributed Digital Architecture

The maturity of digital computing devices suitable for airborne use led to the adoption of digital computers, allowing greater speed of computation, greater accuracy, and removal of bias and drift problems. The digital computers installed on these early systems were a far cry from those available today, being heavy, slow in computing terms, housing very limited memory, and difficult to re-programme, requiring removal from the aircraft in order that modifications could be embodied.

A simplified version of the distributed digital architecture is shown in Figure 5.15. The key characteristics of this system as described below.

Major functional units contained their own digital computer and memory. In the early days of military applications, memory was comprised of magnetic core elements that were

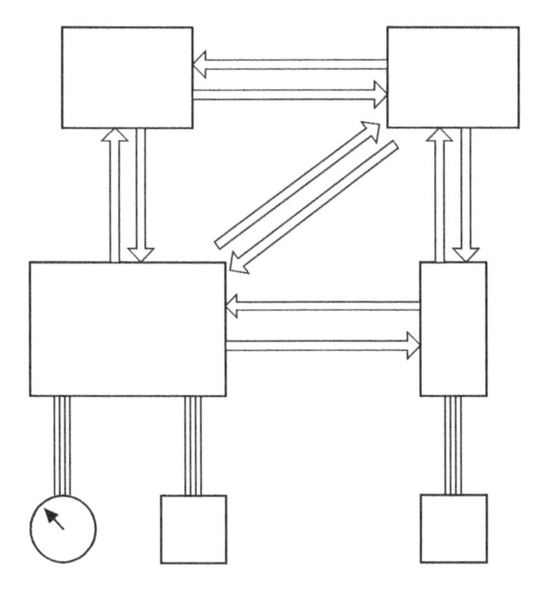

Architecture characteristics:

- Communications between major units via serial data buses - standard data bus
- Dedicated wiring to some sensors/displays
- Digital processing used for control functions
- Software reprogrammable (off-aircraft)
- Dedicated LRUs/ subsystems
- Increased use of standard interfaces
- Increased accuracy & performance

Figure 5.15 Distributed digital architecture.

very heavy and could only re-programmed off-aircraft in a maintenance shop. This, combined with the lack of experience in programming real-time computers with limited memory and the almost total lack of effective software development tools, resulted in heavy maintenance penalties.

At a later stage, as electrically re-programmable memory became available, this was used in preference to magnetic memory, especially for civil applications.

A good feature accompanying digital processing was the adoption of serial half-duplex (uni-directional) digital data buses – ARINC 429 and Tornado serial – which allowed important system data to be passed in digital form between the major processing centres on the aircraft. Although slow by today's standards (110 kbit/s for the ARINC 429 and 64 kbit/s for the Tornado serial data link), this represented a major step forward and navigation and weapon aiming systems secured major performance improvements by adopting this technology.

At this stage systems were still dedicated in function, although clearly the ability to transfer data between the units had significantly improved. The adoption of data buses – particularly the ARINC 429 – spawned a series of ARINC standards which standardised the digital interfaces for different types of equipment. This equipment therefore began to be standardised such that different manufacturers producing an inertial navigation system (INS) would prepare standard interfaces for that system. This eventually led to the standardisation between systems of different manufacturers, potentially easing the prospect of system modification or upgrade.

Displays in the cockpit were dedicated to their function as for the analogue architecture already described. The displays were still the intricate electro-mechanical devices used previously, with the accompanying problems. In later implementations the displays become multi-functional and the following display systems were developed in the civil field:

- electronic flight instrument system (EFIS)
- engine indication and crew alerting system (EICAS), Boeing and others
- electronic checkout and maintenance (ECAM), Airbus.

The Airbus EFIS/ECAM top-level system architecture is shown in Figure 5.12.

The data buses did remove a great deal of aircraft wiring though the question of adding an additional unit to the system at a later stage was still difficult. In ARINC 429 implementations data buses were replicated so that the failure of a single link between equipment did not render the system inoperable.

Overall the adoption of even the early digital technology brought great advantages in system accuracy and performance although the development and maintenance of these early digital systems was far from easy. Aircraft of this system vintage include:

- military: Jaguar, Nimrod MR1 and MR2, Tornado and Sea Harrier
- civil: MD-80 series, Airbus 310 and subsequent models, Boeing 757/767, 747-400, and 737-300/400/500, Avro RJ.

5.6.3 Federated Digital Architecture

The next development, federated digital architecture, is shown in Figure 5.16. The federated architecture – from now on all architectures described are digital – relied principally upon the availability of the extremely widely used MIL-STD-1553B data bus. Originally conceived by the US Air Force Wright Patterson Development Laboratories, as they were called at the time, it evolved through two iterations from a basic standard, finally ending up with the 1553B standard, for which there are also UK Def-Stan equivalents.

The eventual adoption of the 1553B data bus standard offered significant advantages and some drawbacks. The advantage was that this was a standard that could be applied across all North Atlantic Treaty Organisation (NATO) members, offering a data bus standard across a huge military market, and beyond. This has been an exceptionally successful application and the vast electronic component market meant that prices of data bus interface devices could be reduced as the volume could be maintained. It also turned out – as had been the case with previous data bus implementations – that the devices and hence the data buses were far more reliable that anyone could have reasonably expected. Consequently, the resulting system architectures were more robust and reliable than the preceding architectures.

The federated architectures generally use dedicated line replaceable units (LRUs) and sub-systems but the wide availability of such systems data meant that significant advances could be made in the displays and other aircraft systems such as utilities or aircraft systems where avionics technology had not previously been applied.

Although the higher data rates – approximately 10 times that of ARINC 429 and about 15 times that of the Tornado serial data link – this standard was a victim of its own success. The full-duplex (bi-directional), multi-drop protocol meant that it was rapidly seized upon as being a huge advance in terms of digital data transfer (which it was). However, system

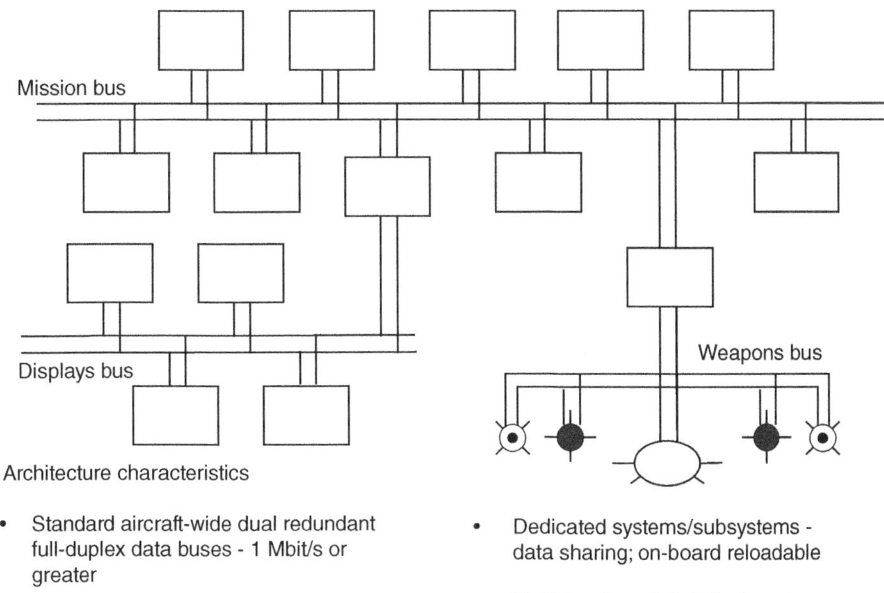

Figure 5.16 Federated architecture.

designers soon began to realise that in a practical system perhaps only 10–12 of the 30 possible remote terminals (RTs) could be used due to data bus loading considerations. At the time, it was the policy of government procurement agencies to insist that at system entry into service for a military system, only 50% of the available bandwidth could be utilised to allow growth for future system expansion. Similar capacity constraints applied to processor throughput and memory, therefore system designers were prevented from using the last ounce of system capability in terms of either data transfer or computing capability.

It was also recognised that it was not necessary to have every single data bus equipment talk to every other across the aircraft. Indeed there were sound system reasons for partitioning systems by data bus to enable all like-minded systems to interchange information with each other and then to provide inter-bus bridges or links between different functional areas. From this premise many architectures similar to the one portrayed in Figure 5.16 were evolved. With minor variations, this architecture is representative of most military avionics systems flying today: F-16 Mid-Life Update, SAAB Gripen, Boeing AH-64 C/D, and so on.

The civil community was less eager to adapt to the federated approach, having collectively invested heavily in the ARINC 429 standard that was already widely established and proving its worth in the civil fleets. Furthermore, this group did not like some of the detailed implementation and protocol issues associated with 1553B and accordingly decided to derive a new civil standard that eventually became ARINC 629.

MIL-STD-1553B utilises a 'command-response' protocol that requires a central control entity called a bus controller (BC) and the civil community voiced concerns regarding this centralised control philosophy. The civil-orientated ARINC 629 is a 2 Mbit/s system that uses a collision-avoidance protocol that provides each terminal with its own time slot during which it may transmit data onto the bus. This represents a distributed control approach. To be fair to both parties in the debate, they operate in differing environments. Military systems are subject to continuous modification as the Armed Forces need to respond to a continually evolving threat scenario requiring new or improved sensors or weapons. In general, the civil operating environment is more stable and requires far fewer system modifications.

ARINC 629 has only been employed on Boeing 777 aircraft, where it is used in a federated architecture. The pace of aerospace and the gestation time required for technology developments to achieve maturity probably mean that the Boeing 777 will be the sole user of the ARINC 629 implementation.

Along with the developing maturity of the electrical memory ICs, in particular non-volatile memory, the federated architecture enabled software re-programming in the various system LRUs and systems via the aircraft level data buses(s). This is a significant improvement in maintainability terms on the constraints that previously applied. For military systems it confers the ability to re-programme essential mission equipment on a mission-by-mission basis. For the civil market it also allows operational improvements/updates to be speedily incorporated.

The more highly integrated federated system provides a huge data capture capability by virtue of the extensive data-handling capability provided by the interconnected data buses.

5.6.4 Integrated Modular Architecture

The commercial pressures of the aerospace industry have resulted in other solutions and perhaps the most impressive is the embracing of COTS by companies such as Honeywell, illustrated in Figure 5.17.

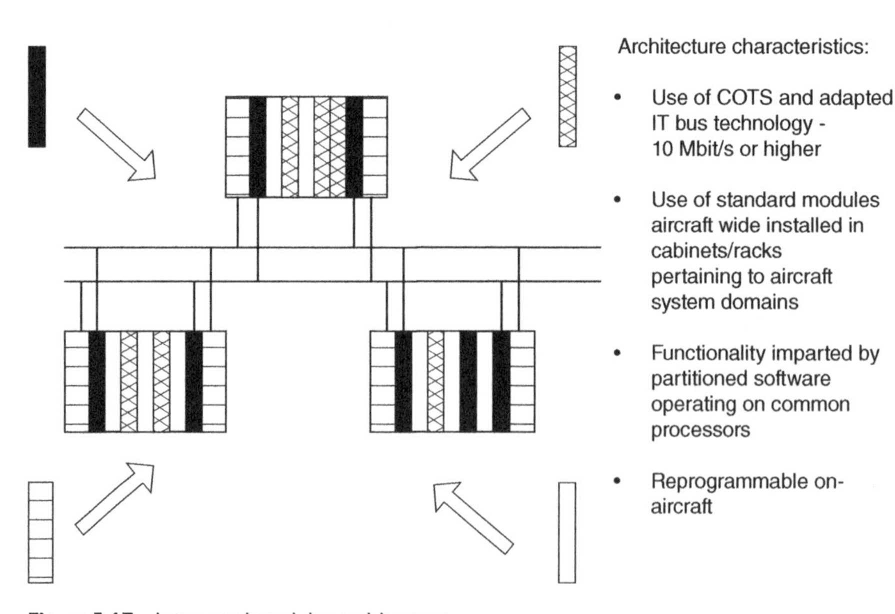

Figure 5.17 Integrated modular architecture.

The resulting architecture used ruggedised commercial technology to provide the data bus interconnections between cabinets. It is interesting that the business jet community has been the initial champion of many of the early developments in this architectural field.

When business jets were first introduced to the market some 30 years ago they were somewhat of an anachronism – generally they were used to represent the status of the chief executive of the company involved and aircraft utilisation was very low in terms of flying hours per annum. In the past decade, fractional ownership – the ability to own a part-share in a business jet – has meant that these assets are continually utilised, often flying in excess of 3000 hours per annum. Furthermore, the capability of the top of the range models such as the Gulfstream GV and Bombardier Global Express provide aircraft capable of flying 6000 miles from continent to continent. This offers huge advantages in transferring government ministers or chief executives across the globe in the fastest possible time. Finally, and perhaps most important in the present environment, this can be done with the utmost security.

In the case of the Primus Epic system used on the Raytheon Horizon business jet an adaptation of 10 Mbit/s Ethernet is used to provide the data bus that connects the modular avionics units (MAUs) that house the system modules. In a typical system a total of four MAUs house all the modules associated with the avionics function as well as those associated with utilities such as fuel, proximity switch interfaces and so on. Many of these are standard modules. In previous architectures the system units or LRUs were dedicated to function. In this architecture functions are spread across common systems modules and the systems' operational functionality imparted purely by software. A high-integrity software executive system provides the ability to partition sub-system software functions.

Other systems, such as the avionics system developed for the Airbus A380 aircraft, use a derivative of 100 Mbit/s Fast Ethernet called Avionics fast switched Ethernet (AFDX), now formalised as ARINC 664. In this system, the cabinets are partitioned by aircraft functional domains: cockpit, cabin management, energy management, and utilities management. These functional domain-related cabinets are populated by standard avionic modules supplied by one supplier. Figure 5.18 illustrates typical architectures from modern in-service aircraft (see Moir et al. (2013) for further explanation of these architectures).

The system functions are embedded in the partitioned software that is downloaded onto common processor modules (CPMs) by means of a dedicated download data bus. This necessitates the ability to partition – with adequate levels of integrity – the various aircraft system control laws within a particular CPM. The ability to implement and assure the integrity of not just single systems but the combined criticality effect of a number of systems hosted within the same functional area (hardware and software) cannot be ignored.

As well as the CPMs, many of the AFDX (now identified as the ARINC 664 data bus) and aircraft systems interface modules input/output (I/O) are standardised and supplied by one supplier. The main characteristics of this architecture are:

- common set of core modules used across all functional domains
- standardised processing elements
- common use of software tools, standards, and languages
- dispenses with a multitude of specialised and dedicated LRUs
- able to accommodate specialised aircraft system interfaces

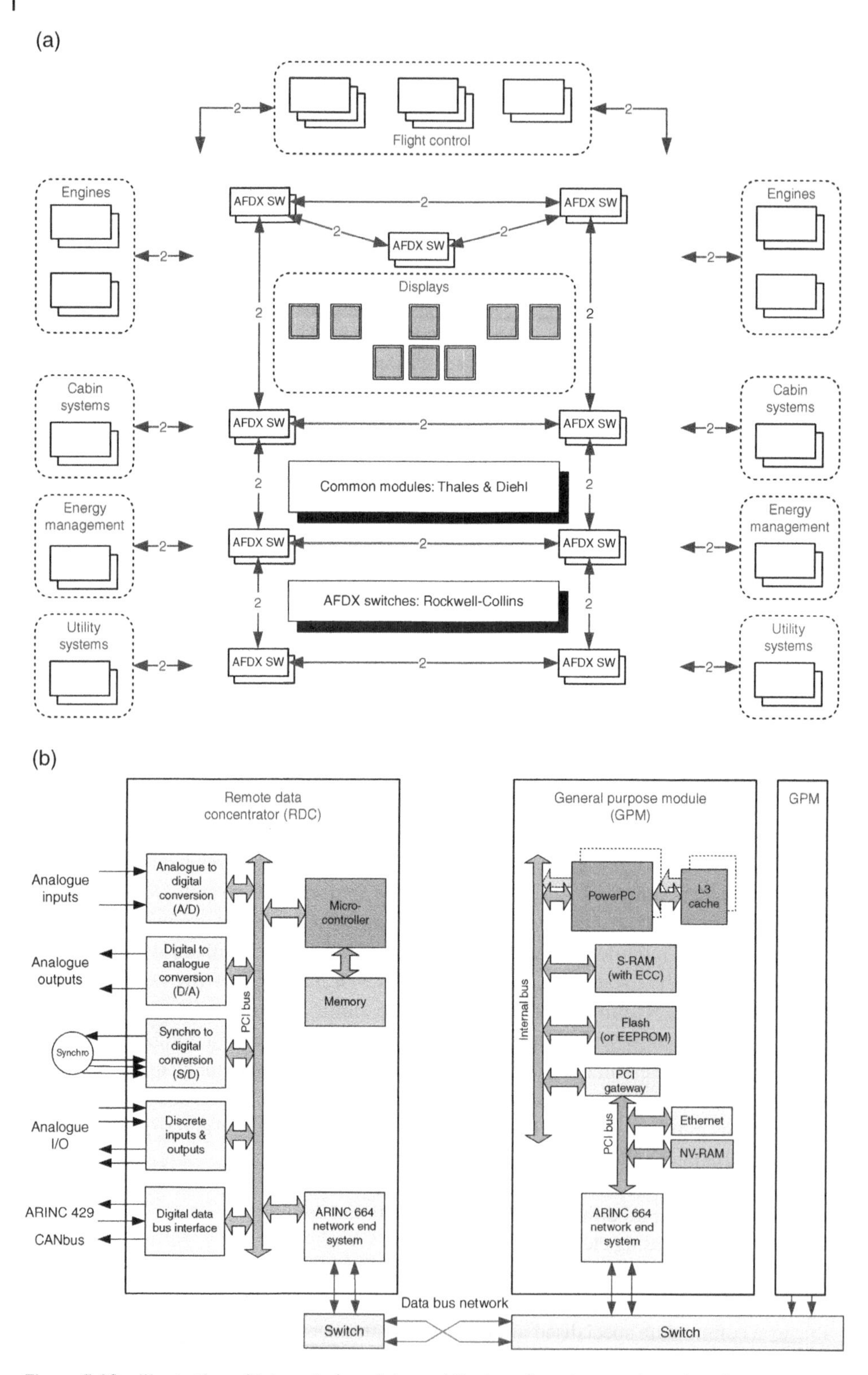

Figure 5.18 Illustration of integrated modular architecture from two modern aircraft.

- offers benefits of scale across the entire aircraft
- improves logistics for the airlines
- provides a scaleable architecture with scope for application to future projects.

The Boeing 787 uses a different approach with remote data concentrator units (equivalent to remote input/output [RIO]) to condition inputs and outputs, and general purpose modules to perform the processing task.

5.7 Example Architectures

This section will make use of the information presented previously in order to show how elements of different system architectures can be used to build architectures for individual and integrated systems. Some examples are given for different systems in the aircraft to illustrate what an architecture represents and how it is used in the design of a system. The architectures shown do not represent any specific aircraft, they are solely for example. No specific formal method is used for their representation and no method is recommended. The systems engineer must decide what is most suitable for their task, but must also ascertain if there is a common tool or method demanded by the project, in which case that must be adopted to ease communication.

5.7.1 Example 1: System Architecture

This architecture was shown previously in Figure 5.4 and is shown again here in Figure 5.19 simply to illustrate where the following examples fit.

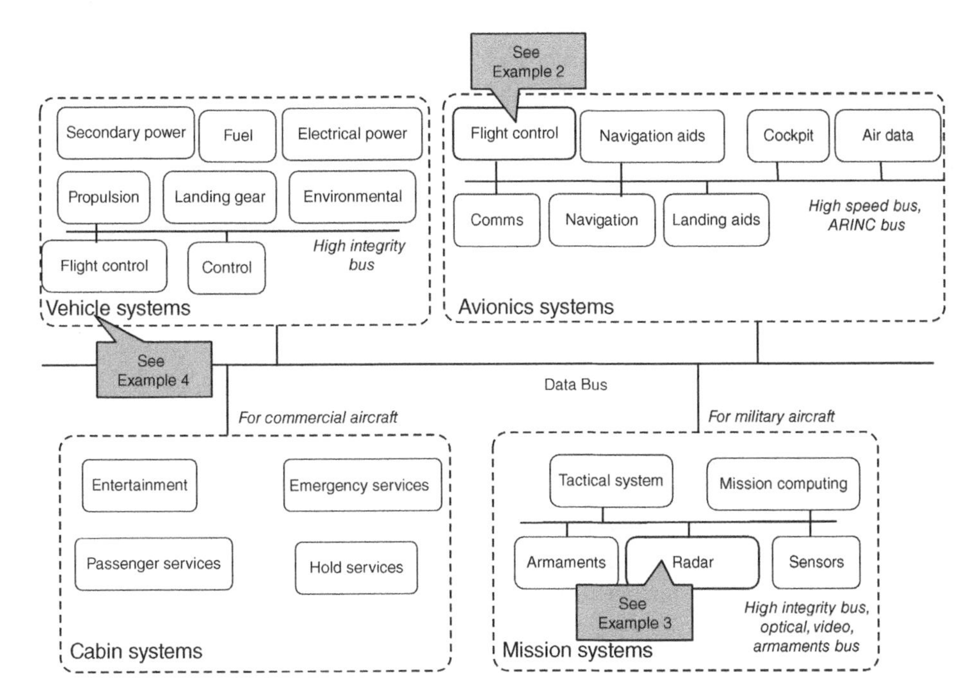

Figure 5.19 A top-level system architecture.

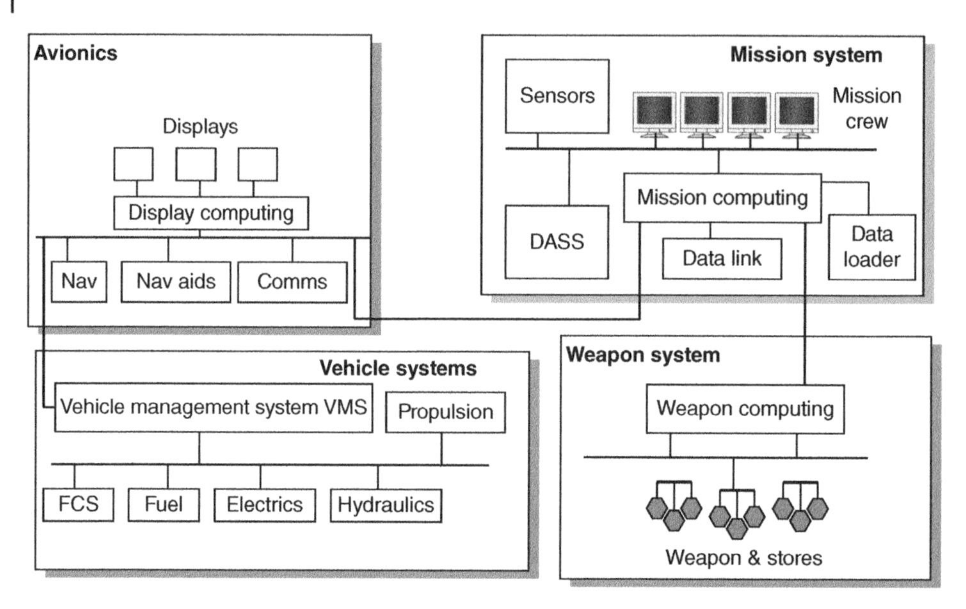

Figure 5.20 Example of a large military aircraft architecture.

The architecture can be developed further to show more detailed definition and to show how it can be used to further define requirements, responsibilities, and functional segregation. Figure 5.20 shows how the architecture can be used to segregate sub-system architectures in a military aircraft. For reasons of integrity it is necessary to separate the main functions of avionics, aircraft systems, the mission system, and the weapon system, as well as segregating their wiring harnesses. The main reasons are to ensure a suitable integrity for the aircraft systems functions and to apply specific armaments standards to the weapon system.

Usually this leads to use of military standards for the data bus, for example MIL-STD-1553B or STANAG 3910 for all functions. However, in the case of converting an existing commercial platform into a military type, for example for surveillance purposes, there may be sound reasons for keeping the commercial avionics and blending that with military standard systems. Figure 5.21 shows an architecture used as the basis for achieving such a system by using the mission computer as a buffer between the two systems.

5.7.2 Example 2: Flight Control System

In Figure 5.19 the flight control system (FCS) is shown as a single box as part of the airframe systems. Figure 5.22 shows the contents of this box as an FCS top-level generic architecture and shows some preliminary interfaces to the aircraft in the form of the displays and controls, the air data system, and the control surfaces.

This has been developed further to the architecture shown in Figure 5.23. This is a historic diagram since it shows a flight control structure developed for an experiment to demonstrate fly-by-wire control in an unstable aircraft, the fly-by-wire (FBW) Jaguar

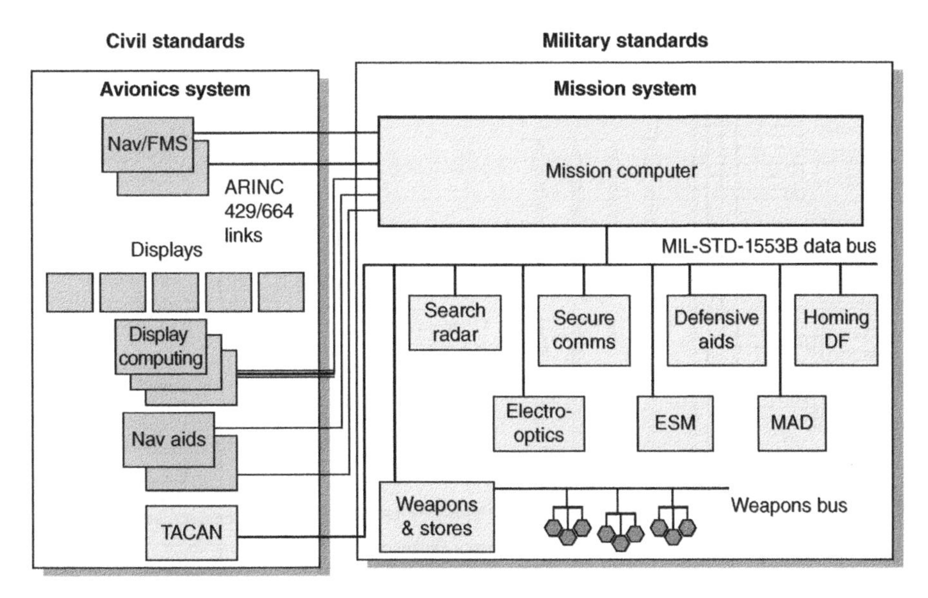

Figure 5.21 Allocating responsibilities between civil and military systems. DF, direction finding; ESM, electronic support measures; Nav Aids, navigation aids; Nav/FMS, navigation/flight management system; MAD, magnetic anomaly detector; TACAN, tactical air navigation.

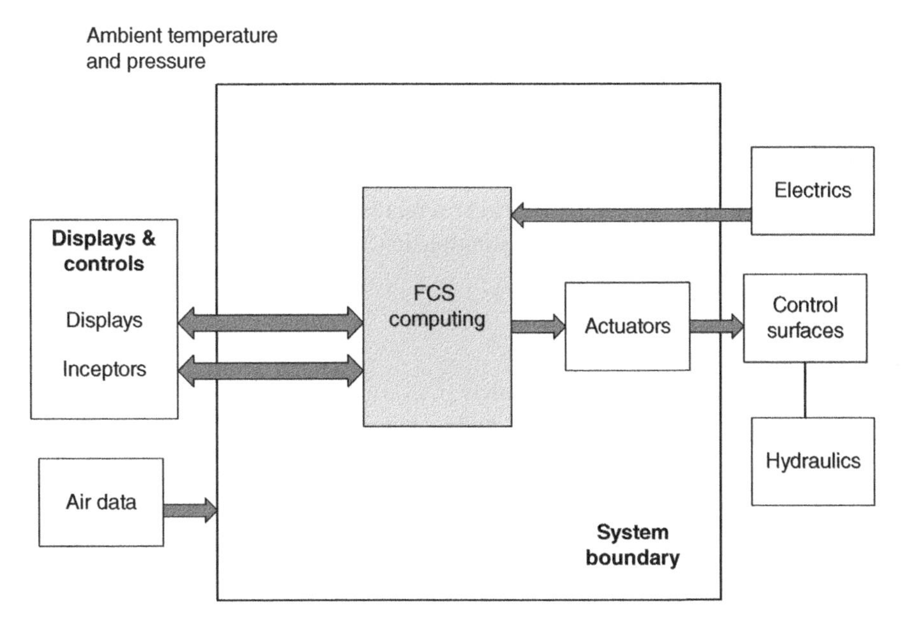

Figure 5.22 A generic FCS architecture.

(Weller 2018). It records some important decisions made on a quadruplex implementation of a software intensive system designed for high integrity. This architecture led on to a complete design that was successfully tested on a Jaguar aircraft and formed the basis for following aircraft in the UK such as EAP and Typhoon.

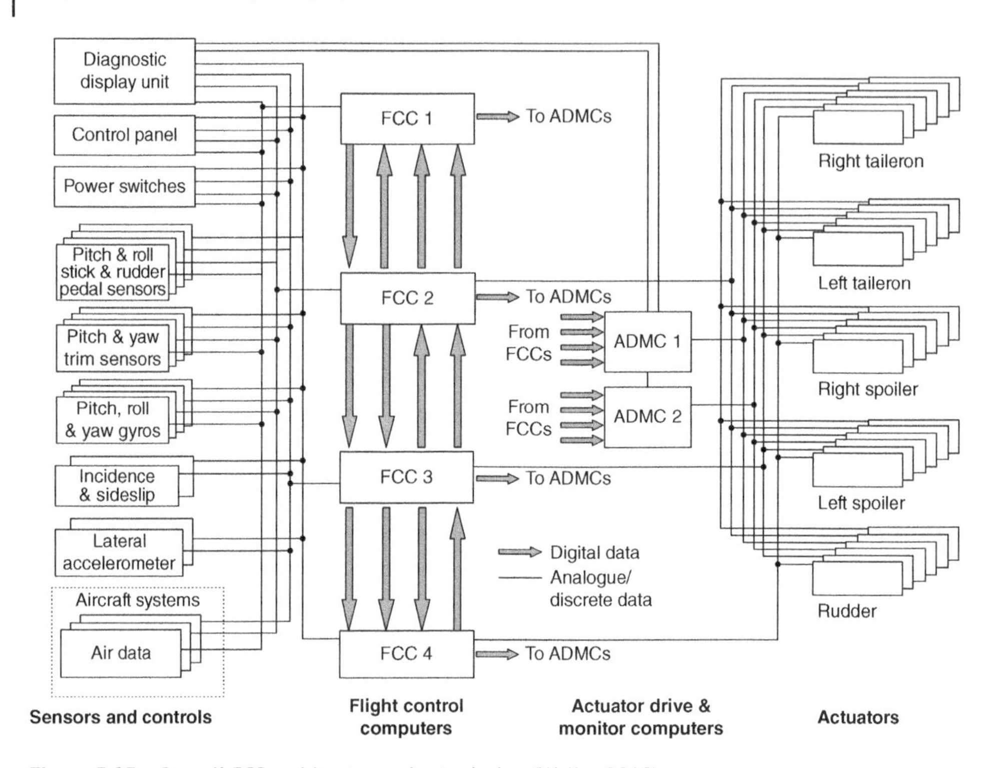

Figure 5.23 Overall FCS architecture prior to design (Weller 2018).

5.7.3 Example 3: Radar System

This example has been chosen as it appears to be a relatively simple, although very high value, system and it is used as a verification example in Chapter 7. The customer's operational requirement will include:

- detecting a particular set of targets
- detecting targets at a particular range
- discriminating target from background clutter
- detecting, classifying, identifying, tracking, and locking on to targets.

Figure 5.24 shows a sketch of a system that will meet these requirements. This is the beginning of an integrated weapon system in which the radar is physically integrated into the airframe and is then integrated with the aircraft avionics, the armaments system, and the cockpit to provide a system that can detect targets, identify, them, track them, and then engage them with missiles with a single pilot to perform these tasks and to fly the aircraft.

Figure 5.25 shows the main elements of the sketch put into an architecture to show the relevant interfaces. This architecture is an interconnection of the basic navigation systems, the radar sensor, and the armaments system – in a military aircraft. Thus there is no need for interconnection of commercial and military standards – all the sensors and data buses are designed to military standards. The resulting system is a comprehensive target acquisition and engagement system in a fast jet military aircraft operated by a single pilot.

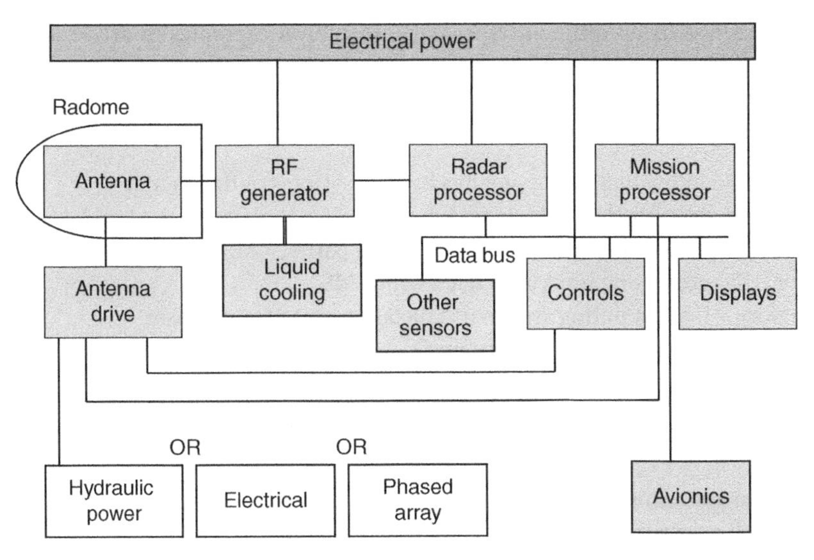

Figure 5.24 Scheme of a radar system.

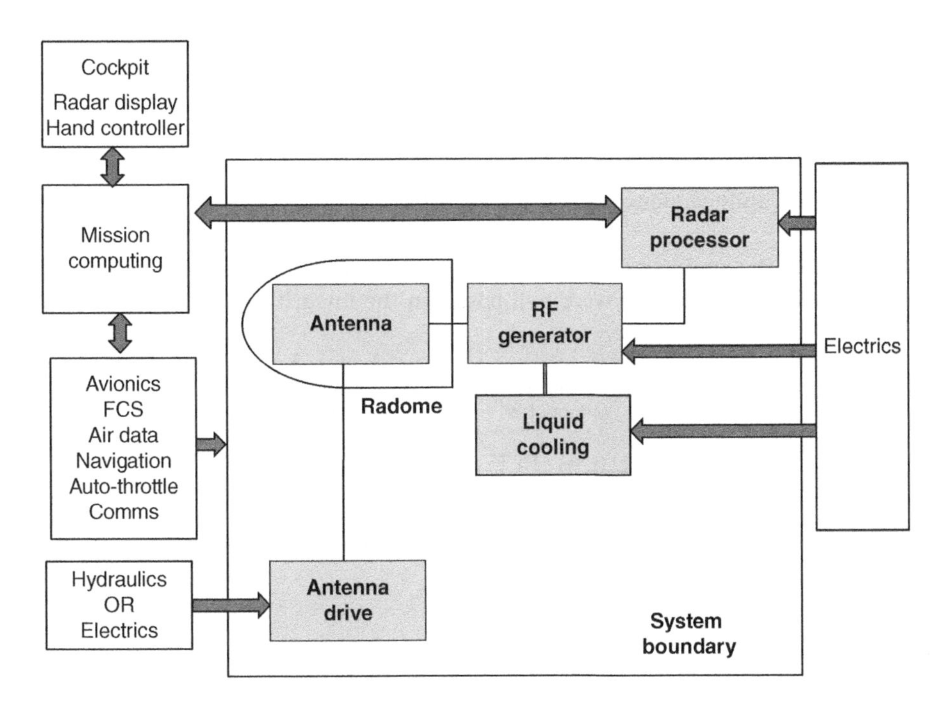

Figure 5.25 Architecture of a radar system.

5.7.4 Example 4: Vehicle Systems Management

This rather lengthy example explains how a computing system for the set of power and mechanical systems, known as vehicle systems, was developed for a range of systems that are described in section 12.2. This is a historical example but it illustrates how a whole system can be developed using different types of architectural examples to arrive at a robust solution. The final system was demonstrated on an experimental aircraft and it formed the basis for systems in use today, albeit in rather more modern implementations. The following terminology has been used:

- vehicle systems – the individual power and mechanical systems of the aircraft
- VMS – the system that controls all vehicle systems
- VMS architecture – the VMS processors and data bus as part of VMS
- VMS processor – the processing units associated with VMS
- processor/memory – the computing system installed in each VMS processor
- I/O – signals associated with the vehicle systems.

The process began with the desire to develop a computing system for the interfacing, control, and management of a number of systems. The nature of the problem is illustrated in Figure 5.26.

These systems were a challenge because they were predominantly mechanical or electrical and usually involved a transfer of significant amounts of energy to perform a function. They were also directly associated with the safe operation of the aircraft. This is in contrast to the avionics systems, which are predominantly managing information. They had a diverse range of input and output characteristics with a wide range of sensor types.

There were some decisions that could be made early in the project based on the integrity of the systems that would determine which system could be entrusted to computer control and which were of a level of integrity that ruled them out of this system. This was decided on the basis of the need for complex functions to be performed balanced against a need for independence for safety reasons, and the result is illustrated in Figure 5.27.

Discussion with the engineers of individual systems led to an agreement on which systems were suitable and which were not, based on the integrity of each system, the

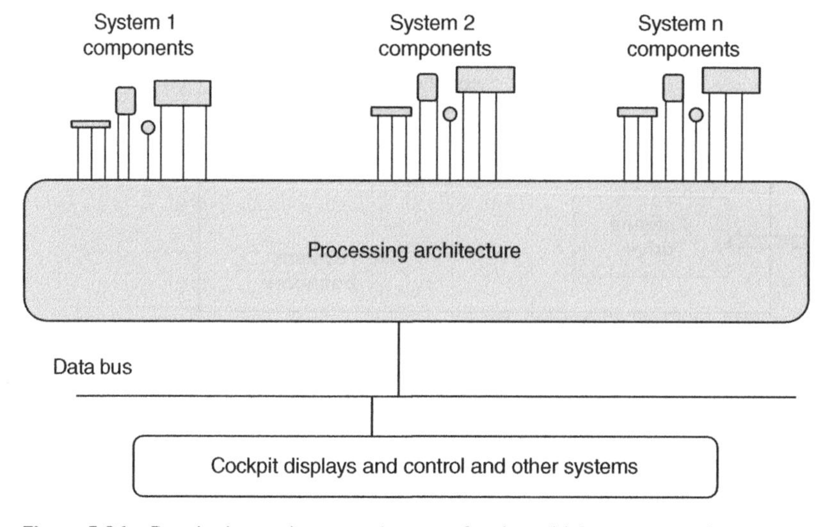

Figure 5.26 Developing an integrated system for the vehicle systems – the start.

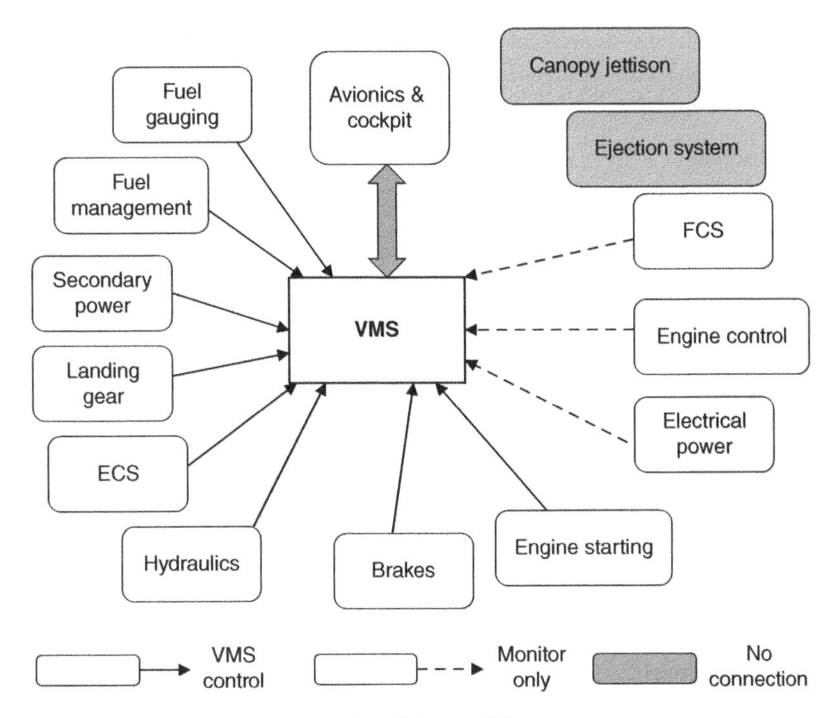

Figure 5.27 An early selection of candidate vehicle systems.

need for closer control of the systems, weight savings, and engineering judgement. The result showed that it was not suitable for the systems associated with crew escape to be considered for inclusion because there was a need for those systems to be simple, to prevent inadvertent ejection, and to work immediately when commanded. The engine control system was traditionally the realm of the engine manufacturer and there was an increasing trend to incorporating the control system on the engine, therefore in this case it should remain independent. The flight control system for this aircraft had already been designed as a quadruplex system and for its use in a full authority fly-by-wire aircraft unstable by design it was considered wise to keep this as a separate system with data interfaces as required.

Each of the remaining systems used a variety of sensors to detect conditions within the system. These formed a set of input types with a diversity of type, range, source impedance, and slewing rate. Typical examples are:

- proximity switch analogue or discrete
- fuel gauge probe AC capacitance
- gearbox speed pulse probe
- actuator position potentiometer, rotary or linear
- actuator position variable transformer, rotary or linear
- temperature thermistor or platinum resistance
- pressure piezo-clcctric
- pressure switch discrete

The outputs from the vehicle systems to perform some form of action in the aircraft also formed a set of output types with a diversity of type, range, load impedance/resistance, and reactance. Typical examples are:

- valve discrete
- DC motor DC drive
- actuator drive low-power analogue
- torque motor low-current servo drive
- fuel pump high-power servo drive
- filament lamp lamp drive

Links to other systems on the aircraft by data bus were required and these included:

- other processors in the system
- avionics and mission systems
- cockpit/flight deck displays and warnings
- prognostics and health warning
- accident data recording
- flight test instrumentation.

All of the physical inputs and outputs were to be connected to a set of input conditioning modules to interface with the aircraft wiring and to convert the input signals into a digital signal capable of being used by a processor, and to convert demands from the processor into a form suitable for operating devices in the aircraft. Links to other systems were made by means of a common type of data bus. This arrangement is shown in Figure 5.28, which shows the structure of a generic VMS processor in terms of acquiring inputs, performing some form of processing function, and generating outputs with appropriate feedback.

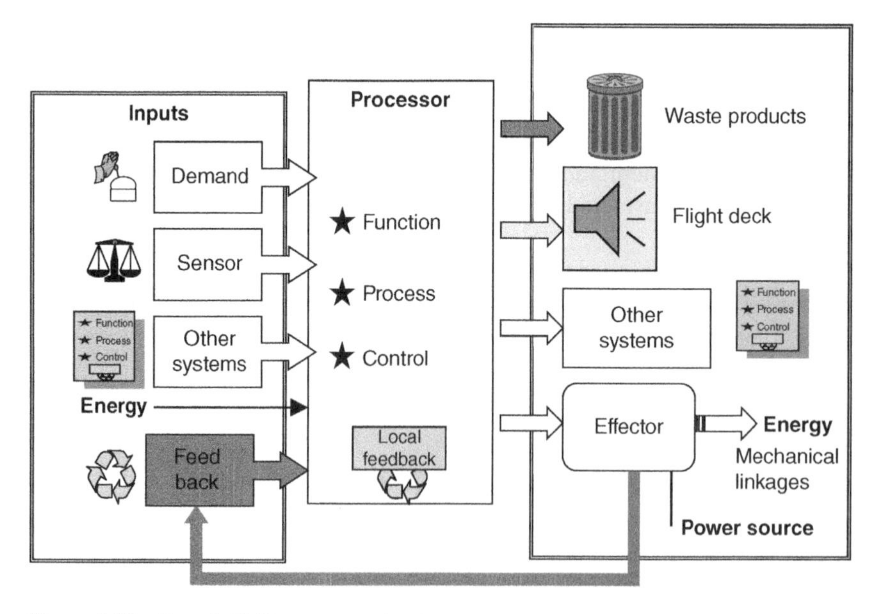

Figure 5.28 Generic VMS processor structure.

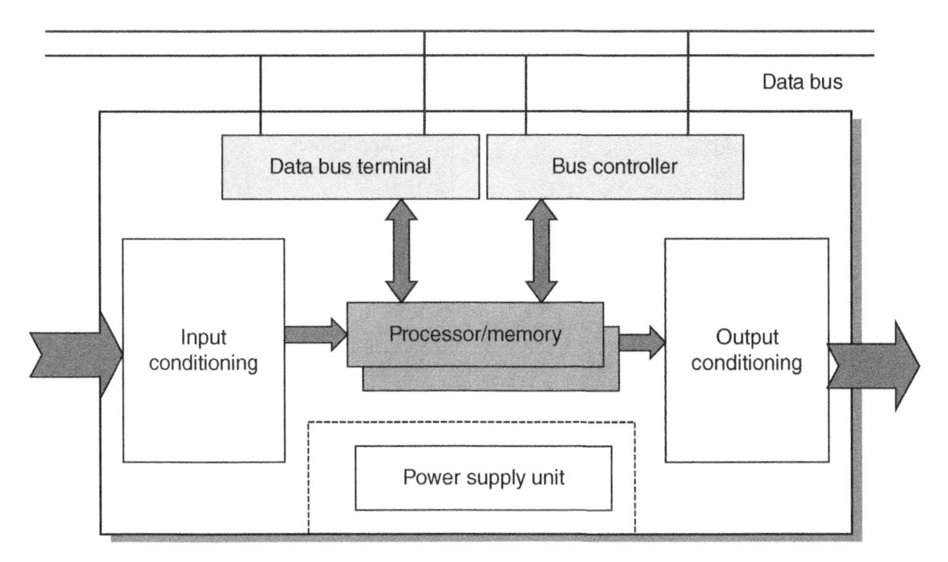

Figure 5.29 A generic VMS processor architecture.

To perform the functional algorithms of the systems a processor and memory combination was required, to send information to the other aircraft systems a data bus was required, and the VMS processor needed its own internal power supplies. All of this information led to discussions with a supplier to arrive at a physical architecture for a VMS processor, as illustrated in Figure 5.29.

A generic architecture of the processor memory module was attempted before any attempt was made to select a processor type. This allowed responsibility to be attached to the suppliers of the software. A brief analysis of the functions to be performed within the system led to a preliminary assessment of the magnitude of the task, the processor loading, memory capacity required, and data bus loading. This led to the preliminary software processing architecture shown in Figure 5.30.

With all this preliminary work completed there was now an understanding of the architecture of the total system, the processors within the system, including the structure of I/O conditioning, and the preliminary software architecture, without yet a selection of any detailed parts. This was sufficient to enable a specification to be issued to the supply industry to start the design of the processors. The following work was undertaken to analyse each system in order to define the overall architecture using the following information from the individual system engineers:

- definition of input characteristics
- definition of the functions to be performed using a functional requirements document
- definition of output characteristics
- understanding of the integrity of the connected systems
- definition of data bus connections to other systems.

One result of this is a detailed design of the processor boxes shown in Figure 5.31 in which each processor has been allocated a set of modules associated with each type of input and output signal.

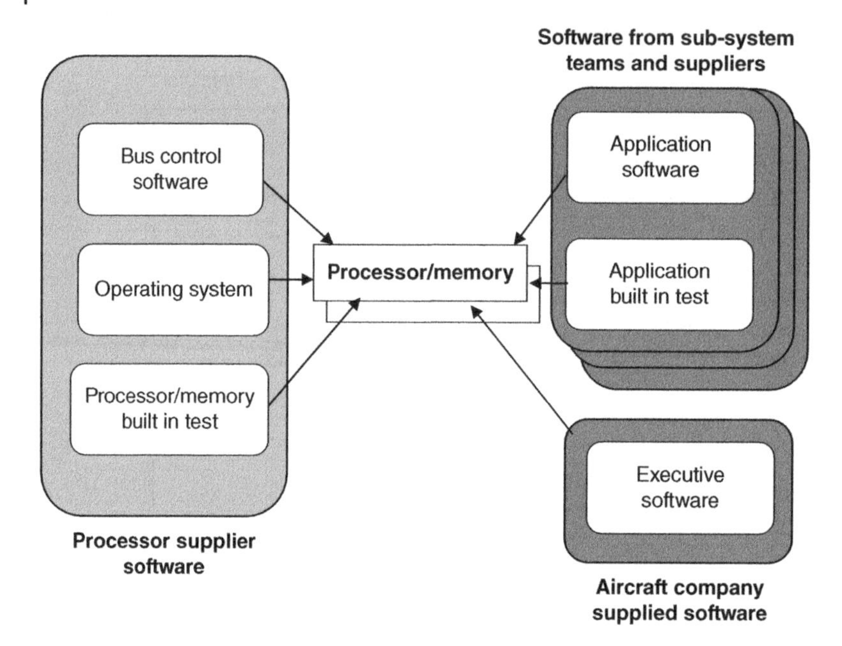

Figure 5.30 A generic processor/memory architecture.

A study was conducted to decide how each vehicle system was to be connected into the VMS architecture and how the functional processing task was to be divided between one or more computers. This was to achieve the appropriate separation between channels and to provide any redundancy that was required. Doing this for all systems led to an optimum number of VMS processors. A location in the aircraft was sought for each of these that would optimise the length of wiring harnesses (and mass) and provide a reasonably benign temperature and vibration environment. This led to a decision to use four VMS processors, as shown in Figure 5.32. The data bus in this architecture was to conform to a common standard for the aircraft, and a common high-order software language was to be used. This was the final design and drawings were prepared to allow the system to be procured and installed into the aircraft after suitable testing.

Architectures played a significant role in the development of the system from its initial concept through to the design stages. Many decisions were made early on in the project with no certain knowledge of the final implementation. The number of stakeholders was high to enable all the vehicle systems to be represented in the design process. Agreements were reached and very few changes were made after commitment to the design. The role of the architecture as a tool in this process is illustrated in Figure 5.33.

Figure 5.34 shows an overall system architecture of the aircraft used for the basis of some of the individual system architectures. This shows how the individual systems fit back up into a top-level architecture to produce an increased level of detail.

After the success of this development other projects followed a similar path. One important aspect of future project systems was the development that enabled the I/O processing

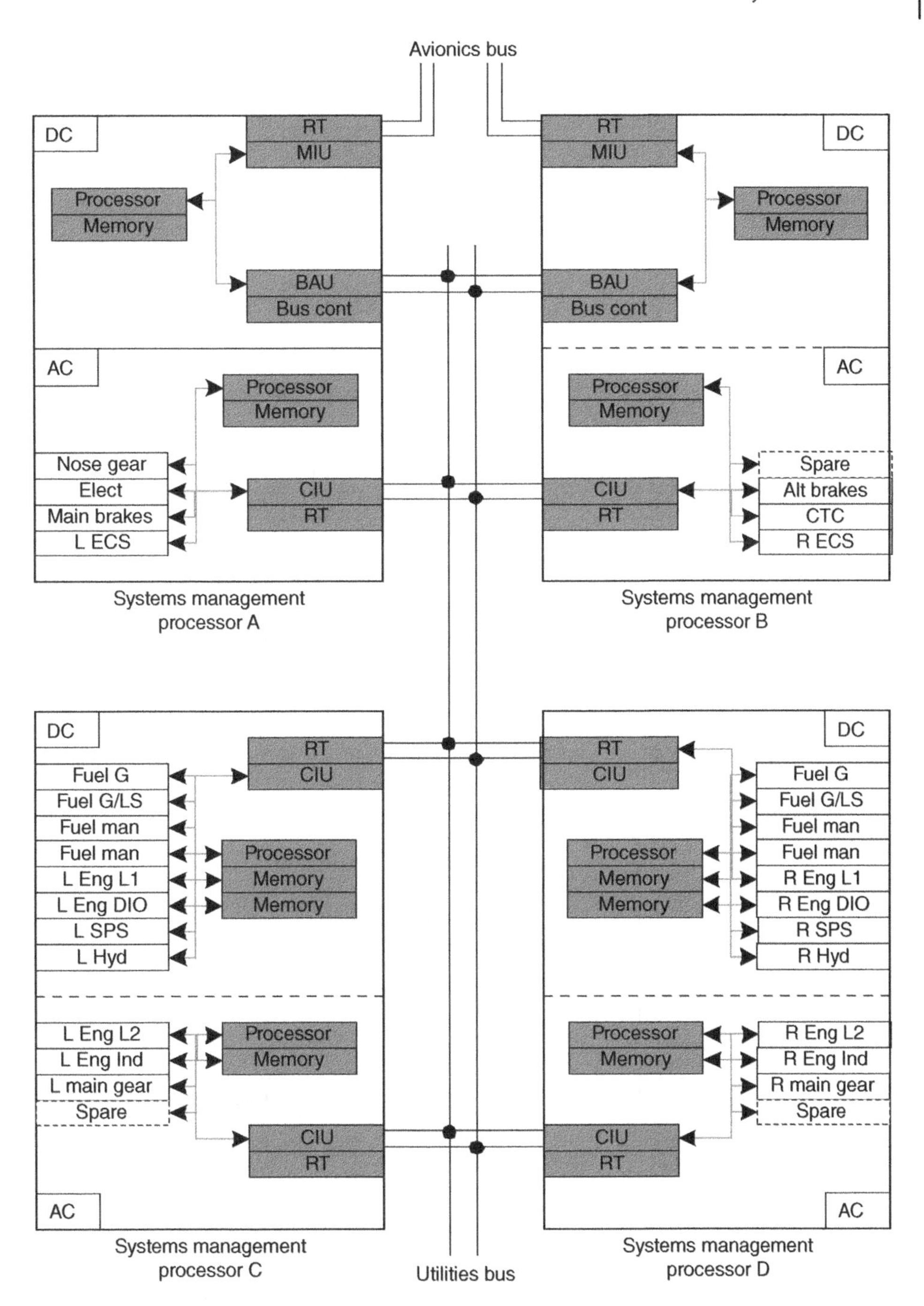

Figure 5.31 Input/output conditioning modules.

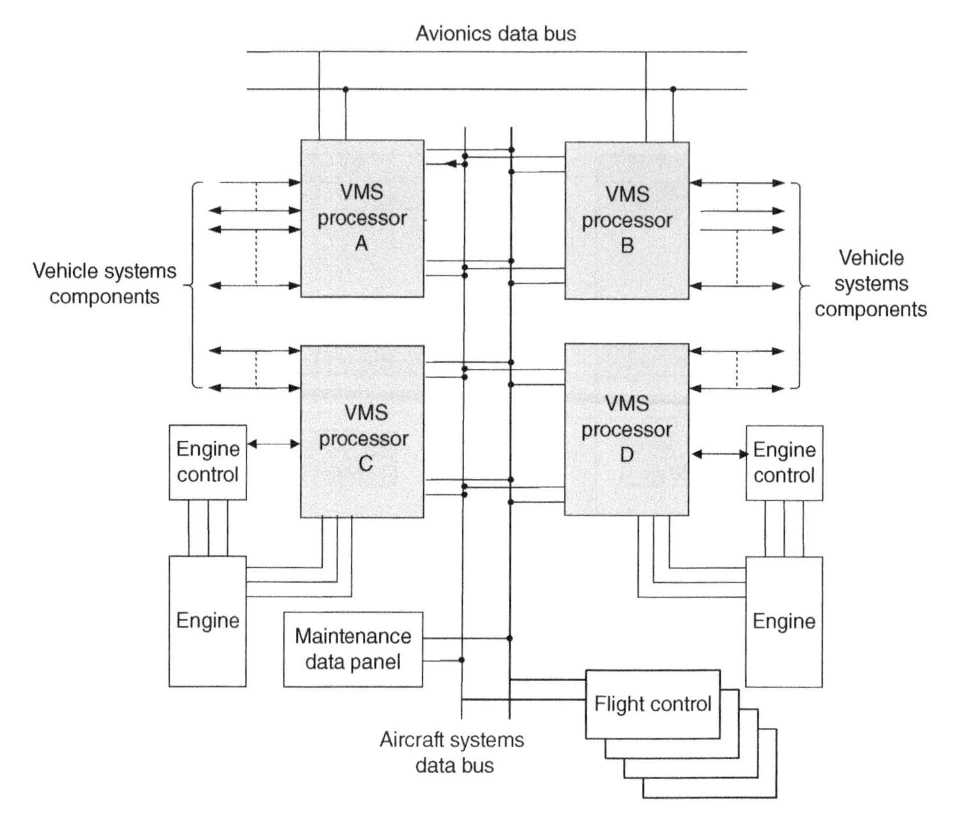

Figure 5.32 The complete VMS.

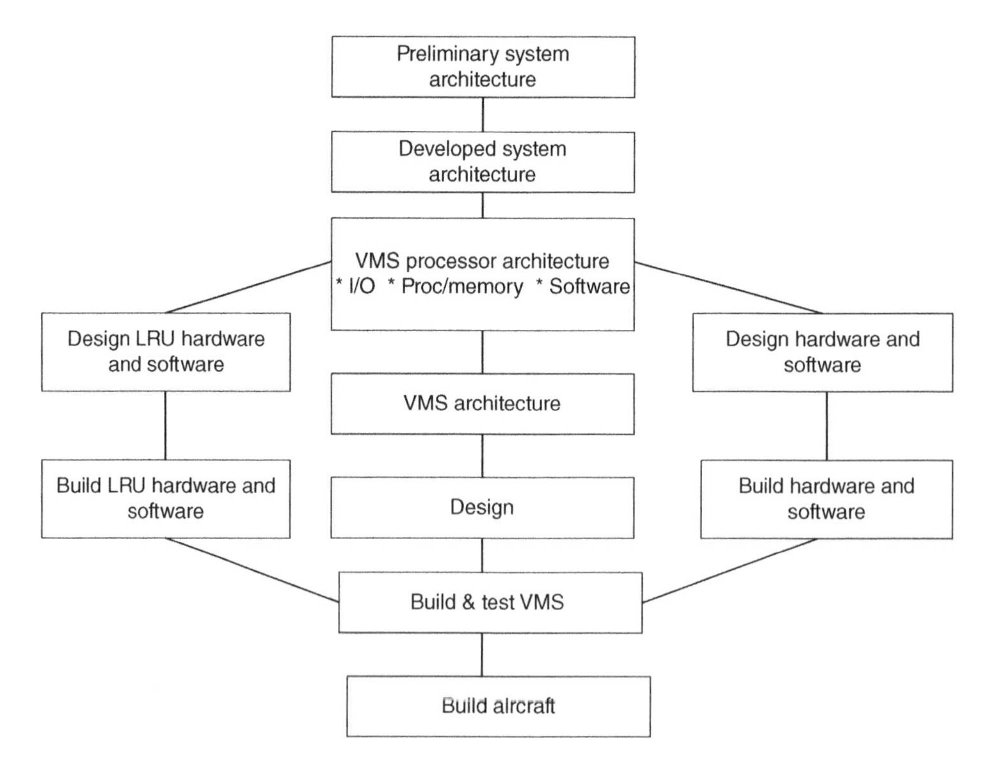

Figure 5.33 The role of architectures in the design process.

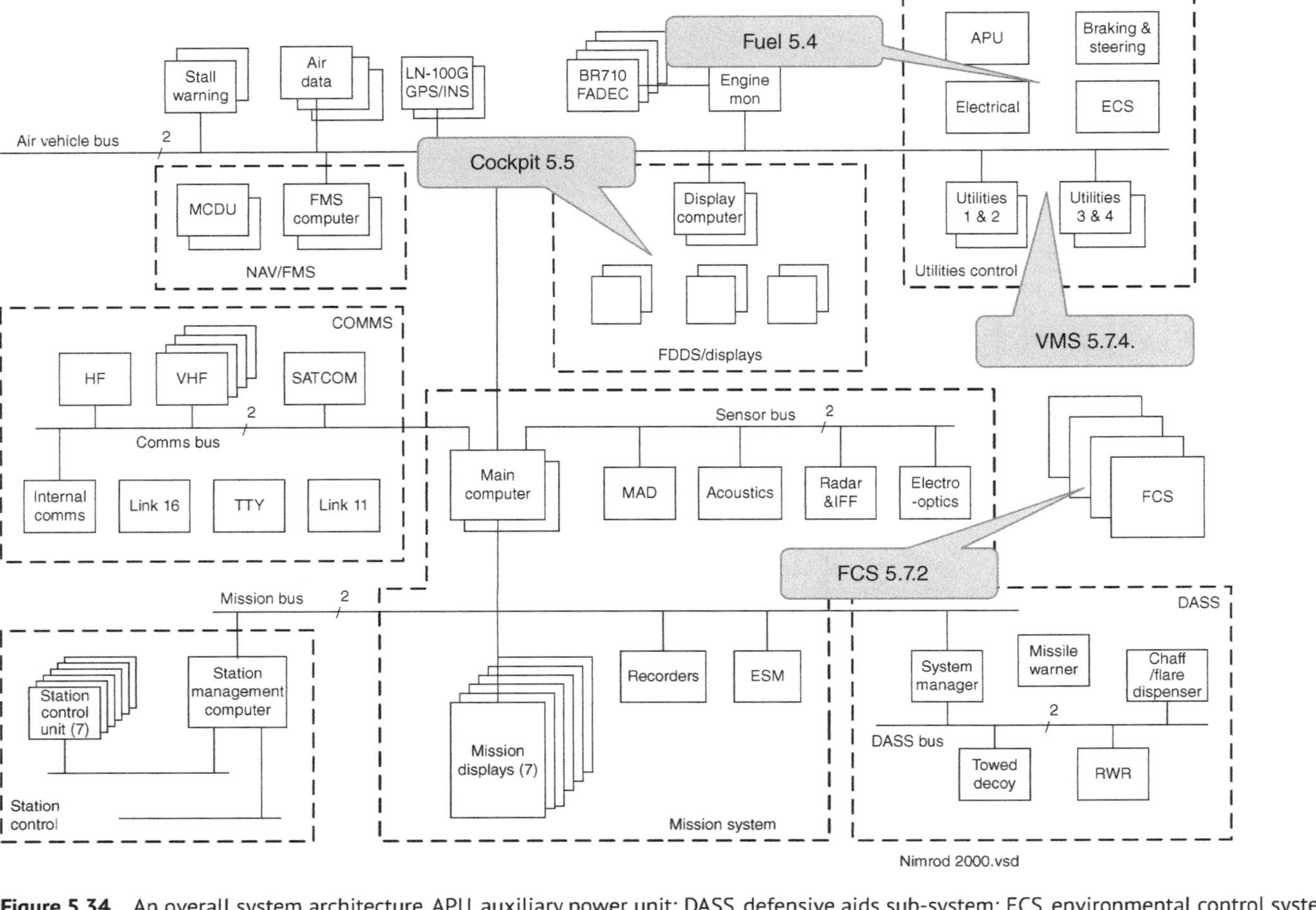

Figure 5.34 An overall system architecture. APU, auxiliary power unit; DASS, defensive aids sub-system; ECS, environmental control system; ESM, electronic support measures; FADEC, full authority digital engine control; FCS, flight control system; FDDS, flight deck display system; FMS, flight management system; GPS/INS, global positioning system/inertial navigation system; HF, high frequency; IFF, identification friend or foe; MAD, magnetic anomaly detector; MCDU, multi-function control and display unit; Nav/FMS, navigation/flight management system; RWR, radar warning receiver; Sat Com, satellite communications; TTY, teletype; UHF, ultra high frequency.

to be separated from the processing function to devices known as RIO. With developing technology it became possible for RIOs to be installed in parts of the aircraft that had a relatively harsh environment, but in remote areas that further reduced the lengths of wiring harnesses and reduced weight, whilst keeping the processing components in a cooled environment. Figure 5.35 show how this separation was devised using a VMS architecture and Figure 5.36 shows a VMS architecture using two RIOs.

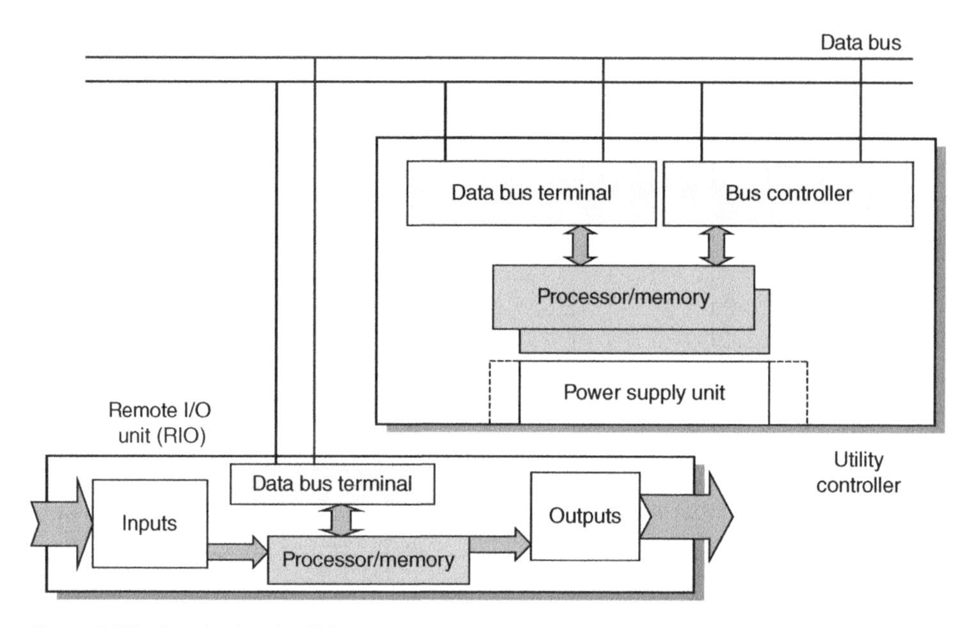

Figure 5.35 Developing the RIO concept.

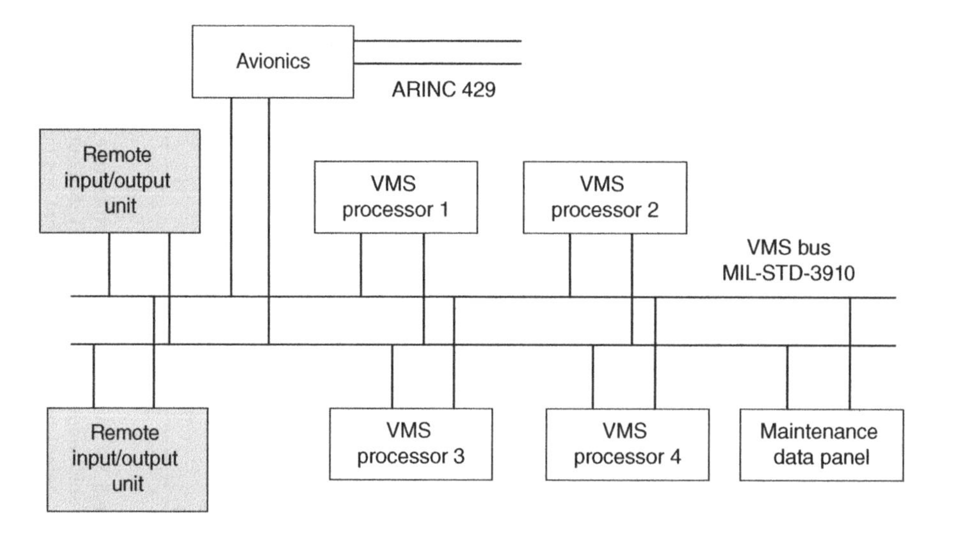

Figure 5.36 Example VMS architecture using RIOs.

Exercises

1 Design an architecture for a remotely piloted aircraft to deliver commercial items in a rural setting. Identify all essential sub-systems on board and identify the ground segment and suitable telemetry links.

2 Repeat the above exercise for an autonomous system.

3 Describe how you would prepare the evidence for certification of these systems.

4 Put yourself in the position of the engineer responsible for the data bus structure for a long-range commercial aircraft. Produce a diagram of your proposed data bus architecture and identify all the bus types you need. Define the key characteristics to enable users of your architecture to integrate to meet their system performance and to meet the safety and performance requirements of the aircraft.

5 Take into account the notion of a human being as an integrated system given in Chapter 6 and devise a system architecture of a human. How does the performance of this differ from an aircraft architecture?

6 Compare and contrast the architecture of a smart phone and an aircraft. Comment on the similarities and differences between the two architectures.

References

Chadwick, J. (1987). *The Decipherment of Linear B*. Cambridge University Press.
Moir, I., Seabridge, A., and Jukes, M. (2013). *Civil Avionic Systems*, 2e. Wiley.
Pallet, E.H.J. (1987). *Aircraft Instruments & Integrated Systems*. Longman, ISBN: 0-582-08627-2[1].
Weller, B. (2018). *A History of the Fly-by-Wire Jaguar*. BAE Systems Heritage Department.

Further Reading

Chau, S., Dang, V., Xu, J., and Lu, J. (2006). *An Automatic Technique to Synthesize Avionics Architecture. Proceedings of the First NASA/ESA Conference on Adaptive Hardware and Systems (AHS'06)*. IEEE Computer Society, Washington, DC.
Maier, M.W. and Eberhardt, R. (2002). *The Art of Systems Architecting*, 2e. CRC Press.
Moir, I., Seabridge, A., and Jukes, M. (2013). *Civil Avionic Systems*, 2e. Wiley.
Salomon, U. and Reichel, R.. (2013). *Automatic Design of IMA Systems*. IEEEAC Paper #2018, Version 3, Updated 19/01/2013.
Seabridge, A. and Leon, S. (2016). *EAP – The Experimental Aircraft Programme: Britain's Last Manned Aircraft Demonstrator*. BAE Systems Heritage Department.
Stevens, R., Brook, P., Jackson, K., and Arnold, S. (1998). *Systems Engineering – Coping with Complexity*. Prentice Hall.

6

System Integration

6.1 Introduction

The term 'system integration' is understood to mean different things by different people and by many different organisations. This chapter will examine some aspects of system integration and offer the reader some of the potential down sides to encourage a level of caution and scrutiny of the design to ensure safe solutions. This is discussed further in Chapter 11.

Integration arises because engineers want to pursue solutions that are efficient in operation and in their use of equipment. To achieve this latter desire, engineers often seek to incorporate many functions into single devices such as hardware components, line replaceable items (LRIs) or software packages. The design drivers for this are to do with cost, weight, installation volume in the aircraft, reliability, and in some cases technological challenge. In addition to this most computing devices encourage multi-tasking solutions, as illustrated by personal computers.

Whilst the results have many tangible benefits, there are some drawbacks. Some integrated solutions appear on the surface to be simple and to offer a simple man–machine interface, rather like the Apple iPod. This is achieved by quite high levels of sophistication and complexity within the device. In a complex aircraft system encompassing all of the avionic and aircraft systems the result is hugely complex with interactions of hardware, software, data, and functions occurring throughout the system.

Perhaps the easiest way to grasp the concept is to provide an example of a familiar integrated system such as the human being.

The human being is a good example in terms of understanding system integration as it embodies all the attributes that an engineer would like to incorporate into a perfect system. An aircraft can be considered as a complex set of interacting sub-systems, not dissimilar to a human being. Indeed, with the emergence of unmanned air systems, and especially autonomous unmanned air systems, this ideal of an integrated system behaving with minimal operator intervention is fast becoming the target of future systems designers. Figure 6.1 shows the similarities that can be observed between the human being and an aircraft in systems terms.

The human being has a frame consisting of a skeleton with a surrounding structure. Into this structure are incorporated a source of energy converted from an appropriate fuel, a

Design and Development of Aircraft Systems, Third Edition. Allan Seabridge and Ian Moir.

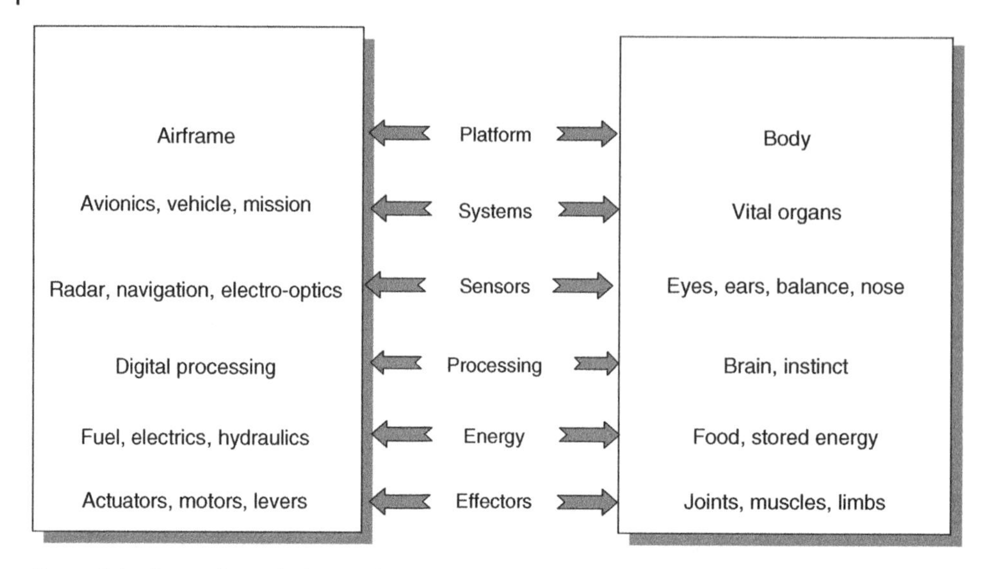

Figure 6.1 Comparison of a human being and an aircraft as a system.

complex processing system, mechanisms for converting energy into movement, the ability to sense the conditions of the surrounding environment, mechanisms for reacting to and compensating for climatic conditions, and finally a means of undertaking purposeful activity or providing motive power. The human being can perform a number of complex tasks simultaneously:

- acquire and process information from various sensors
- think, analyse, calculate, judge
- perform vital functions – breathing, blood circulation, balance, movement, digestion, etc.
- react to information received in a purposeful manner
- react instinctively to external stimuli
- exercise moral and ethical considerations in decision making
- prioritise all these functions for maximum effectiveness.

This can be expressed as the merging or integration of individual sources of data, information, and knowledge to perform a function. The human being is an effective and innate integrator of information. For example, knowing where one is and where one needs to move to is a simple, intuitive function: *Our sense of orientation is derived from integrating information from balance organs in the inner ears with that from muscle and joint receptors that signal the position of the limbs, together with visual cues* (Ashcroft 2000). This ability to assimilate and integrate information and to react in many different ways is true for many human functions.

The human being, therefore, represents the successful interaction and integration of complex information acquisition and processing together with a means of energy transformation that enables it to achieve an outcome. This, together with the ability to compensate for changes in the physical environment and the ability to assess threats and react to them is a central attribute. This ability has assured the position of mankind as the ‘supreme’ animal in the global environment. This is also a key requirement of aircraft weapon systems. In the aircraft this use of information from sensors, knowledge of the outside world, and transfer of energy into action must be processed by some form of computing system to mimic the intuitive processing of the human brain. Bringing together all these actions to achieve a desired outcome using mechanical and electrical/electronic means in a machine with a human operator is a key skill of system integration. It is the genetic influences and received wisdom that shape the intellectual development of the human being over a long period of evolution according to Darwin. An aircraft, on the other hand, must be designed to meet a particular set of requirements in a relatively short time period. A diverse range of skills and processes is needed to do this to produce a vehicle capable of achieving a wide range of activities in a diverse range of operating scenarios. This task is system integration.

6.2 Definitions

The term system integration can be interpreted in a number of ways, and the following interpretations are commonly used in the aircraft industry:

- *Integration at the component level*: The ability of a component or LRI to ensure that the discrete function it offers contributes to the overall system in which it resides.
- *Integration at the system level*: The merging of discrete functions and characteristics previously performed by discrete control items into common areas of control.
- *Integration at the process level*: The progressive build-up of product components into a single, working and tested product.
- *Integration at the functional level*: The identification of integrated functions that are an amalgamation of many individual functions to form a demonstrable measure of performance.
- *Integration at the information level*: the recording and authorisation of information to define, design, document, and certify fitness for purpose of the complete system.
- *Integration at the prime contract level*: the ability to design, develop and manufacture a complex product that precisely meets the customer's requirement throughout the product lifecycle.
- *Integration from emergent properties*: a phenomenon of interactions between sub-systems that may not have been purposely designed but arise as a result of emergent properties of the constituent systems.

6.3 Examples of System Integration

An explanation with examples will provide an understanding of each of these instances of integration.

6.3.1 Integration at the Component Level

Integration at the component level is important as this provides the building blocks from which a sub-system or system is constructed. A number of electronic components, when assembled together on an electronic circuit board, provide a module that forms a building block for an LRI or system. Similarly, an electric motor, rotary valve, associated pipework, mounting flanges, and connectors may be assembled to form a motorised valve to be used in an aircraft fuel system. In a large aircraft there may be 30 or 40 such valves used in various ways to provide all the fuel systems functions such as refuel, defuel, engine feed, and fuel transfer.

At a smaller level such component integration takes place in specially designed electronic devices designed to meet specific customer specifications. This may require devices to be programmed or substrates to be designed to incorporate logical functions. This results in a device designed to perform a specific function, often referred to as 'firmware', and requires a software programme as part of the design process. Such devices may also be known as application-specific integrated circuits (ASICs). An early example of an ASIC designed for an integrated system was the MIL-STD-1553B combined remote terminal and bus controller chip designed and manufactured by a division of Smiths Industries (now GE Aviation) for the application described in the next section.

Each component will have its own specific requirements in terms of operating environment, location in the aircraft, orientation, mounting, etc. The same components may perform in different ways when installed in different positions on the aircraft or different parts of a system.

An example of an integrated mechanical component is the electro-hydrostatic actuator (EHA), a device for moving flight control surfaces in an energy-efficient manner. It consists of a single enclosure containing an electric motor, a hydraulic pump, and an actuator ram that provides energy without being connected to the main hydraulic system (Moir and Seabridge 2008).

6.3.2 Integration at the System Level

Examples of integration at the system level have been described as integration in the domains of:

- *Avionics Integration – on the basis of the reduction of discrete control units and the performance of functions in general purpose computing systems and data bus interconnections* (Warwick 1989). An example of this can be seen in the development of a system for controlling general systems in the experimental aircraft programme (EAP), a UK programme which first flew in 1986. This system, known as the utility systems management system (Moir and Seabridge 2008), performed the functions previously hosted in 20–25 individual items of equipment in four general purpose computing modules as shown in Figure 6.2. This not only reduced the number of items of equipment in the aircraft, but also reduced the bulk of wiring with an overall reduction in weight. This has since been developed further to the modern vehicle management system (VMS) to be found on many new projects, and can also be seen in propulsion systems with the many separate items of engine control being integrated into a single, engine-mounted control unit (Moir and Seabridge 2008). In the field of avionics, functions are becoming integrated into a small number of open architecture computing units. Based upon general purpose computing, memory, and interfacing modules with standard back plane interconnections,

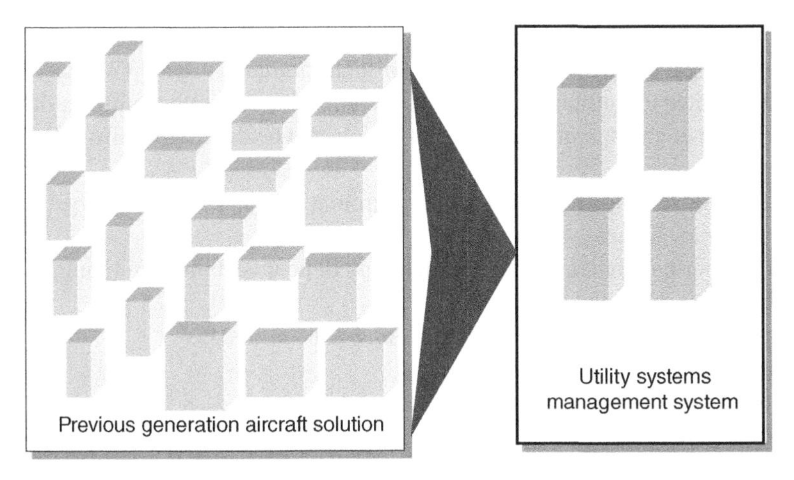

Figure 6.2 Utility systems management in the EAP.

such systems allow functions to be distributed throughout the system architecture. Examples of this form of integration can be found in the Boeing 777 aircraft information management system (AIMS) and electrical load management system (ELMS), and the aircraft systems controllers for Boeing 787, Airbus A350 and A380. In the military field the EuroFighter Typhoon and the Lockheed Martin F-35 also have VMSs based on this principle.

- *Cockpit Integration – on the basis of the reduction of discrete, single purpose displays and the emergence of multi-function displays and voice based systems* (Warwick 1989). Cockpits and flight decks were once designed or evolved as a layout of individual switches, control knobs, indicators, and lamps. These were grouped in such a manner that the pilot instinctively knew where to look or reach. Nevertheless the overall impression was of a mass of items providing information in different formats and methods of presentation. This may have led to accidents from the misreading of instruments and incorrect selection of controls (Moir and Seabridge 2013). Most modern aircraft have flight decks or cockpits that present information to the crew on multi-function displays based on flat, liquid crystal display (LCD) screens. These are able to present information to the crew in colour using graphics and text in 'pages' that can be selected as required. Sound and synthetic voices are also used to draw attention to critical conditions.

One important aspect of cockpit design is to achieve an overall consistency in the design. The cockpit or flight deck comprises a number of different sub-systems and it receives information from a variety of sources in the aircraft systems. It is important to establish a common set of principles that must be applied to display formats, fonts and font sizes, colours, lighting levels, and warning tones. This is to ensure a consistency of presentation to reduce the potential for misunderstanding of the information and reduce potential mistakes.

Integration at this level produces a clean and uncluttered cockpit environment in which information about the status of systems is presented only when it is required. Whilst this is good during normal operation there is a potential for information to be 'buried' deep in the display pages, requiring a number of key presses to access it. It is vital that care and attention are paid to moding of the displays to minimise this risk.

- *Sensor Integration – on the basis of multi-role sensors and the processing and fusion of data from sensors into a single comprehensive and recognisable situation display* (Warwick 1989). An aircraft designed for military surveillance operations incorporates a number of different sensors which enable targets of interest to be detected by different means. This set of sensors includes:
  - radar of various types for 'electronic' detection of ground, air, and seaborne targets, and also for weather avoidance
  - electro-optical for thermal imaging for use at night and in poor visibility conditions
  - TV and digital cameras for visual data capture
  - electronic support measures for detection of radar and radio emissions
  - acoustic sensors
  - ultra-violet and infra-red detectors for missile motor detection
  - a magnetic anomaly detector for detecting large magnetic masses beneath the surface of the sea for anti-submarine warfare.

The information received from these sensors can be integrated to provide a tactical 'picture' on a screen of the sea or land surface and the surrounding air space, sometimes referred to as the recognised air surface picture (RASP). This will be used by a team in the aircraft to locate, identify, and track contacts, to discriminate between hostile and friendly contacts, and to prosecute an attack. Very often further integration is provided by interrogation of on-board intelligence databases and by information received from external sources and other forces (Figure 6.3).

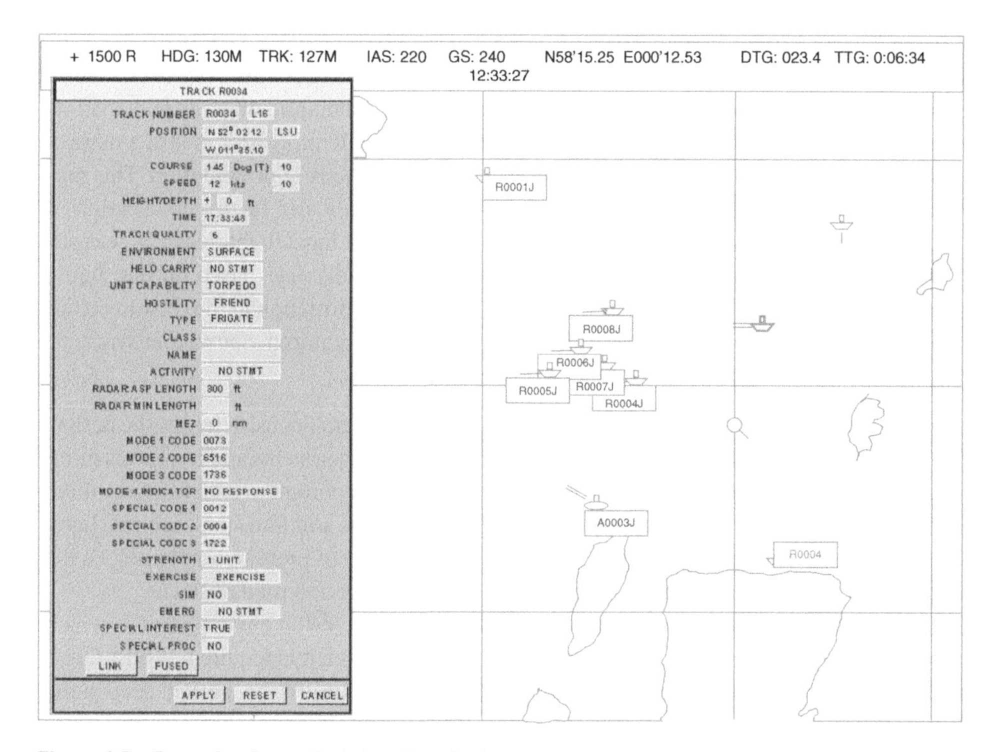

Figure 6.3 Example of a tactical situation display.

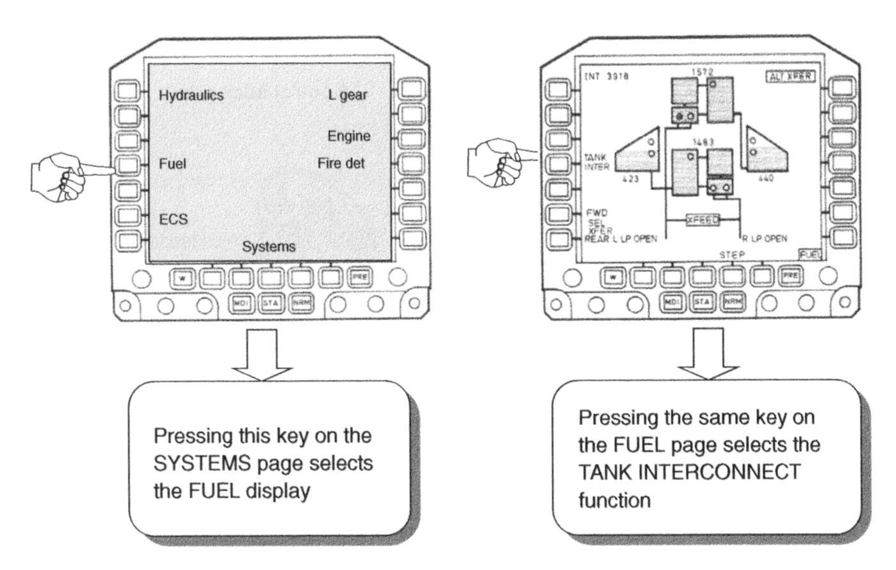

Figure 6.4 An example of 'soft' keys.

- *Control Integration – on the basis of the reduction of discrete, single purpose controls and the use of multi-purpose and soft-key controls* (Warwick 1989). An example of this can be seen in the use of 'soft' or programmable keys, often associated with multi-function displays. The function that the key or switch performs when it is activated depends on the legend ascribed to it by the aircraft processing system. For example, when the key bears the legend 'FUEL' on the SYSTEMS page, the action of pressing the key will select the FUEL page of the display; on this page the legend next to the same switch may be 'TANK INTER' and pressing it will activate the tank interconnect sequence. An example of this can be seen in the multi-function display shown in Figure 6.4. Careful design and location of keys and the key caption will help to reduce the risk of inappropriate selection. An alternative approach to the use of physical soft keys is to use touch screen capability.

A further example of the integration of controls to ensure an optimum human factors layout can be seen in the HOTAS (Hands on Throttle and Stick) concept adopted by many combat aircraft. In this example all the controls and switches necessary to fly the aircraft in a combat situation are located so that the pilot can conduct the mission without moving his hands from the throttle and control stick.

There are many fixed-wing and rotary-wing applications of HOTAS in military projects. Figure 6.5 shows the F/A-18C/D Hornet in which all critical controls are located on the throttle and stick to ensure effective one-person performance in all combat missions. This allows control of weapons, sensors, and avionics in both air-to-air and air-to-ground modes. The adoption of HOTAS is almost universally pursued in modern military cockpits since it allows more immediate and effective operation during the most critical phases in the mission. In combat the pilot cannot afford to look into the cockpit for the correct switch and take his hands off the throttle or the stick without incurring the risk of an uncommanded action on the controls. The combination of functions into a collection of switches and controls is a challenging ergonomic task (AGARD 1996).

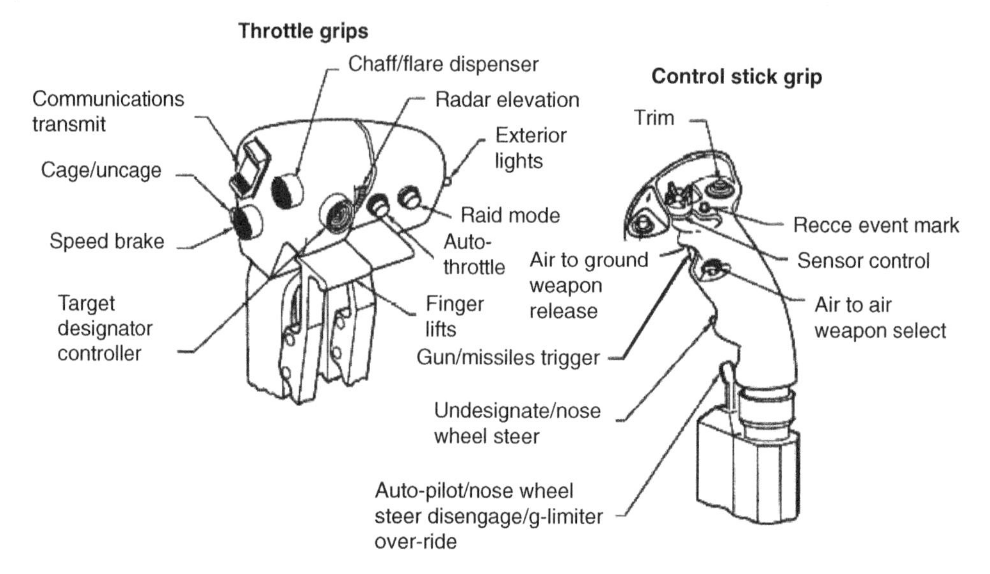

Figure 6.5 The HOTAS (hands on throttle and stick) concept in the F/A-18 Hornet (AGARD 1996).

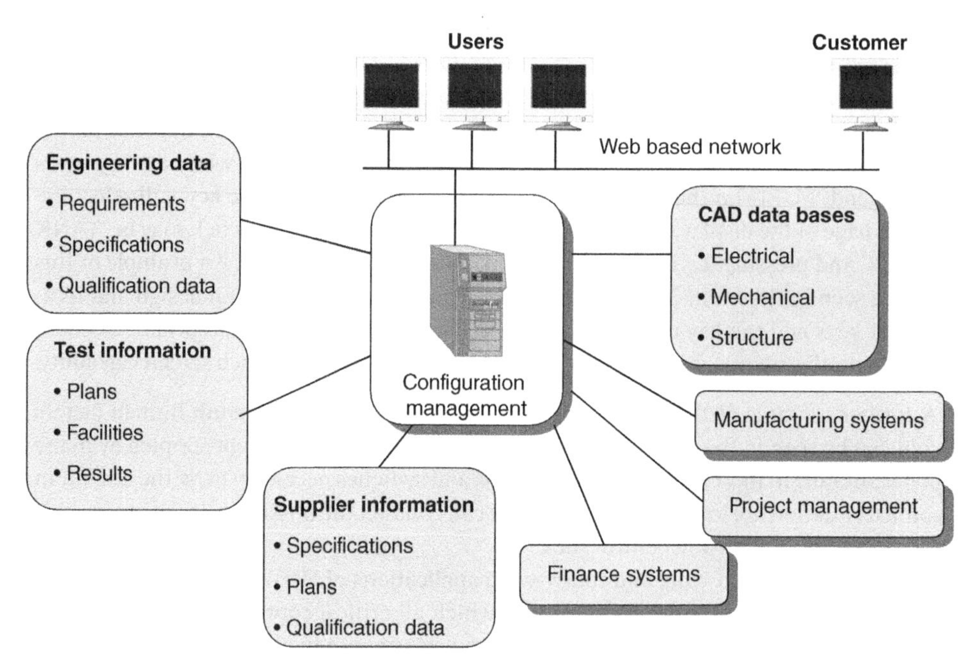

Figure 6.6 Contributors to the design database.

- *Data base integration – on the basis of shared data access from numerous systems to common areas of data* (Warwick 1989). The design process of a complex system produces many databases that define the product design as three-dimensional model data, interfaces, software design, hardware record, etc. This captures the design baseline and any changes to it. Figure 6.6 shows the typical contributors to the design database.

There is an increasing trend in industry to develop products in a collaborative environment with a number of partners, associates or suppliers. Inevitably this association tends to be widely distributed geographically, making communication and sharing of information an issue. The need for a process of securely and selectively exchanging, reviewing, and managing the change of product information with other internal and external participants such as customers, design partners, suppliers, and distributed manufacturing organisations demands a mechanism for managing vast amounts of information and data. Easy, but secure, access is required for all authorised parties into a shared data environment.

An example of such a mechanism is PTC's Windchill (www.ptc.com). This is a set of integral, modular solutions for rapid distributed collaborative development of products which removes the traditional boundaries that exist within organisations. Windchill creates a single system-of-record for a variety of digital product information such as computer-aided design (CAD) tools, often from different tool-sets, design data, specifications, test plans, information and results, supplier data, etc. This information is made available to users at their desktops by means of an internet-based distribution system.

An example of on-board database integration is the use of multiple source intelligence data to provide a composite view of threats to the crews of frontline aircraft. Various tactical and strategic databases will be used to provide information on the nature, location, and deployment of threats such as:

- surface-to-air missile sites
- anti-aircraft artillery sites
- surveillance and threat radar types
- electronic warfare capability.

This information allows the mission to be planned and for the aircraft to be routed to avoid major threat concentrations, and also to be equipped with the appropriate countermeasures. Contributors to the tactical and strategic databases are illustrated in Figure 6.7.

- *Knowledge integration – on the basis of knowledge based systems providing information and assistance to aircrew and ground crew* (Warwick 1989) Information is collected by most military aircraft as a main product or a by-product of their principal role. This data is received from various sources, such as communications, radio frequency signals, photographic images, human observation, etc. This data can be combined with historical data and analysed to form a source of intelligence that is used to provide a strategic or tactical picture of the battle space, as shown in Figure 6.7.

Knowledge databases are used to complement human operators to enable patterns of signals to be interpreted and used to identify classes of target. Modern techniques allow targets to be identified to a class of ship, a type of aircraft or land vehicle, or a human being. This knowledge is continually refined to form a continually changing record of intelligence used during peace time and war time to gain an understanding of the state of readiness of a potential adversary.

Knowledge about the state of the aircraft systems is of value to the ground crew to ensure rapid turn-around and to plan for repairs and servicing at the optimum base location. Maintenance data-recording systems have been available for many years on aircraft, but they mainly collected failure status information which was interrogated on the ground and

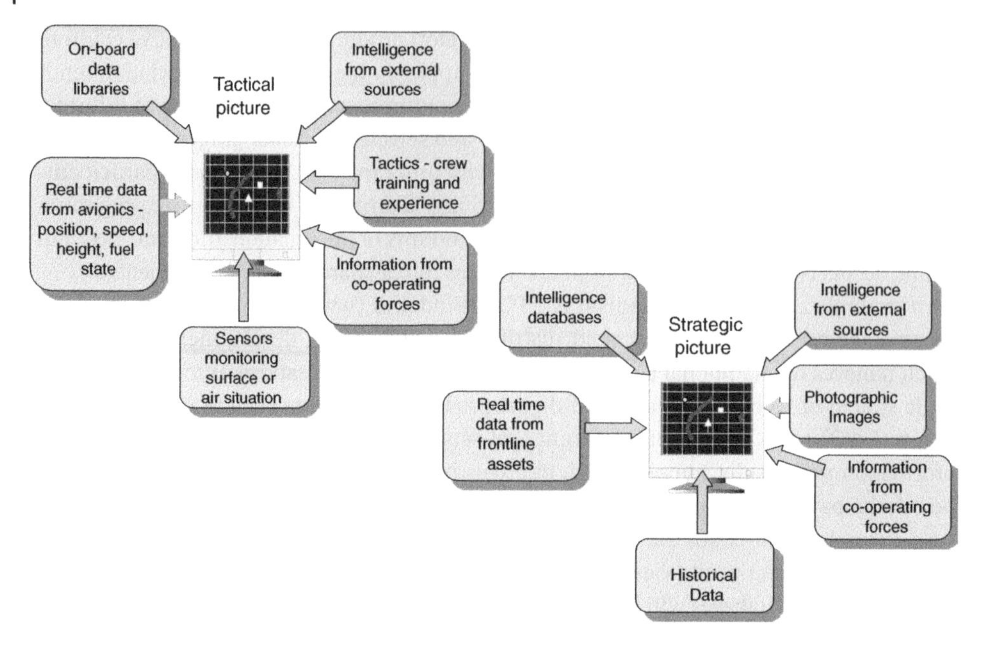

Figure 6.7 Tactical and strategic pictures.

used to identify faulty components to enable repairs to be carried out. Modern systems collect failure status information and also look for trends that indicate a potential failure; examples are increasing amounts of debris in oil or a gradual decline in pump pressure over time. Combined with knowledge databases and algorithms to determine from the trend what is the most likely source of failure this system can be more precise in determining a fault and its location to a single component. This sort of information is much more helpful in providing readiness for repair and replacement action rather than waiting for a failure to occur. Such systems are often known as prognostics and health management systems (PHM). Information can be stored in a removable cassette for connection to a ground database or directly transmitted to the ground during flight via a data link system such as the ARINC communications and reporting system (ACARS).

Another collection of knowledge about the operation of the aircraft is the accident data recording system (ADR). A pre-selected set of mandatory and optional system parameters is continuously recorded throughout flight for interrogation after an accident.

6.3.3 Integration at the Process Level

The progressive testing from sub-system or module level through system to complete product is often referred to as integration. This is essentially the build-up of the system shown in the right-hand leg of the V diagram in Figure 6.8. Integration in this case involves the progressive build-up of fully tested functions, modules, and interfaces, and their eventual progression to final testing on the completed product. Much of this activity takes place in a test laboratory, eventually transferring to the aircraft during build and then to flight testing.

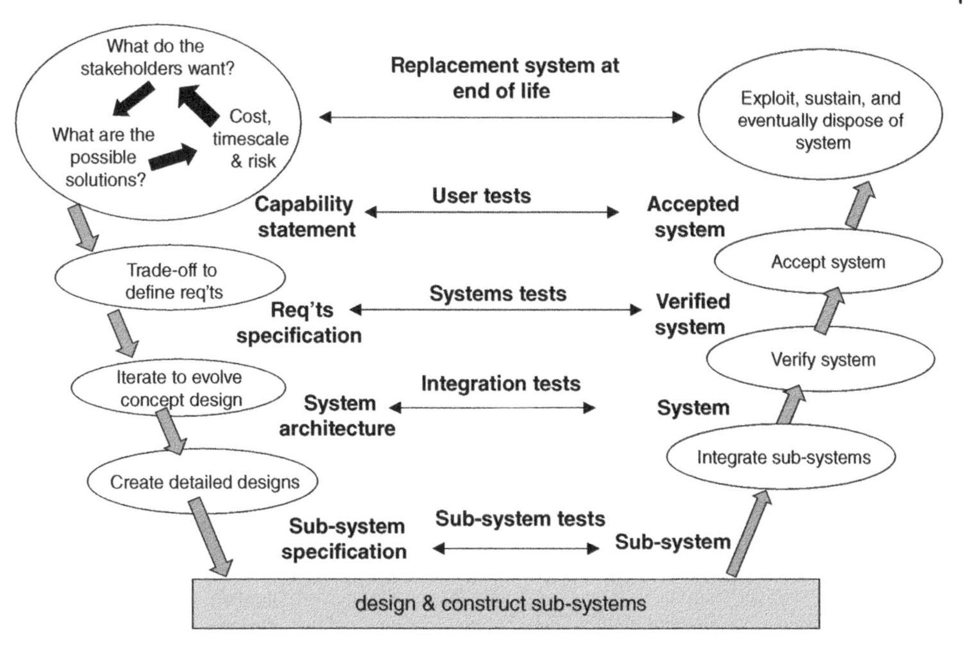

Figure 6.8 The 'classical' V diagram. *Source:* adapted from the Royal Academy of Engineering (2007).

The right-hand part of the V portrays a breakdown and validation of the top-level system requirements so that they flow down towards a module design. The left-hand branch shows a progressive procedure by which module integration, hardware and software integration, and system test are achieved – this is the verification process. The first step involves the application of a traceability matrix to confirm that all of the original requirements have been satisfied and fully met.

This activity enables each element of a system to be thoroughly tested and the test results validated prior to connection with other systems and subsequent testing as a whole. This process is not intended to find faults and rectify them. Its primary purpose is to obtain documentary evidence that a system fulfils its working requirement and that all evidence from progressive testing represents the entire system.

These processes are linked to major systems development milestones, as will be explained in Section 6.5.2.

> An alternative representation of the process is available in the spiral model. *The spiral development model is a risk-driven process model generator. It is used to guide multi-stakeholder concurrent engineering of software-intensive systems. It has two main distinguishing features. One is a cyclic approach for incrementally growing a system's degree of definition and implementation while decreasing its degree of risk. The other is a set of anchor point milestones for ensuring stakeholder commitment to feasible and mutually satisfactory system solutions.* (Dr Barry W. Boehm)

The spiral model combines the idea of an iterative development or prototyping with the systematic, controlled aspects of the waterfall model. It allows for incremental releases of

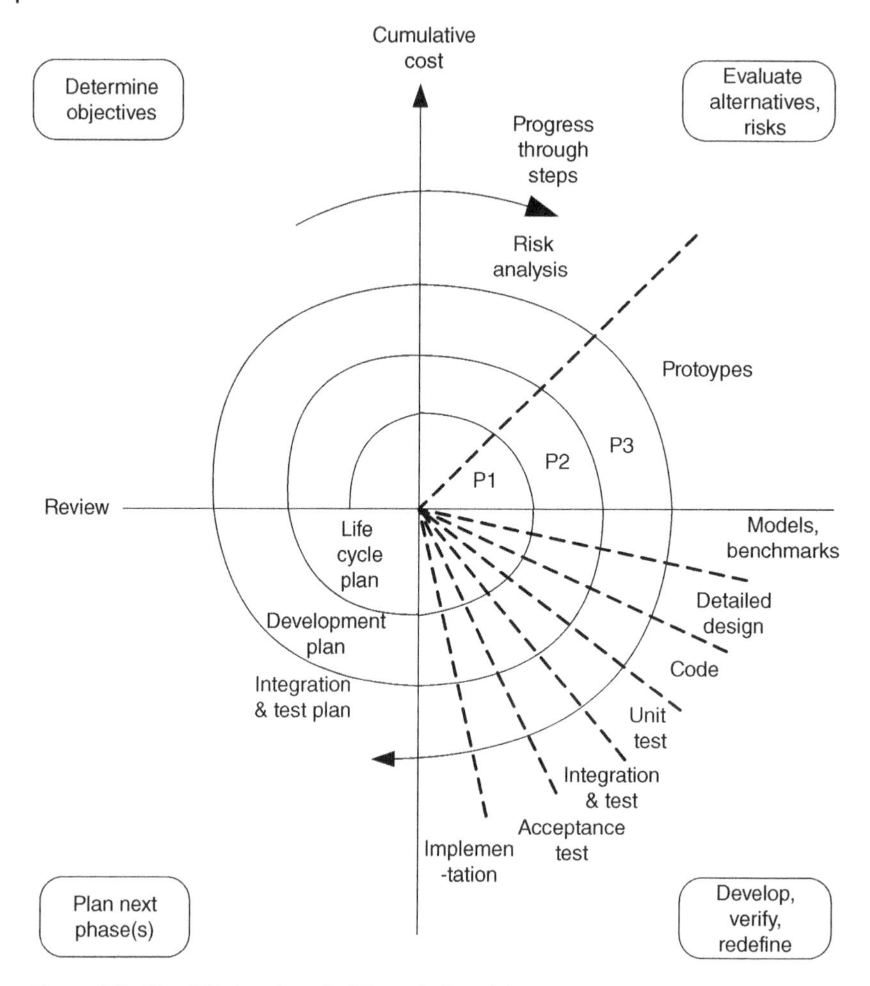

Figure 6.9 Simplified portrayal of the spiral model.

the product, or incremental refinement through each time around the spiral. The spiral model also explicitly includes risk assessment as the spiral unfolds and the financial impact increases. Identifying major risks, both technical and managerial, and determining how to lessen the risk helps keep the development risk – particularly the development of software – under control.

The simplified portrayal of the spiral model in Figure 6.9 shows the major phases of development, in the diagram the start point is the 9 o'clock position:

- determining objectives
- evaluation of alternatives and risk
- developing verifying and re-defining the product
- planning.

The example shown would be typical of the development of a small unmanned air vehicle (UAV) which is destined to carry and deploy a small sensor package. At the outset the designers have a vision of what it is that they intend the vehicle to provide. As the first spiral unfolds, rapid prototyping using commercial-off-the-shelf (COTS) technology would enable the basic characteristics of the system to be evaluated and trade-off and risk assessments to be undertaken. A prototype system or emulation could then be produced to establish 'proof of concept' before moving to the next stage of development.

Having satisfactorily established the system's key characteristics a development plan would then be constructed to provide the framework for more serious and costly development. The second cycle of the spiral would refine the process, still taking account of alternative solutions and risk prior to building a second prototype, possibly using laboratory hardware.

A third spiral could move the development forward to initial flight test and development. As the spirals progressively unwind so the cost of the venture increases proportionately. The spiral model permits an orderly development process with recursive review of the programme objectives such that cost and risk are balanced at each stage to determine an acceptable outcome.

Further spirals could involve refinement of the sensor package(s) and of the mission envelope. Further developments and extensions of the basic platform capabilities could also be embraced.

The model lends itself to the development of relatively small, self-contained projects. It has less application to larger and more complex 'system-of-systems' developments.

6.3.4 Integration at the Functional Level

Requirements for the functions that the aircraft must perform are drawn from a number of sources. Some of these requirements are explicitly stated by the customer, whereas others are derived from experience, from performance requirements or by an understanding of standards, regulatory standards, processes and technology – all extracted with a degree of engineering judgement. This process is described in Chapter 2.

These requirements are 'flowed down' into a work breakdown structure (WBS) that reflects the constituent systems and sub-systems of the aircraft. The requirements then flow down into specifications for sub-systems and equipment. Very often the organisation required to develop this work is structured as teams with responsibility for delivering the products. Figure 6.10 illustrates this flow down of requirements.

In order to ensure that the functional definitions that arise out of this separate development of products are 'integrated', a view is taken across the product lines, as shown in Figure 6.11. This view takes account of physical and functional interfacing, as well as ensuring that common standards and conventions are established and used by the product teams. This task is often performed by a separate team known as the engineering integration team, whose task is to ensure that the individual products combine to form an integrated functional whole.

A simple example of this is a system concept that is envisaged as an integrated system. It will continue to be seen this way in the early stages of definition and preliminary design. For practical reasons, however, when individual major functions have been identified and boundaries established it becomes convenient for each of these to be developed into detailed design by different groups or departments. Throughout the remainder of the

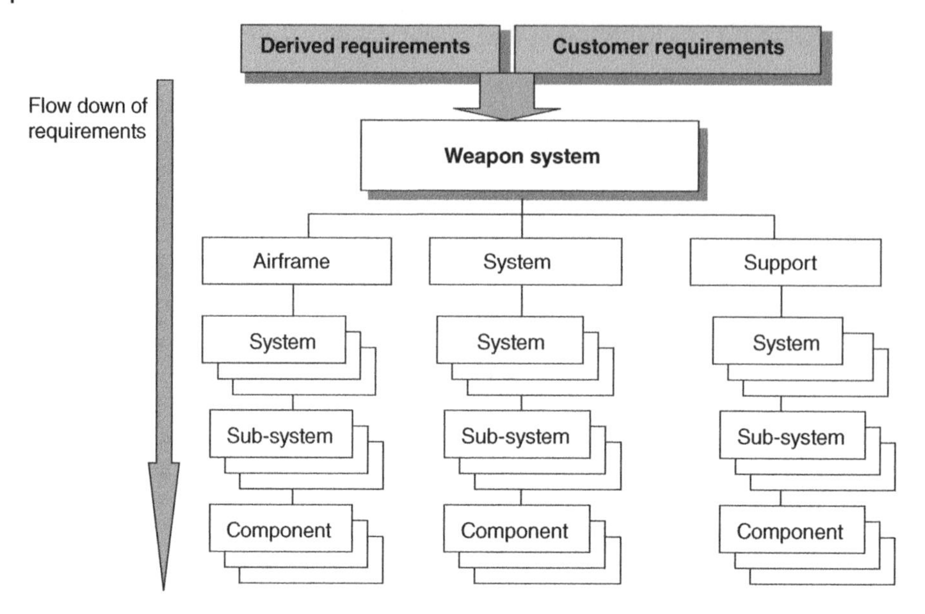

Figure 6.10 Flow down of requirements.

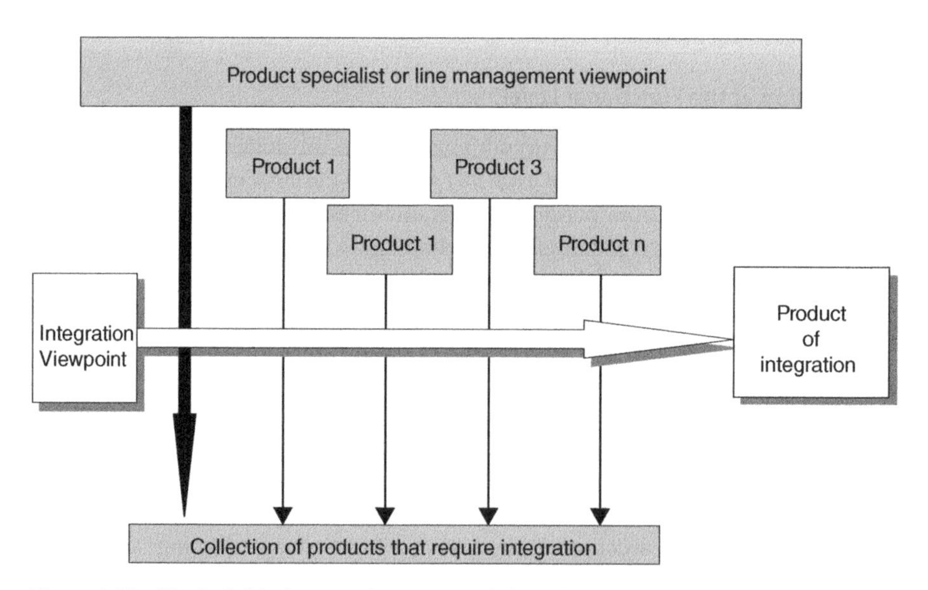

Figure 6.11 The individual versus the integrated view of the product.

lifecycle these responsibilities remain for procurement and test. The task of the system integrator is to ensure that the original concept of an integrated system is maintained so that when the total system is built up during the test phase it is fit for purpose. This means that all agreements made at the concept phase, which include functional separation, functional interfaces, accuracy and resolution of data, timing of data, redundancy, etc., are maintained at all times.

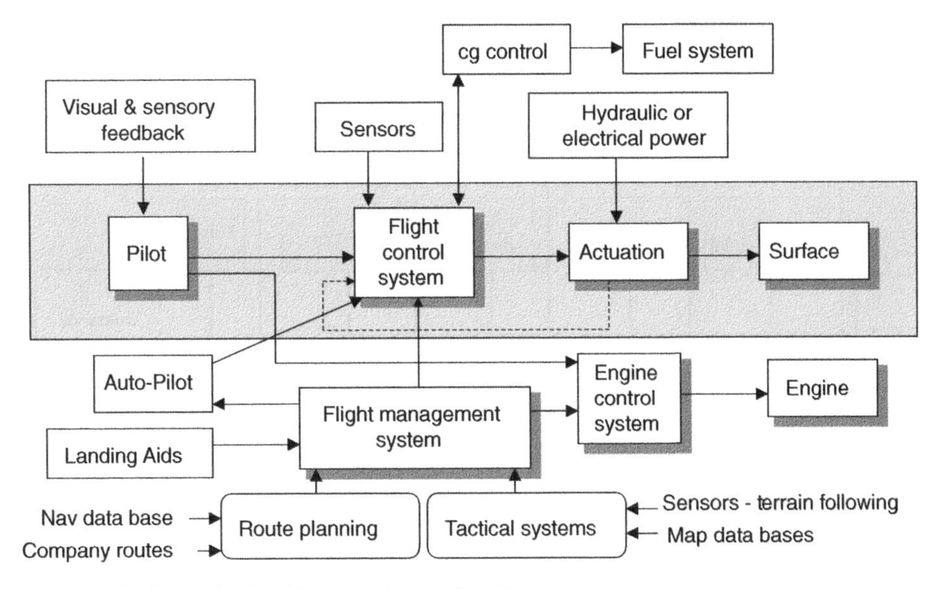

Figure 6.12 Example of guidance and control architecture.

There are whole aircraft functions that are a combination of many individual functions. It is often convenient to view the combined function as a whole and to identify the individual functions that contribute to the whole.

A system architecture example is illustrated in Figure 6.12 in which the function a traditional flight control system is shown running through the centre of the diagram from pilot input through flight control computing to actuator to control surface. In a guidance and control system, other systems which may make a demand on the direction, speed or attitude of the aircraft have been included, and the flight control system may need to influence the centre of gravity (cg) of the aircraft to perform a manoeuvre efficiently. There is now an increased choice of sub-systems to perform different functions of the control of the aircraft flight path. For example, in an extreme implementation all functions could be integrated into the flight management system.

Figure 6.13 shows how this combination of sub-systems should be viewed in order to form an integrated function – guidance and control – and to establish the individual products that are needed to form it. This ensures that the integrated function is established early in the product lifecycle and the criteria for its qualification are understood by all parties in the project.

This thinking can be developed to establish other functions such as the following:

- Information management: the collection of information for presentation to the crew, the best means of presenting that information, and a thorough understanding of the human engineering aspects.
- Target acquisition and prosecution: the selection and moding of sensors to identify, track, and select targets, selection of the appropriate course of action to prosecute the target, and the provision of information to the crew and other participants. Targets in this context means hostile targets, dropping zones, search and rescue targets or destinations.

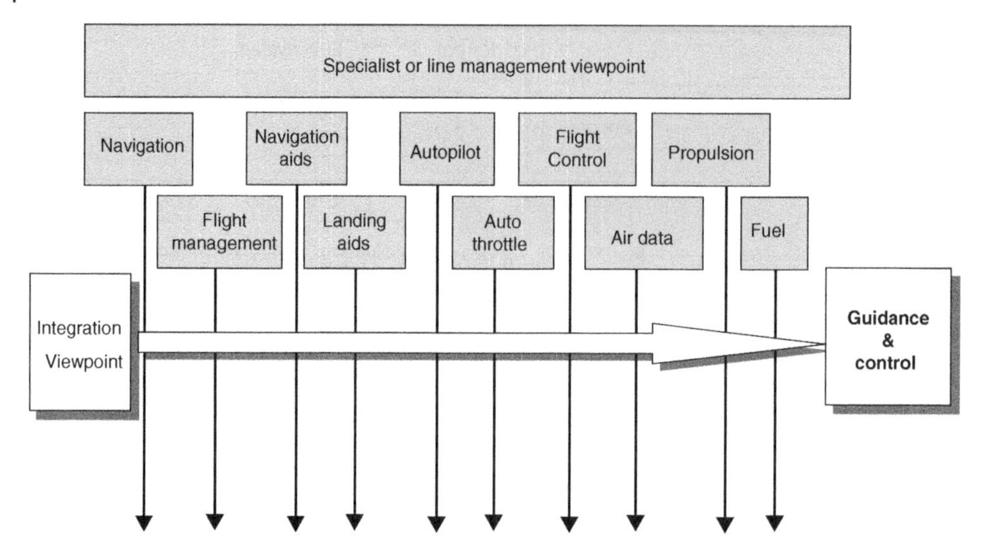

Figure 6.13 Example of guidance and control.

- Communications management: the management of all internal and external communications.
- Displays and controls management in which the design of the cockpit or flight deck and all its constituent systems is integrated to provide an ideal crew working environment.

6.3.5 Integration at the Information Level

The products of the lifecycle are controlled by documenting every stage of the aircraft development. Each stage of development during the process shown in Figure 6.10 must be recorded to show the flow down of requirements, the links to the design, and the evidence gained by testing and modelling to prove that the final product is safe, robust, and fit for purpose.

The information collected in this way is essential in demonstrating to the customer and the regulatory bodies that the aircraft is safe to fly without danger to the operators and the over-flown population. Control of the product is exercised by the application of configuration control. This means that the issue of all models, drawings, reports, analyses, and parts is recorded for the aircraft type. Any deviations or modifications to the type record are introduced in a controlled manner. This task is usually the responsibility of the chief engineer or chief designer.

6.3.6 Integration at the Prime Contractor Level

The prime contractor's interpretation of system integration encompasses all the above and much more. It is concerned with the management of all aspects of providing a product that will meet the customer's requirement throughout the entire lifecycle.

A systems integrator takes responsibility for the whole product and the way in which all the parts work together.

The critical factor in the success of an aircraft is how well it meets the demands of its operational role and environment. This cannot be achieved by focusing on any one attribute of the vehicle – it depends on the characteristics of the complete systems package, including the crew, the vehicle and its internal sub-systems, training and support systems, and in the case of a military aircraft the weapons and the supporting military infrastructure.

Military customers are interested in characteristics such as survivability, lethality, and low lifecycle cost. Commercial customers are interested in availability, purchase price, and operating costs. These characteristics are largely determined by the detailed way in which systems elements are brought together. These tasks include:

- tracking, understanding, and influencing the customer requirement from its earliest conceptual stage
- capturing the requirement in a structured manner and flowing down that requirement through and across all aspects of the product definition, manufacture, and operation
- ensuring that the requirements are correctly interpreted and traceable to the design solution
- ensuring that the design is consistent across all constituent sub-systems and their hardware, software, firmware, and human engineering solutions
- conducting the testing and proving of the design at unit, component, sub-system, and system levels, including models and simulations, to gather evidence that the design is sound, robust, safe, and fit for purpose
- compiling and controlling a complete record of the design, including all assumptions and calculations.

6.3.7 Integration Arising from Emergent Properties

One example of an emergent property is described here to illustrate how the sub-systems of the aircraft become effectively 'integrated' by interactions that may not have been envisaged in the initial design (Moir and Seabridge 2008). Figure 6.14 shows the heat flows in an aircraft arising from energy dissipated in the aircraft systems. The key to the figure is:

1) Air extracted from the engine fan casing is used to cool engine bleed air.
2) Ram air is used to cool engine oil in a primary oil cooler heat exchanger.
3) Fuel is used to cool engine oil in a secondary oil cooler heat exchanger.
4) The electrical integrated drive generator (IDG) oil is cooled by ram air.
5) Hydraulic return line fluid is cooled by fuel before being returned to the reservoir.
6) Fuel is cooled by an air/fuel heat exchanger.
7) Ram air is used in primary heat exchangers in the air-conditioning packs.
8) Ram air is used in secondary heat exchangers in the air-conditioning packs.

In some extreme cases, such as that exemplified by military stealth aircraft, the ultimate disposal of this heat energy may compromise the thermal signature of the aircraft. Alternative mechanisms for dumping the heat overboard have to be sought to reduce the threat of heat-seeking missile detection.

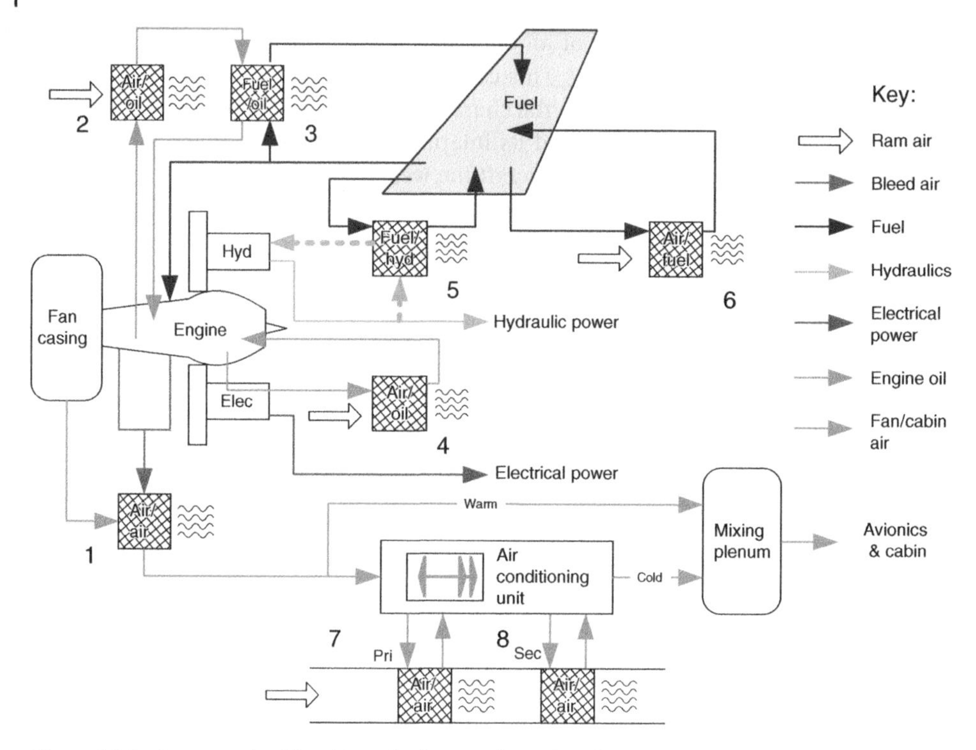

Figure 6.14 An example of heat transfer in a modern aircraft system.

Similar interactions can be observed by analysing current flow in the airframe structure and skin during a lightning strike, by analysing current flows in the airframe in the bonding and earthing of system components, and by analysing structural loads. These effects are especially interesting in structures that are a mix of metallic and composite materials.

Another example concerns a fuel system and the number of computers and data bus paths through which demands and indications must flow. Figure 6.15 shows an example of a fuel system sensor sending information to a cockpit display and a subsequent path for a crew action to reach an actuator, for example an indication of low fuel in a tank and a pilot demand to transfer fuel.

The integration of data flows across a hierarchical system of data buses (federated systems) can provide some interesting results in terms of data latency or 'stale data'. On each occasion that data is exchanged across a data bus there will be occasions where the data is stale because the data can only be refreshed on a cycle-by-cycle basis. This data 'staleness' can be addressed within the system design but the issue needs to be recognised.

Three typical types of transactions are shown in the top-level portrayal in Figure 6.15. For a fuel system example these may be characterised as the following:

- Intra-system data transfers: these are shown as transfers 1 to 4 on Figure 6.15 where data is shared on a system basis without the pilot being involved. Such data transfers could involve a fuel transfer pump being commanded to automatically top-up certain fuel tanks as fuel is expended – an interchange between the fuel gauging and transfer systems.

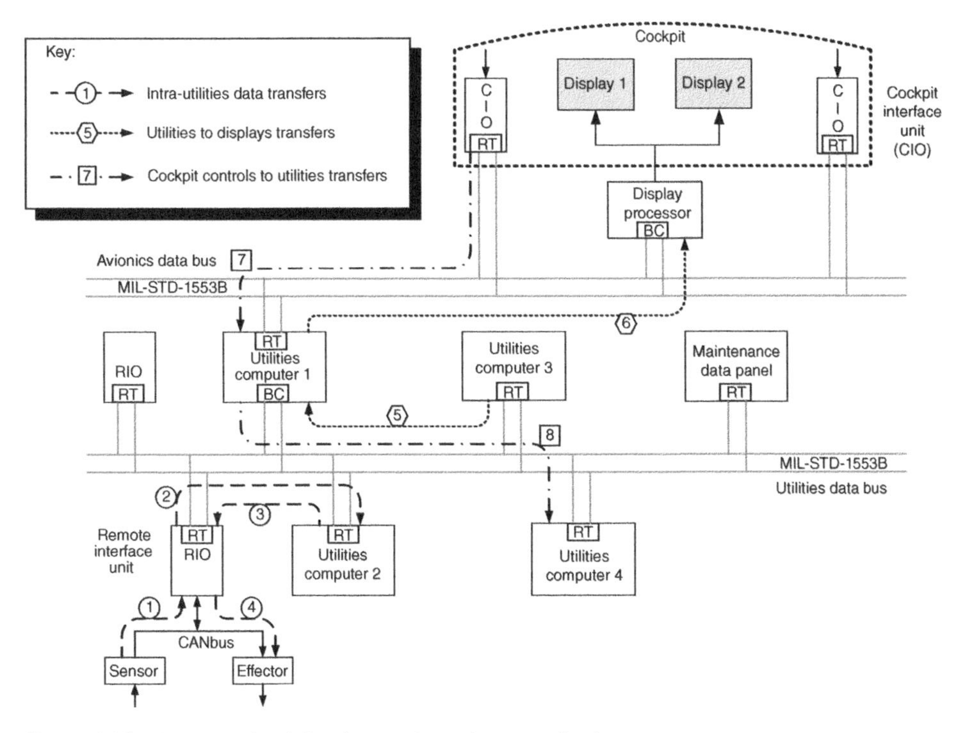

Figure 6.15 An example of data latency in an integrated sub-system.

- System to cockpit: transfers 5 and 6 relate to the transfer of data from the system to the cockpit, perhaps relating to fuel on-board (FOB) and the contents of individual fuel tanks when demanded by the pilot.
- Cockpit to system transfers: pilot selections such as demanded fuel transfers 7 and 8 are input into the system via cockpit interface units whereby discrete pilot fuel mode selections are fed into the system computer.

It can readily be seen that complex operations relating pilot and system interactions could involve all of the above transaction types sequentially. As an example of the impact of system time delays, if each delay is 10 ms on average then a potential total delay from sensor to warning to demand arriving back at the transfer pump could be as much as 180 ms. To this must be added the time taken for the crew to react to the warning, decide what to do, and then to select transfer. There are some instances where this delay may pose a safety hazard. There are other types of system where such a latency of data will be totally unacceptable and may lead to system instability.

6.3.8 Further Examples of Integrated Systems

6.3.8.1 The Airframe

Not surprisingly the airframe is a nest of interconnecting and integrated threads. The departmental structure of many aerospace organisations tends to treat these threads as

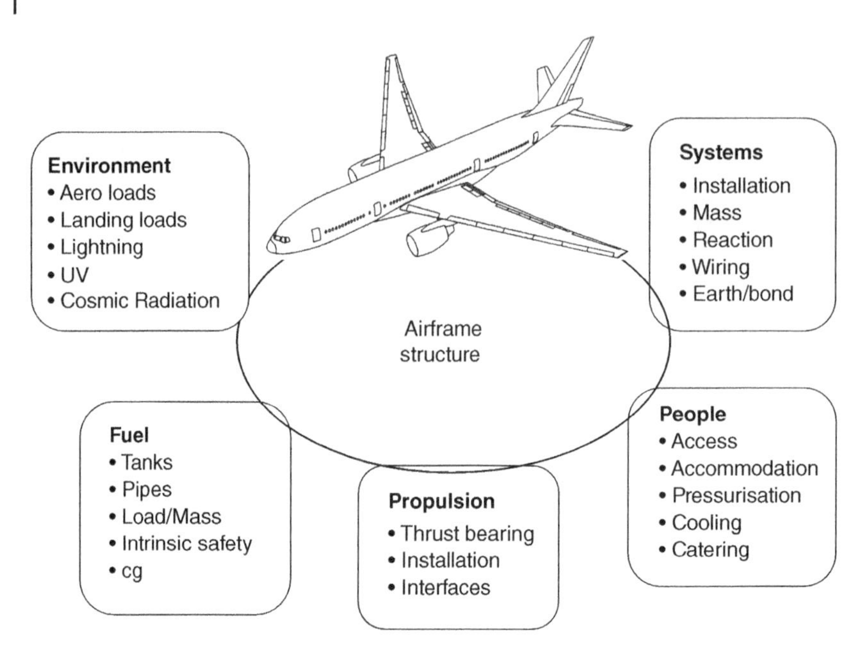

Figure 6.16 The airframe as an integrated system.

single independent disciplines. However, there is much to be gained from treating them as an integrated entity (Figure 6.16).

The structure is intended to withstand all the loads associated with flight, with landing and decelerating, with turbulence as well as vibration, climatic conditions, and age. It may be constructed of different materials, including aluminium alloys, composites, titanium, and plastics, all with different characteristics of strength, corrosion resistance, fire resistance, and electrical conductivity.

These are combined to form a pressure vessel that is the life-support system for the occupants, providing a pressurised environment to allow absorption of oxygen in suitable temperature and humidity conditions.

The airframe is an integral portion of all electrically powered systems since the structure provides the zero potential or 'earth' for all equipment and cable screens. It is also the 'bond' connection for all metallic components such as equipment housings, fuel, and hydraulic pipes, which are vital to preserve intrinsic safety.

The structure forms a Faraday cage to protect the internal equipment from radio interference effects, and also to restrict radio interference escaping from the aircraft to the outside world, an important aspect of some military operations.

Finally, the airframe is the storage location for fuel in wing and fuselage tanks. The continuous movement of fuel and its temperature exert loads on some airframe components. To minimise some load changes, the way in which fuel is moved from tank to tank is scheduled to limit the range of movement of the centre of gravity of fuel, and indeed to maintain the aircraft centre of gravity within precise limits.

6.3.8.2 Propulsion

The powerplant is itself a complex integrated system of mechanical and hydro-mechanical assemblies. On modern engines there is an engine-mounted full authority digital engine controller and a gear box to allow the connection of electrical and hydraulic power generation devices as well as a bleed air off-take for aircraft pneumatic systems. The powerplant is integrated with the intake, which can be quite sophisticated on supersonic aircraft such as Concorde and some military fast jets in which the intake has variable surfaces to allow efficient engine performance.

6.3.8.3 Air Systems

The aircraft cabin is usually pressurised by air bled from the intermediate or high pressure stages of the engine and passed through heat exchangers and refrigeration systems before being used to pressurise and condition the cockpit or cabin. In passenger aircraft the air is recirculated for economy reasons, passing through filters to remove biological contaminations sources. This system is shown in simplified form in Figure 6.17, which shows an integration of the engine and its lubricants, the auxiliary power unit (APU), the aircraft environmental conditioning, the cabin, and the occupants. It also shows some areas where there is a known potential health risk which the system is designed to minimise.

However, it has been found that engine anti-wear additives in the engine oil can lead to contamination of the cabin air (*Telegraph Travel*, 14 February 2009, www.aerotoxic.org). Although the air passes through various stages in the environmental conditioning system

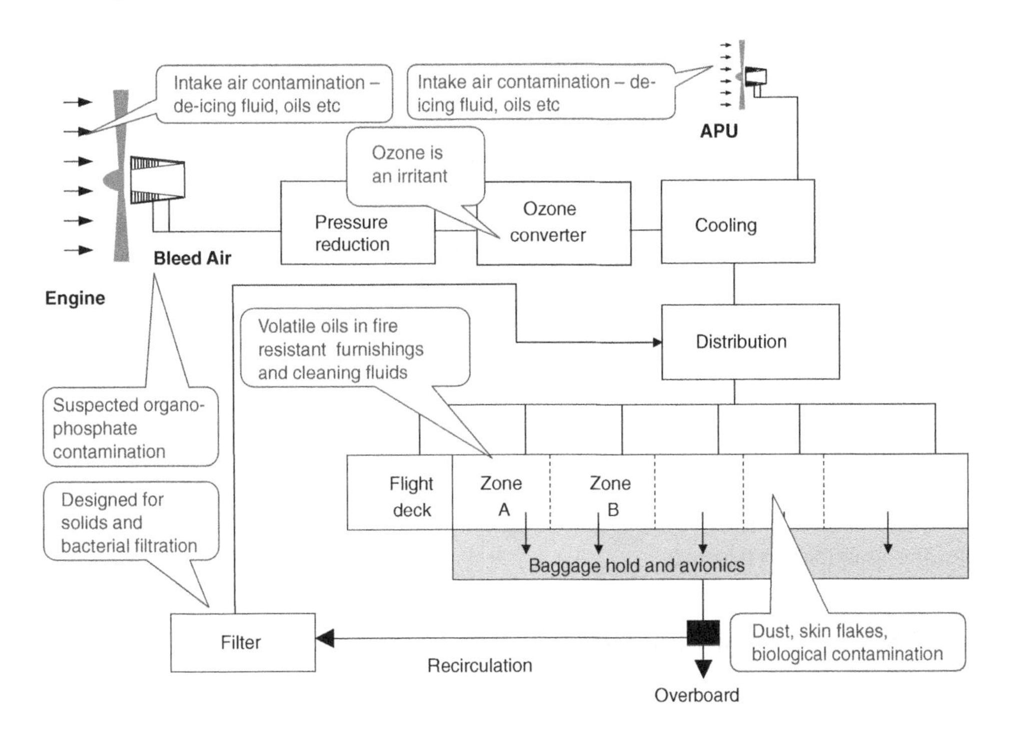

Figure 6.17 The cabin air system.

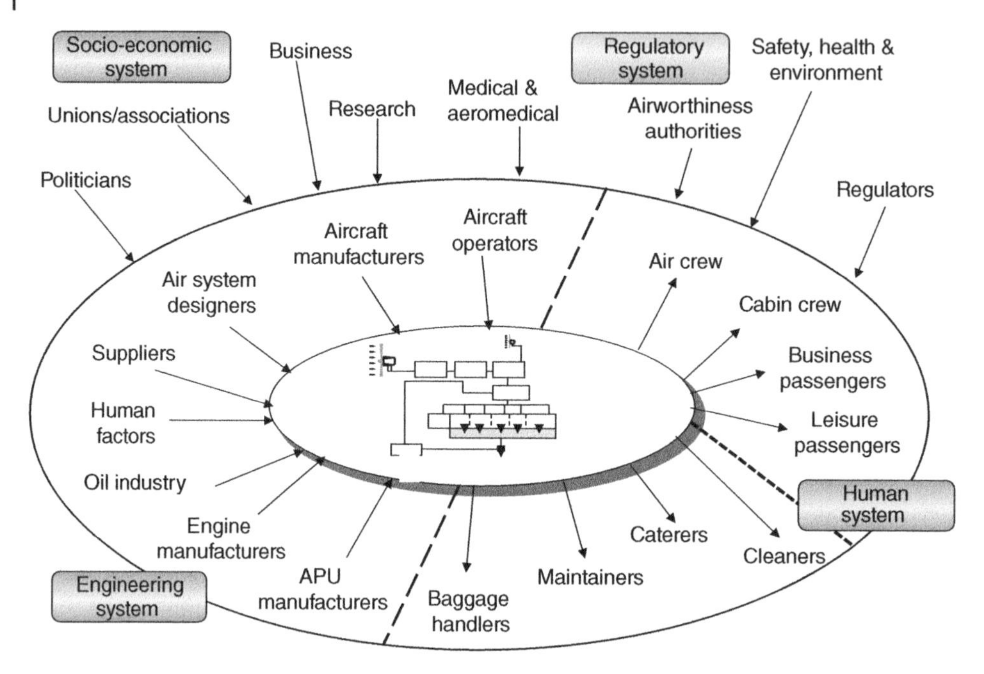

Figure 6.18 Stakeholders in contaminated cabin air.

there is still the possibility that it contains contaminants such as oil vapour or combustion products. This may have a detrimental effect on the respiratory tract, particularly for aircrew with temporary or permanent respiratory ailments. Recent research taking samples of cabin residue found high levels of tricresyl phosphate (TCP) (Michaelis 2011). Exposure to this can lead to symptoms of drowsiness, respiratory problems, and neurological illness. There have also been incidents of smoke and fumes in the cabin air as a result of oil or hydraulic fluid entering the engine bleed air as a result of leakages in the lubrication system. This situation has been reported widely on the internet (aerotoxic.org) and there have been incidents of pilots reporting loss of concentration and neurological disorders, and even some aircrew deaths that have been attributed to this (*Sunday Express* 2013; Learmount 2014). The term aerotoxic syndrome has been used to describe this condition. This is an example of an emergent property or an unexpected consequence of providing a comfortable cabin environment. The example is a relatively simple technical integration, but the emergent property is a complex state of affairs, as can be seen by the lack of agreement as to whether or not it even exists.

The situation has been in debate for some considerable time with no confirmation that the condition does exist. The stakeholder diagram shown in Figure 6.18 shows how complex the situation has become.

6.4 System Integration Skills

System integration skills are related to understanding the total system in all the respective descriptions of the term described above. Having a grasp of all the interfaces and interactions is vital to designing, developing, and certifying the system and guiding the various parties involved in the process. At the conceptual phases of a project a key skill is the ability to understand and develop a requirement to the stage at which it can be sub-divided into more easily manageable blocks. A grasp of the connectivity and dependencies between blocks is essential for the subsequent identification of requirements for each block.

The prime contractor must preserve a view of the system functional performance – how it will meet the customer's requirements and can be demonstrated to do so. The certification and qualification view is that of planning and reviewing all the testing to ensure that the sum of test evidence proves the compliance of the system to its performance and safety criteria.

The sum of this skill set is the ability to manage an end-to-end development of a complex system.

Some of the complexity of the system is illustrated in Figure 6.19, which shows how functions are implemented in a large complex system.

The system of interest is formed of functions derived from the customer's requirement. These functions may be performed in software contained in systems processors, in hardware such as actuators or by the crew operating the aircraft. The aircraft items of equipment in which these functions are implemented are interconnected by hard wiring with discrete signals or by means of a suitable data bus. The whole system is installed in the aircraft and subject to environmental conditions which vary throughout the flight envelope and throughout the world. Some of the sub-systems will interact with each other in the form of emergent properties. Superimposed on this functioning whole are changes resulting from use in development or day-to-day operations and also from maintenance actions.

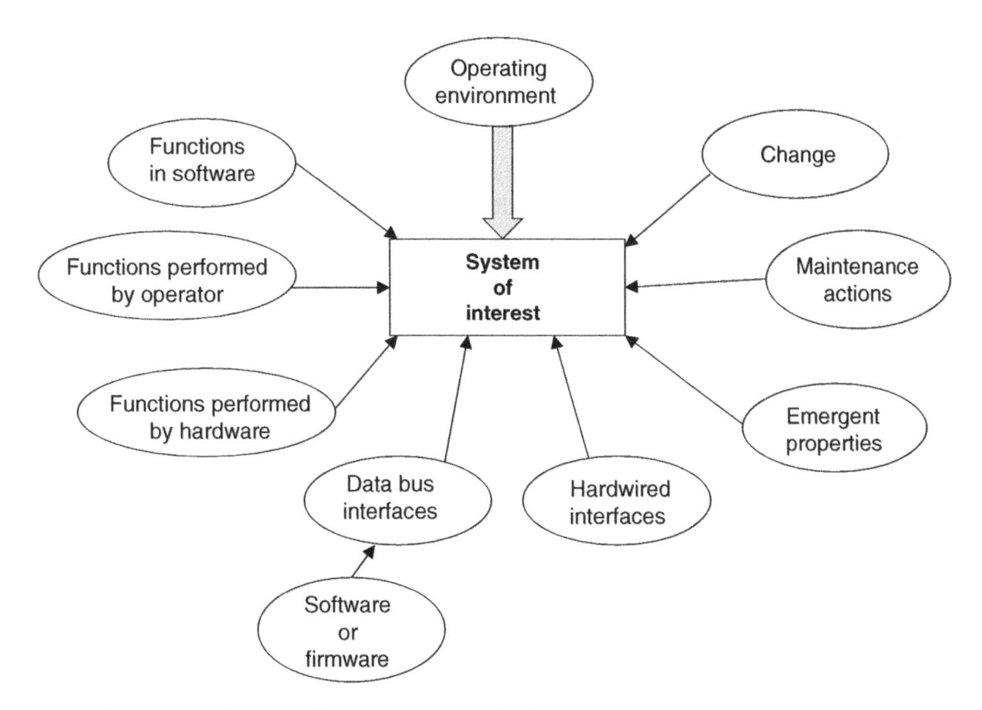

Figure 6.19 An indication of large system complexity.

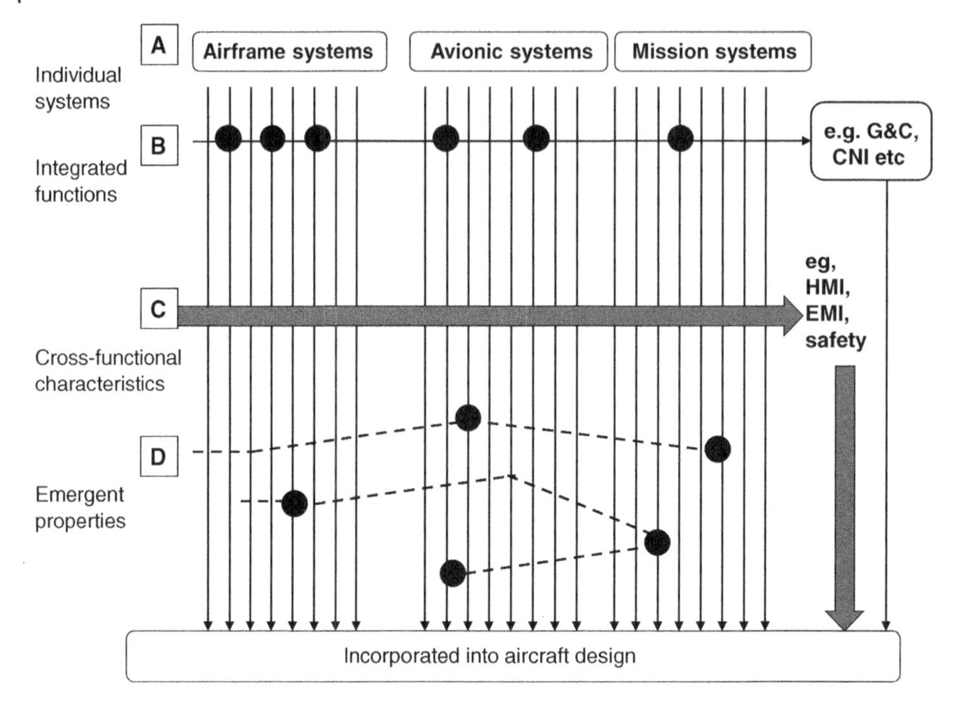

Figure 6.20 Line management and integration viewpoints. CNI, communications, navigation, and identification; EMI, electro-magnetic interference; G&C, guidance and control; HMI, human–machine interface.

Thus an understanding of the normal operation of a large and complex system such as a whole aircraft can only be obtained through an understanding of all of its sub-systems and the impact of their integration. An understanding is required for the chief engineer to sign off the aircraft with total confidence in its safe and effective performance and for the customer to accept the aircraft. This requires all systems individually and in an integrated system to be exhaustively tested to produce evidence for certification. Figure 6.20 shows some of the integration viewpoints that need to be taken in order to manage this process.

- A: the understanding of individual systems as produced by individual engineering teams in the organisation who will produce design and test evidence of satisfactory system performance.
- B: the evaluation of known integrated functions built up from a combination of the individual systems. This will include guidance and control, integrated communications, weapon systems management, etc.
- C: a viewpoint taken by organisational teams to ensure that their discipline has been incorporated throughout the design in a consistent manner. This includes disciplines such as safety, reliability, availability, testability, human factors, and electro-magnetic compatibility.
- D: a view of emergent properties that may arise and an evaluation of their risk. This is no easy task because identification of emergent properties by inspection is difficult; it may become apparent with suitable modelling but this implies modelling of integrated systems rather than simply individual systems.

6.5 Management of System Integration

Figure 6.21 portrays the system development process associated with the production of a system. Various programme milestones are shown across the top of the diagram as the process moves from contract award to the production phase.

6.5.1 Major Activities

The key activities associated with the development process are:

- concept and associated studies
- definition
- design
- build
- test
- operate.

These activities may be aligned with parts of the process described in Chapter 3.

6.5.2 Major Milestones

The major milestones are illustrated in Figure 6.22 for the hardware and software development process. Virtually every sub-system in a modern aircraft will include software embedded in a microprocessor or micro-controller to perform the functions of the system. Both the development of the hardware and the software functionality must be coordinated during development. In Figure 6.22 hardware development follows the upper branch, whilst software development follows the lower branch.

- *Contract award*: down selection of the system supplier who is to take responsibility for developing the system
- *Master program plan (MPP)*: planning of system development activities such that the development timescales are consistent with those for the overall aircraft.
- *System requirements review (SRR)*: collection and review of all system requirements. The SRR is the first top-level multidisciplinary review of the perceived system requirements. It is effectively a sanity check on what the system is required to achieve; a top-level overview of requirements and review against the original objectives. Successful attainment of this milestone leads to a preliminary system design, leading in turn to the parallel development of hardware and software requirements analysis.
- *Software specification review (SSR)*: performs a similar function for software development. As has been painfully discovered over the years the key to a good software design is to spend a lot of time ensuring that the software requirements are fully understood before progressing to software coding and test.
- *System design review (SDR)*: conducted during the requirements analysis phase, ensuring that the design will meet the design objectives as then currently understood.
- *Preliminary design review (PDR)*: preliminary review of the system design, presentation of trade studies, and selection of the preferred system design. The PDR process is the first

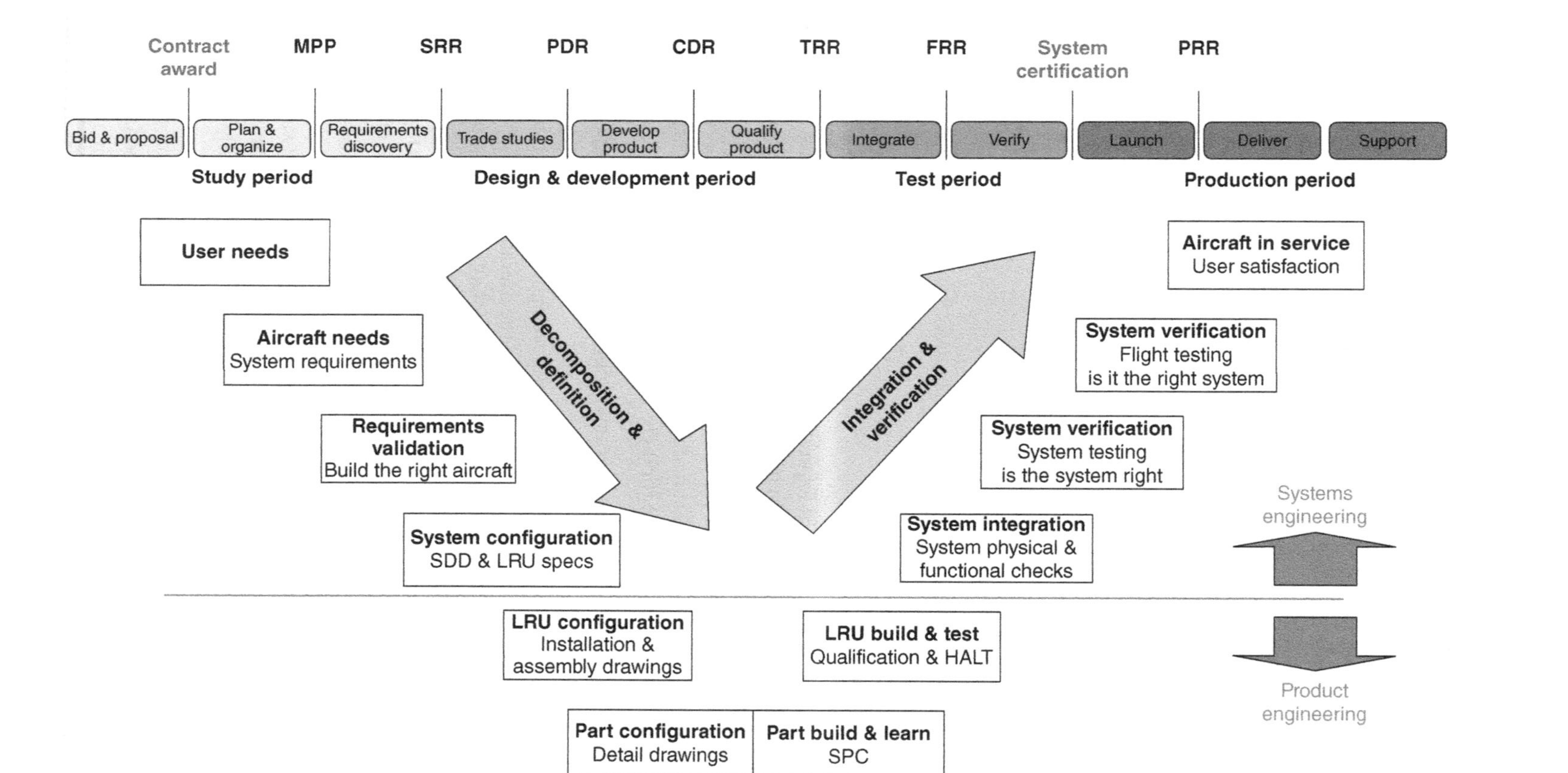

Figure 6.21 The system development process. CDR, critical design review; FRR, final readiness review; MPP, master programmes plan; PDR, preliminary design review; PRR, production readiness review; SRR, system requirements review; TRR, test readiness review.

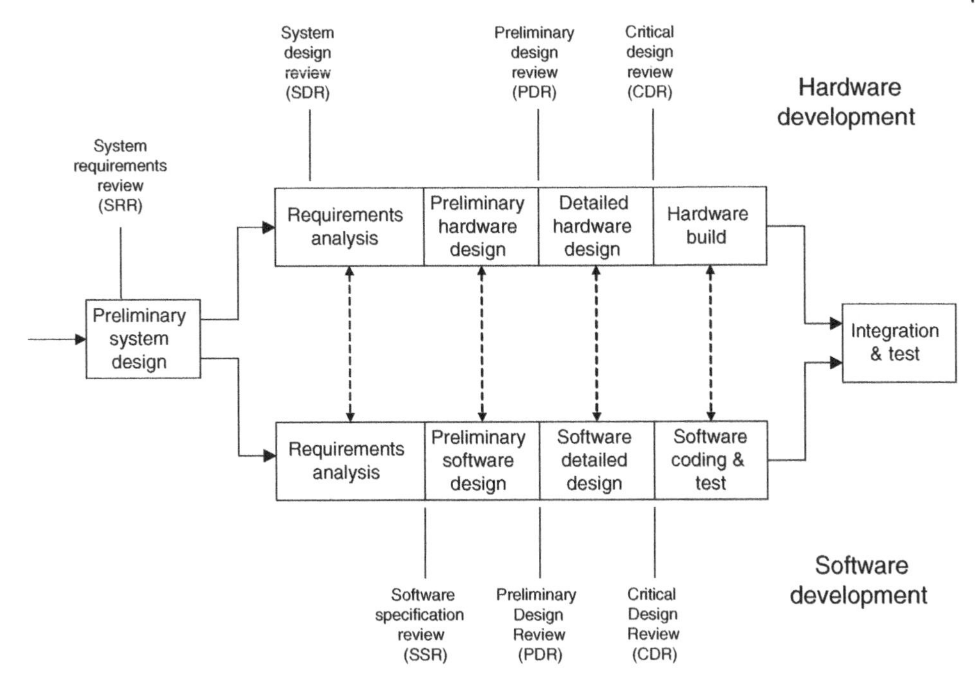

Figure 6.22 Hardware and software process major milestones.

detailed review of an initial design (both hardware and software) versus the derived requirements. This is usually the last review before committing major design resources to the detailed design process. This stage in the design process is the last before major commitment to providing the necessary programme resources and investment.

- *Critical design review (CDR)*: critical review of the design before commitment to the building of development hardware.
- *Test readiness review (TRR)*: review of the test procedures and equipment required for development to confirm that both product and test facility are ready for the test phase to begin.
- *Final readiness review (FRR)*: final scrutiny of the test procedures and equipment before commencing the verification process. By the time of the CDR, major effort will have been committed to the programme design. The CDR offers the possibility of identifying final design flaws or, more likely, trading the risks of one implementation path over another. The CDR represents the last opportunity to review and alter the direction of design before very large commitment and final design decisions are taken. Major changes in design after the CDR will be very costly in terms of financial and schedule loss.
- *System certification*: the culmination of the process of conducting and documenting the system performance and test results such that the appropriate certification authorities are provided with all the necessary documentation to certify the system.
- *Production readiness review (PRR)*: review of all of the processes necessary to ensure smooth and timely production of the system.

The main body of Figure 6.22 shows the decomposition and definition of system requirements and the integration and verification processes as shown in the earlier V diagrams (Figures 6.8 and 6.21).

Processes above the horizontal line are associated with systems engineering. Those below the line are associated with product or component engineering.

6.5.3 Decomposition and Definition Process

Key steps in this process are:

- identification of the user needs
- identification of the aircraft needs – system requirements
- validation of the requirements – are we going to build the right aircraft?
- establishing the system configuration, developing the system description document (SDD) and line replaceable unit (LRU) specifications.

6.5.4 Integration and Verification Process

As the system integration and verification process proceeds the following tasks are undertaken:

- System integration, including system physical and functional checks.
- System verification on the ground – is the system right?
- System verification in flight test – is it the right system for the aircraft?
- Aircraft in service – does the system perform and is the user happy?

6.5.5 Component Engineering

At the component level:

- establishing the LRU configuration; develop installation and assembly drawings
- component configuration and detailed drawings
- component build and learning process, statistical process control (SPC)
- LRU build and test, qualification testing and hardware accelerated life test.

6.6 Highly Integrated Systems

The design rules and methodology have evolved by best practice over the years. Seasoned industry professionals worked together to develop the design rules that are prevalent today for the design of integrated aircraft systems. The design guidelines are illustrated in Figure 6.23.

Within the UK the legal foundations for aircraft are embraced by the Air Navigation Orders, which are a British Civil Airworthiness Requirement (BCAR). In the USA the Federal airworthiness requirement (FARs) and within Europe Joint Airworthiness Requirement (JARs) lead to a set of specifications (now superseded by Certification Specifications [CS] documents) governing the design of specific aircraft types:

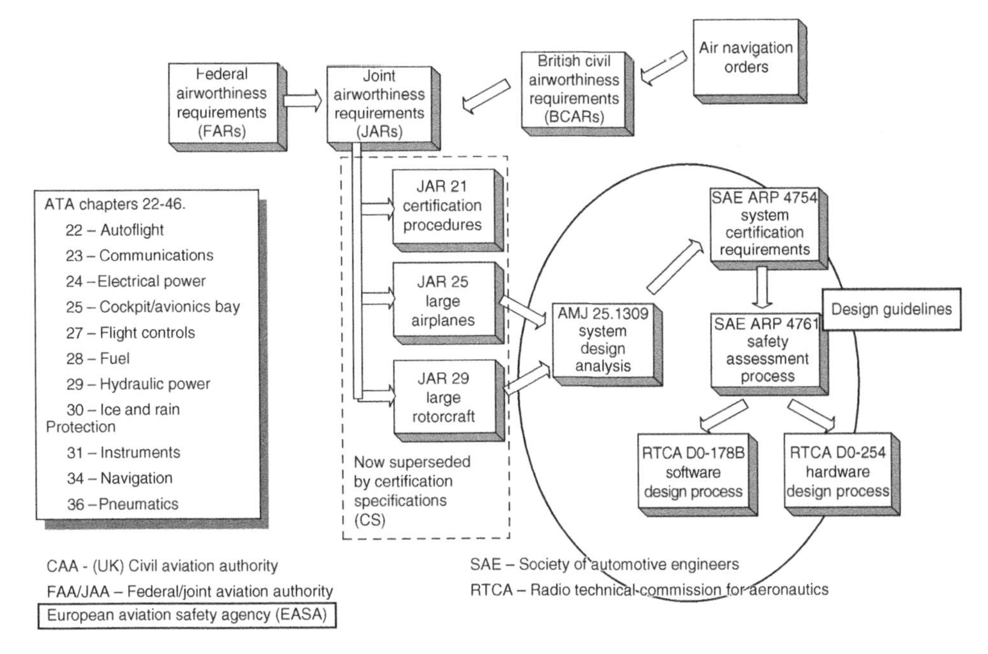

Figure 6.23 Aircraft system design guidelines.

- JAR 21 governing certification procedures
- JAR 25 governing the design of large aircraft
- JAR 29 governing the design of large rotorcraft.

Key design guidelines are contained within a series of documents that represent industry best practice but which are not mandatory, and these are illustrated in Figure 6.25. Designs do not have to follow these guidelines but a system designer who invented his own rules rather than adopt these guidelines would do so at his peril. The key documents are shown in Figure 6.24:

- system assessment process guidelines and methods – SAE ARP 4761
- system development processes – SAE ARP 4754
- hardware development lifecycle – DO-254
- software development lifecycle – DO-178B/C.

The equivalence of other documents is shown in Table 6.1.

6.6.1 Integration of Primary Flight Control Systems

The highly integrated nature of flight control systems is sometimes difficult to comprehend. Figure 6.25 shows a three-level nested control loop in a very simplified form with the following attributes:

- an inner loop controlling aircraft attitude using a high integrity fly-by-wire (FBW) system with triplex implementation
- a secondary loop controlling the aircraft trajectory by means of a dual–dual autopilot system
- an outer loop using a dual flight management system (FMS) to control the aircraft mission from take-off to arrival at the destination airfield.

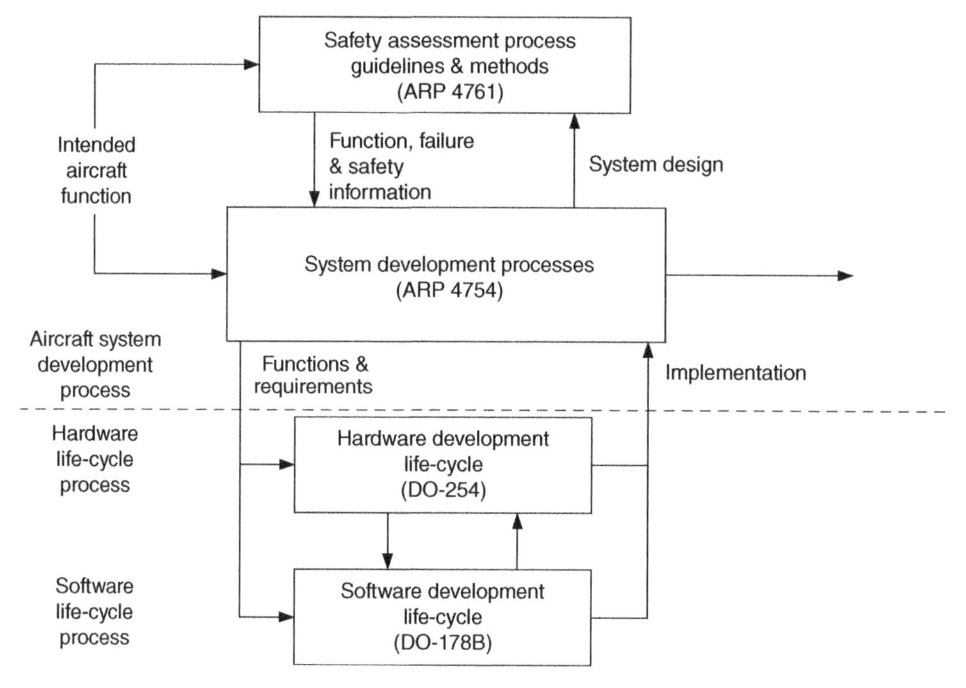

Figure 6.24 Complex system design methodologies.

Table 6.1 Equivalence of documents.

Specification topic	**US RTCA[a] specification**	**European EUROCAE specification**
Systems development processes	SAE 4754	ED-79
Safety assessment process guidelines and methods	SAE 4761	
Software design	DO-178B	ED-12
Hardware design	DO-254	ED-80
Environmental test	DO-160	ED-14

EUROCAE, European Organisation for Civil Aviation Equipment.
[a] RTCA Inc.

As the functions migrate from inner to outer loops the functionality increases as the integrity decreases, as depicted by the arrow. The figure shows inputs from automatic and manual systems, and the functions now typically performed by the flight control system.

Returning to the Air Transport Association (ATA) chapters illustrated in Figure 2.5 and emphasising those functional areas associated with the provision of the mission management yields Figure 6.26. Note that ATA has now been superseded by the JASC 4 digit codes hence:

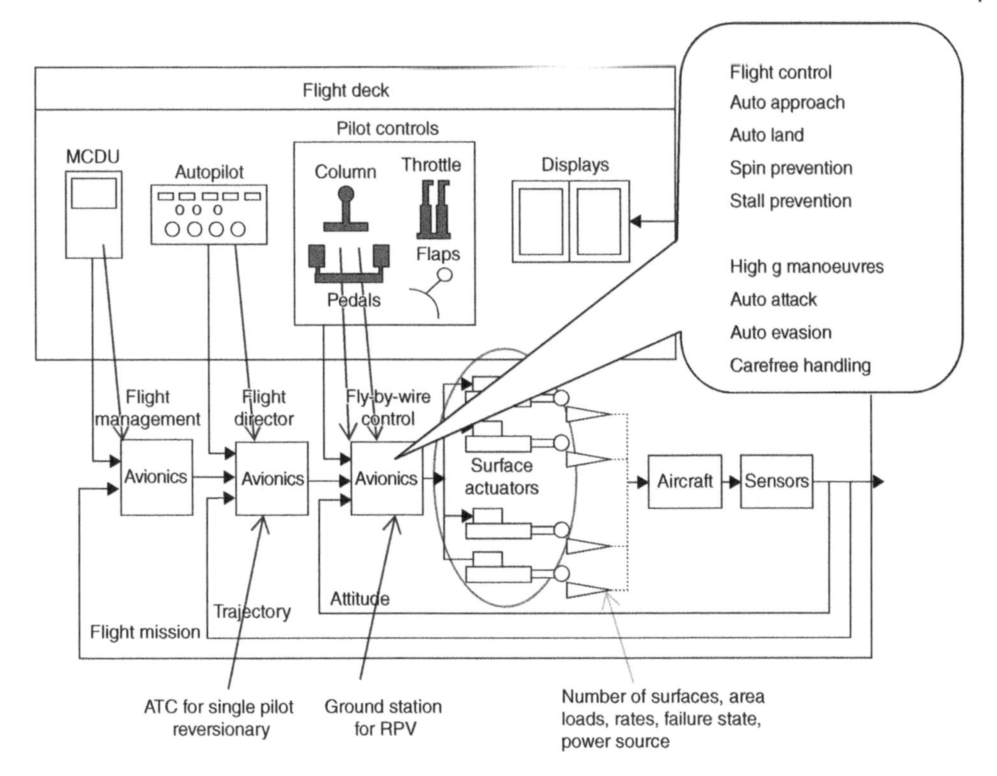

Figure 6.25 Three-level nested control loop as a complex system example.

- Auto-Flight is 2200
- Comms is 2300
- Indications is 3100
- Navigation is 3400
- Electrics is 2400
- Flight control is 2700
- Hydraulics is 2900
- Fuel is 2800
- Engine control is 7600
- Powerplant is 7100.

This highlights all the functional areas associated with providing the overall mission management function:

- avionics functions such as auto-flight, communications, recording and indications, and navigation
- electrical power
- flight controls and hydraulic power
- fuel system, powerplant, and power control.

Without the necessary contributions from all these elements the system will not function.

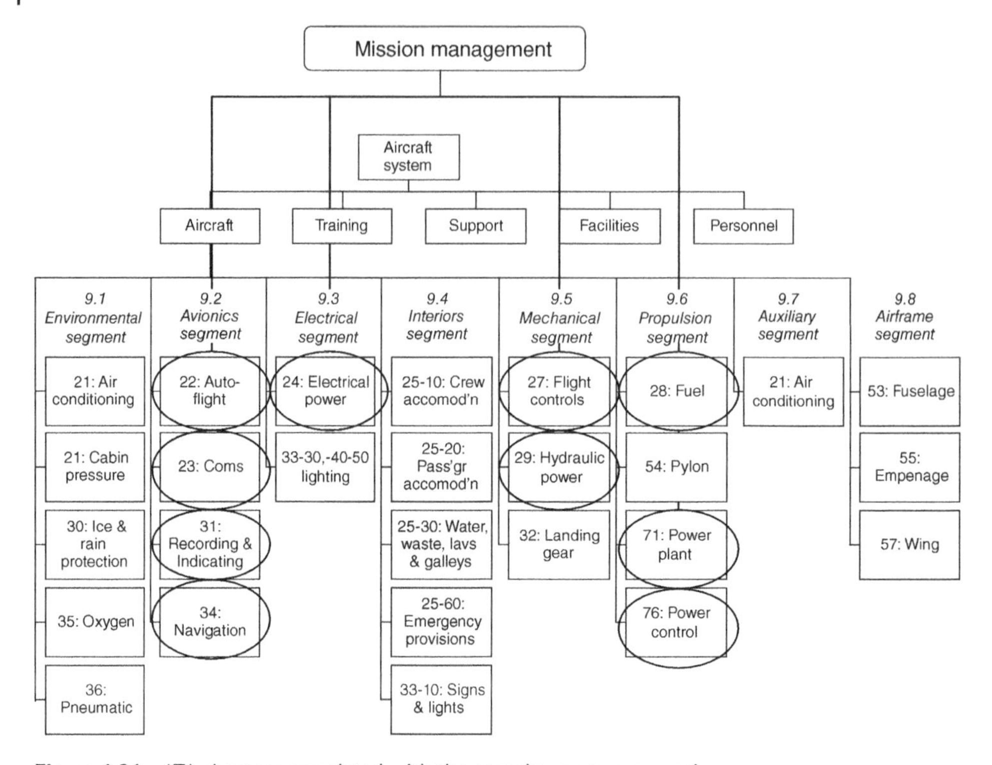

Figure 6.26 ATA chapters associated with the complex system example.

6.7 Discussion

It is clear that integration at a number of levels is here to stay in modern aircraft design. The drive towards more automation in the systems to reduce crew workload means that many more functions must be performed in the systems. A move towards unmanned vehicles, especially those with high levels of autonomy or independence from ground-based control, is a further driver towards more automation. In order to achieve these levels of automation more integration of functions into systems with software-based functional design is required.

The result of this is to produce architectures that are complex to the extent that there is no way that system behaviour can be explained by simple inspection of diagrams. The 'hidden' nature of functions in software and firmware, and the exchange of information as streams of digital data words by various data bus types accentuates the difficulty. The vast amount of design and test data produced for a complete aircraft makes it difficult for anyone to fully comprehend the behaviour of the system and to analyse the test evidence to demonstrate that the system has been tested exhaustively. However, someone must do so in order that the product can be signed off and accepted by the customer.

There is a potential impact on safety that needs to be considered. The circumstances described above make it difficult for a comprehensive safety analysis to be performed. The system architecture shown in Figure 6.27 illustrates the point of complexity. The diagram

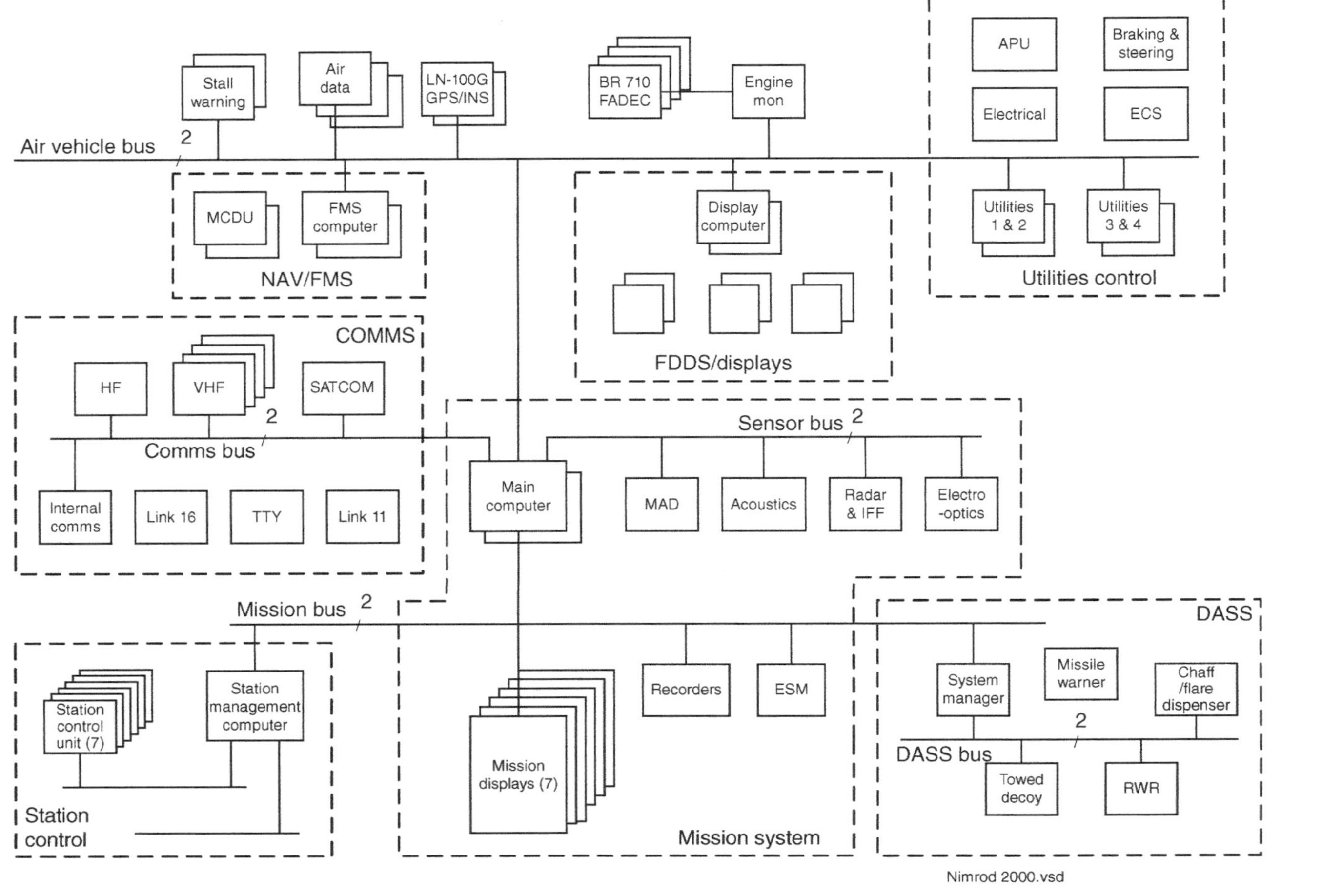

Figure 6.27 Simplified military surveillance aircraft architecture. APU, auxiliary power unit; DASS, defensive aids sub-system; ECS, environmental control system; ESM, electronic support measures; FADEC, full authority digital engine control; FCS, flight control system; FDDS, flight deck display system; FMS, flight management system; GPS/INS, global positioning system/inertial navigation system; HF, high frequency; IFF, identification friend or foe; MAD, magnetic anomaly detector; MCDU, multi-function control and display unit; Nav/FMS, navigation/flight management system; RWR, radar warning receiver; Sat Com, satellite communications; TTY, teletype; UHF, ultra high frequency.

has been much simplified – even so it will be illegible to the reader – and in reality each of the blocks on the diagram contains even more blocks and more interconnections.

Chapter 11 examines the impact of integration and its potential impact on flight safety in more detail in order to stress the importance of understanding the behaviour of the total system and of devising suitable tests to verify its performance.

Exercises

1 A lawn maintenance system.

This exercise is intended to take people out of their comfort zone by looking at a system that is not related to aircraft systems. It can be conducted by individuals or by groups. The exercise can be conducted over an extended period to simulate a real project.

Top-level requirement: To design a lawn maintenance system for domestic use.

Key performance parameters:

- to provide a lawn with a pleasing appearance all year round
- to be safe in operation
- to require minimum user effort
- to present minimum nuisance to neighbours
- to present no hazard to family, pets and neighbours
- to have a life of at least 15 years
- to be capable of recycling with no environmental impact
- to meet all requisite legislation.

A Brainstorm a number of solutions ranging from conventional to innovative. Consider at least the following:
- a mechanical system
- an animal-based system
- a service provision
- genetic modification.

B For each solution:
- provide a brief description of the solution you have chosen
- provide a stakeholder diagram
- discuss and record the major issues you envisage in qualification using a verification cross reference document (VCRM) to explain how you will plan to qualify the system. How will you prove to the regulatory authorities and the customer that is fit for purpose? Are there any residual risks?

C Perform a trade-off to determine what you believe is the optimum solution that:
- meets the requirement in terms of finess for purpose
- is most beneficial in environmental terms.

Things to consider: animal-based solution

- What subsidiary costs have been considered – vet bills, feed, shearing, etc.?
- Will local by-laws allow this?
- Have you considered disease threats and possible limitations – foot and mouth, blue tongue, etc.?
- Are there any noise issues?
- What about interaction with pets – dogs in particular?
- How will you avoid incursions into neighbours' gardens and possible damage?
- How will quality of cut be demonstrated for fussy gardeners or lawn anoraks?
- How will you deal with edges?
- How will the solution discriminate grass from flowers and vegetables?
- What are the costs of stabling/shelter?
- Are there any useful by-products?

Things to consider: technical-based solution

- Have you considered noise and pollution?
- How will the solution deal with irregular shapes?
- How will it deal with edges?
- What about hazards from stones, pets, ornaments?
- What storage is needed?
- What are the costs of servicing?
- What are the estimates of mean time to repair, life, obsolescence, etc.?

Things to consider: service provision solution

- Is the service provider experienced?
- What is their local reputation?
- Is the service provider reliable?
- Have you agreed a price escalation guarantee?

Things to consider: genetic modification solution

- Is the technology available?
- What is the long-term impact?
- What are the implications of contaminating neighbouring pasture?

D What does the term 'system architecture' mean in the context of your system? Explain and then construct an architecture with a meaningful system boundary. Identify stakeholders external to that boundary and explain the mutual dependence.

2 Power line inspection system.

Design a system to monitor the status and health of insulators and conductor connections of electricity powerline pylons.

Understanding the requirement

What conditions need to be monitored?

What sensors are available to do this?

Over what time period is the monitoring required to be performed?
What weather conditions need to be withstood?
How is the sensor information to be stored or relayed?

Trade-off
Determine the most appropriate parameters for comparison.
Perform a trade-off study.
Select the optimum candidate.
Draw a system architecture.
Describe how the system is to be tested and qualified.
Describe how the system is to be certificated for UK and European use.
Examine how to extend that certification for use in other areas of the world.

Performance
Determine the most appropriate performance parameters.
Describe how you would market the product.

References

AGARD (1996). *Advisory Group Report 349*. Flight Vehicle Integration Panel Working Group 21 on Glass Cockpit Operational Effectiveness.

Ashcroft, F. (2000). *Life at the Extremes – The Science of Survival*. Harper-Colllins.

Sunday Express (2013). Dead BA pilots 'victims of toxic cabin fumes'. Sunday 27 January.

Learmount, D. (2014) *BA crew autopsies show organophosphate poisoning*. www.flightglobal.com. 31.07.2014

Michaelis, S. (2011). *Health and Flight Safety Implications from Exposure to Contaminated Air in Aircraft*, PhD thesis. http://handle.unsw.edu.au/1959.4/50342 (online)

Moir, I. and Seabridge, A. (2008). *Aircraft Systems*, 3e. Wiley.

Moir, I. and Seabridge, A. (2013). *Civil Avionic Systems*, 2e. Wiley.

Royal Academy of Engineering. (2007). *Creating systems that work: Principles of engineering systems for the 21st century*.

Warwick, G. (1989). Future trends and developments" from. In: *Avionic Systems* (ed. D.H. Middleton), 247–256. Longman.

Further Reading

Brookes, A. (1996). *Flights to Disaster*. Ian Allen.

Elliott, C. and Easley, P.D. (eds.) (2007). *Creating Systems that Work: Principles of Engineering Systems for the 21st Century*. Royal Academy of Engineering ISBN1-9033496.

7

Verification of System Requirements

7.1 Introduction

In a large and complex system such as an aircraft or a warship it is impractical to build the entire product without first conducting some form of analysis to provide a high degree of confidence that the completed product will work to specification. The cost and the time involved in so doing would have a large impact on any programme, and yet the risk of proceeding without some kind of confidence that the system will work to specification is unacceptable.

In order to demonstrate to the customer that their requirement has been met, it is necessary to test the product throughout the lifecycle. The test results are evidence that the requirements can be demonstrably met. This testing is often performed on physical products such as prototypes. However, it is time-consuming and costly to build experimental systems and prototypes, particularly if the design contains errors that must be corrected or if the operation is highly influenced by factors in the environment.

Figure 7.1 shows the V diagram and its review points as discussed in Chapter 6. In this version of the diagram the left-hand side has been obscured to illustrate the fact that testing is usually thought to be conducted on the right-hand side of the diagram, once hardware is available to test.

However, it should be acknowledged that there is a lot of evidence of correct understanding of requirements and solutions to the requirements that emerges during the design of the system. This evidence should not be overlooked but should be gathered and used to contribute to the qualification of the product.

Engineers need rigorous methods for analysing and observing the performance and behaviour of their systems that allow decisions to be made on continuing along a development path without committing expensive resources or incurring excessive risk. An ideal method of doing this is to experiment with the system under planned experimental conditions that can be repeated or modified in a controlled manner. This can be used to explore the limits of operation of a system, and can also be used to obtain test results that can be used as evidence that the requirement has been met.

Design and Development of Aircraft Systems, Third Edition. Allan Seabridge and Ian Moir.

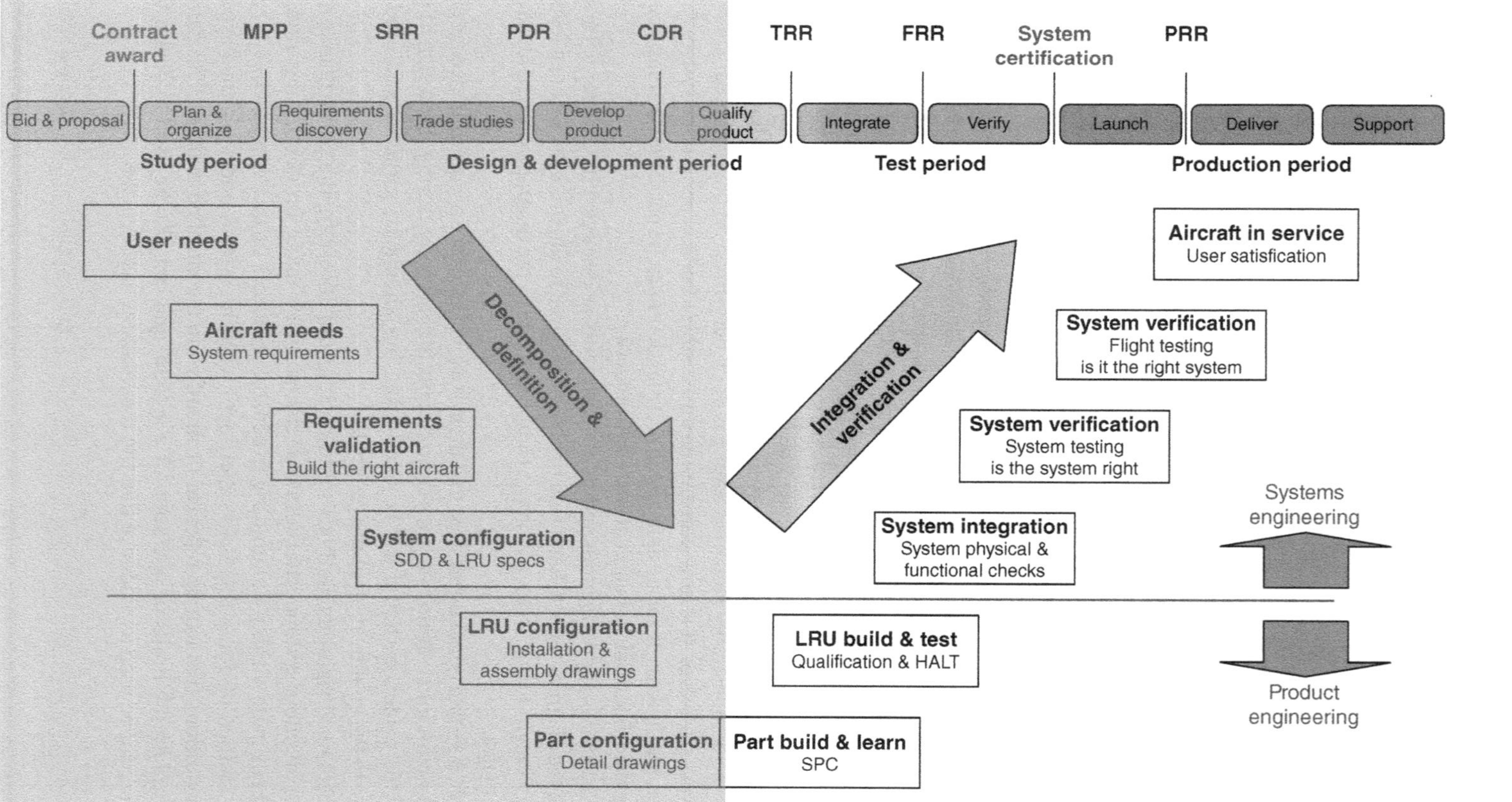

Figure 7.1 Traditional view of testing in the V diagram. CDR, critical design review; FRR, final readiness review; MPP, master programmes plan; PDR, preliminary design review; PRR, production readiness review; SRR, system requirements review; TRR, test readiness review.

7.2 Gathering Qualification Evidence in the Lifecycle

Modelling, simulation test rigs, and prototypes are tools used to provide a sound basis for examining the behaviours and performance of aircraft systems under a variety of conditions that give a high degree of confidence in the design. The gathering of test evidence is performed by a combination of all of these tools. This includes all supplier evidence gathered from the test rigs for sub-systems and components provided by the supply chain. For the evidence to be relevant it is essential that every source of evidence is controlled and the relationship between sources is clearly understood. This process requires a robust approach to identifying every source and exercising configuration control.

Evidence is gathered from a number of different sources and processes. From the test process alone evidence is gathered from the sources shown in Figure 7.2.

This evidence is supplemented by other information gathered from the design process which can be presented to the customer as suitable evidence of qualification at early stages in the lifecycle. The evidence is then used to gather up and submit evidence at stages in the programme to support clearance for engine ground running, prototype flying, preliminary customer acceptance, and full customer acceptance, as shown in Figure 7.3.

It is important to plan how this evidence is to be gathered and also to record the successful completion of each stage of testing. The verification cross reference matrix (VCRM) should start life as a planning tool for the test regime. A simple Excel spreadsheet can be constructed to divide the flight into as many phases as are thought to be necessary. Then decide what type of test is to be used to validate each phase of the mission, what tools are required, what facilities, what test procedures, and what instrumentation.

This will be used to form the basis of a test and qualification plan so that the whole process can be managed effectively. A simple spreadsheet-based cross-verification matrix is shown in Figure 7.4 in a generic form. The columns can be expanded to show detailed test

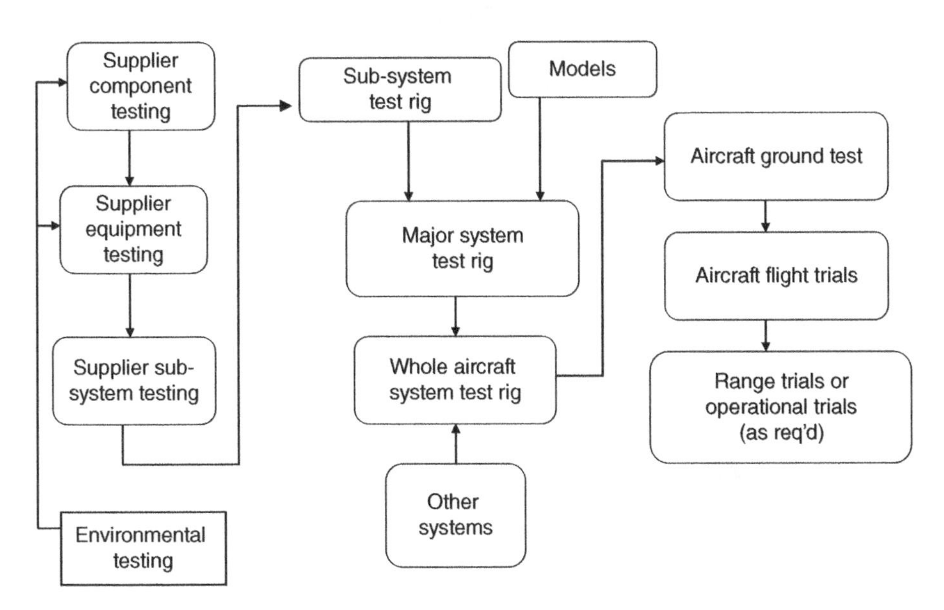

Figure 7.2 Sources of test evidence.

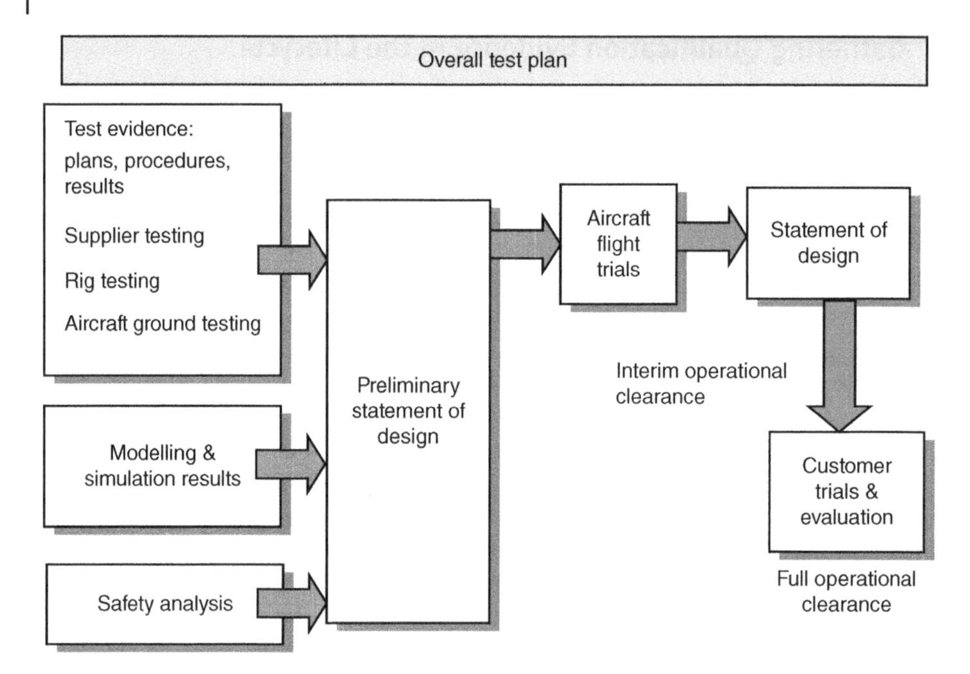

Figure 7.3 The path to full product acceptance.

Requirement	Design	Calc	Analogy	Model	Test	Operation
R001.01						
R001.02						
R001.03						
R002.01						
R002.02						
R00n.0n						

Each requirement has one or more verification methods declared, and dates can be established to form a plan. This VRCM can be used to form the basis of a plan for demonstrating compliance and agreeing early payment plans with the customer. Progress can be indicated using colours, e.g.:

Planned

Achieved to plan

Risk to plan

Late to plan

Figure 7.4 Verification cross-reference matrix.

events and the matrix used to record simple completion results or to record test procedure numbers/issue, dates planned, and achieved authorising signatures etc. to form a complete record of test achievements. Ideally this form of record should be part of the requirements analysis tool so that an exact correspondence between test evidence and requirements can be recorded.

This has an added benefit of forming a suitable tool to negotiate with the acceptance authority or the customer to arrive at a payment plan. It is always of benefit in this costly phase to try to achieve test targets early and to use them as the basis for an early demonstration of design completion. This allows the customer to get involved very early on the demonstration process, and allows the contractor access to early payments. An example payment plan is shown in Figure 7.5.

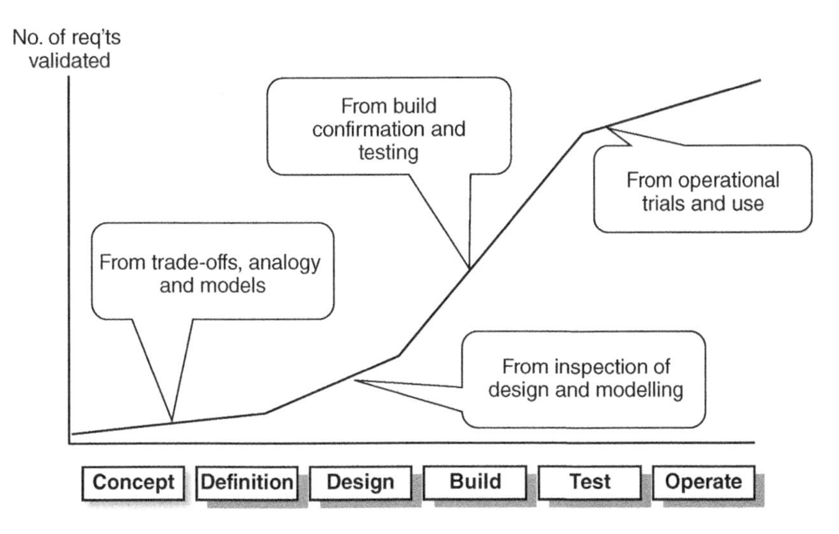

Figure 7.5 Using early test evidence to improve the payment plan.

7.3 Test Methods

In this chapter the following test methods are used as examples of providing qualification evidence, and each of these methods will be explained:

- inspection of design
- calculation
- analogy
- modelling and simulation
- test rigs
- environmental testing
- integration test rigs
- flight resting

- trials
- operational testing
- demonstrations.

7.3.1 Inspection of Design

Design information is available in a number of different forms ranging from preliminary schemes through to detailed drawings or three-dimensional (3D) computer-aided design (CAD) models, specifications, interface control documents, and software design documents. At some stage in the design process these documents will become more and more definitive as they pass each review stage. A point can be agreed at which they can be said to contribute to qualification and accepted as a defining point which merits an agreement of completion.

An example is the 3D model of a radar scanner which sweeps in azimuth and elevation to provide a forward-looking search pattern for air-to-air target detection, and can also be tilted down for ground surveillance. A Catia model can be demanded to replicate the antenna sweep envelope and to allow the profile of a radome to be designed which will not foul the antenna throughput its sweep. Figure 7.6 shows the model from which this process will start. Once the model has been completed it can be animated to demonstrate the appropriate clearances before the design is frozen and a commitment made to manufacture. The scanner will be animated throughout its entire range of azimuth, elevation, and tilt to provide an envelope for the design of a radome with appropriate clearances. (Figure 7.6 shows for interest a heritage air intercept radar, AI 23B, and a CAD model of a more modern radar).

7.3.2 Calculation

In the early stages of design the evolution of a system progresses by way of calculations to determine such aspects as pressures, electrical loads flight envelopes etc., usually assisted by computer programmes or even spreadsheets. It is vital that these calculations are formalised and preserved. The ready access to calculating and computing facilities on laptops means that some engineers are tempted to perform a calculation without treating the

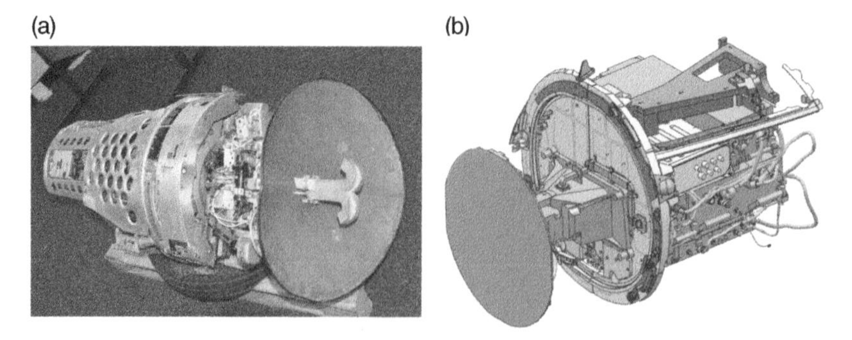

Figure 7.6 Catia model of an example radar scanner. (a) A Heritage AI 23B radar (photo: Leon Skorczewski) and (b) CAD image of a typical radar (photo: Allan Seabridge collection).

exercise as a design tool. In other words, the algorithm or method should be stored, tested under a range of variables, and then frozen in much the same way as a drawing. This ensures that the calculation can be repeated at any subsequent review or re-design. Used with caution a calculation can be performed with different ranges of variables to explore the limits of the design.

7.3.3 Analogy

Quite often a design solution is found that has been used previously and found to be a sound solution. The same design can be transported to a new project provided the designer is confident that the engineering and environmental conditions are the same. The test evidence for the previous design must be scrutinised to ensure that the installed conditions are identical and that the evidence can be used. Aircraft seats are an example since they are designed to meet a well-understood security of attachment case and can be purchased as commercial off-the-shelf (COTS) items. Many small components can be used in this way and may form the basis of stock items suitable for a project.

7.3.4 Modelling and Simulation

Modelling can be used throughout the product lifecycle to represent many design options and to help to define the metrics necessary to assess these options. Models are the foundation of all analysis, whether for the purpose of quick calculations or to provide qualification evidence. Simulation is a computational framework whereby it is possible to predict and/or replicate the performance of systems over time. This allows systems to be observed under conditions that exceed the operational limitations of test rigs. It can provide access to performance trends that would be very difficult or expensive to experience or measure. In some instances simulation may actually replace certain aspects of rig testing altogether.

The term 'modelling' is used in this chapter to refer to a number of different techniques used to describe the behaviour and performance of a system using means other than actual operation of the system in a real-life environment. This definition includes the use of computer-aided modelling so that 3D representations of system components can be used for analysis of installation, human machine interface, and access, pure mathematical modelling or state analysis, and simulation.

The distinction between the last two terms is nicely explained as: *If the relationships that compose the model are simple enough, it may be possible to use mathematical modelling methods (such as algebra, calculus, or probability theory) to obtain exact information on questions of interest; this is called an analytic solution. However, most real-world systems are too complex to allow realistic models to be evaluated analytically, and these models must be studied by means of simulation. In simulation we use a computer to evaluate a model numerically, and data are gathered in order to estimate the desired characteristics of the model.* (Law and Kelton 1991). This simulation can be tested by running it under realistic conditions in a controlled environment.

A means of overcoming these difficulties is to use tools to emulate or simulate system operation. Preferably these tools should provide a 'soft' representation of a system that can be modelled and re-modelled without incurring excessive cost. Tools and techniques are available to allow this to happen. Alternative system designs can be compared and contrasted using simulation to see which best meets a specific requirement.

An additional advantage of producing robust and validated models is that they can be provided to other teams or to suppliers for them to develop their own functions to mutually agreed interface and performance definitions.

Chapter 3 illustrated how the cost of change increases as the product advances through its lifecycle. One advantage of modelling is to increase confidence in the correctness of the solutions that make up the complete system so that the likelihood of change is diminished. In addition, the results obtained can be used as evidence that the design is compliant with the requirement. It can be an advantage to gather this evidence early in the lifecycle in order to gain the customer's confidence that the design is maturing and converging to a solution that can be qualified.

Typical types of models used in systems engineering are:

- *Simple diagrammatic model*: the thought process or intellectual process in which a systems engineer 'imagines' or envisages the structure and behaviour of a system during the mental process of perception, intuition, and reasoning during the evolution of a system concept.
- *Simple scale model*: a physical, scaled representation of the system or the components of a system.
- *Mathematical model*: a simple model of system behaviour described as changes of state with defined probability.
- *Simulation*: most complex, real-world systems with stochasitic elements cannot be accurately described by a mathematical model that can be evaluated analytically. Thus a simulation is often the only type of investigation possible. Simulation allows one to estimate the performance of an existing system under some projected set of operating conditions.
- *Test rig*: a test facility that mimics part or the whole of the product, allowing experiments to be conducted under instrumented conditions.
- *Prototype*: a full-scale representation of a product or system that can be tested exhaustively to establish performance before manufacture.

It is often not practical to attempt to model a product in its entirety, and it may not even be necessary. What is most useful is to model those aspects of a system that are least understood and whose incorrect operation later in the lifecycle may pose a significant cost or time risk. It is important to use models correctly and generate simple models: *Models don't have to be an exact representation of reality to be useful. In fact making them less realistic generally makes them more useful as long as they still provide useful insights. A model that is as complex as the thing it represents is likely to be too complex to be useful. A simple model is easier to work with.* (Stewart 2007).

Generating simple models can be accomplished by analysing the system and breaking it down into elements that can be modelled, and by understanding their interfaces and dependencies upon other elements of the system. Figure 7.7 shows how a number of different modelling techniques can be used in a complex system.

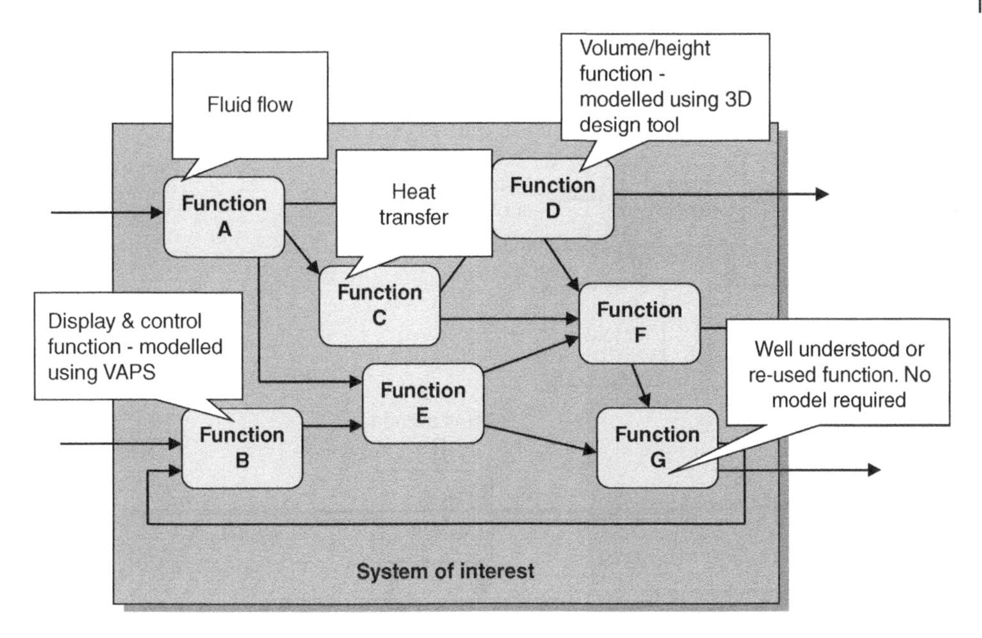

Figure 7.7 Example of the use of different modelling techniques.

Using different techniques in this way allows each function to be tested and refined as necessary until its performance is judged to be satisfactory. The results of each tested function can then be assembled manually to determine whether or not the combination also performs satisfactorily. An alternative, and better, method would be for the different models to be assembled into a framework that enables them to be connected to form a complete, or near-complete, simulation of the system. An illustration of the interconnection of models is shown in Figure 7.8, with an example of application to a fuel system.

The tank model in this example is important because it is used to translate the level of fuel determined by a number of probes under all conditions of roll and pitch, and translates this into fuel volume and then mass. This model application then in theory enables a fuel model to be designed which could replace a physical fuel rig. In fact the Nimrod MRA4 fuel system was qualified in this way.

This model can be extrapolated to show the complexity of a modern fuel system, as illustrated in Figure 7.9. The fuel system model must be designed to represent the following sub-systems and their interactions to complete a total simulation of a fuel system:

- A1 is the fuel system, the collection of tanks, fuel gauge probes, pumps, and valves that measure the quantity of the fuel in the tanks and ensure that it is moved from tank to tank and to the engine under the control of the fuel management section of the vehicle management system. For accurate measurement of fuel quantity the properties of fuel must be understood since fuel is known to stratify in the tanks in layers of density and temperature.
- A2 is the action of the aircraft in terms of pitch, roll and yaw position, and rates of change. This can cause the fuel attitude to change, thereby demanding careful selection

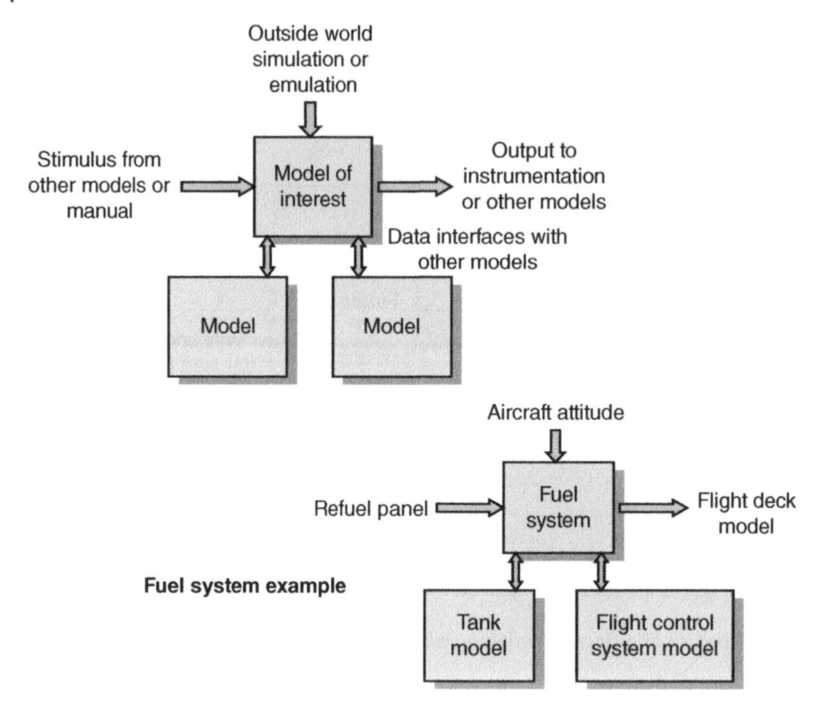

Figure 7.8 Examples of interconnected models.

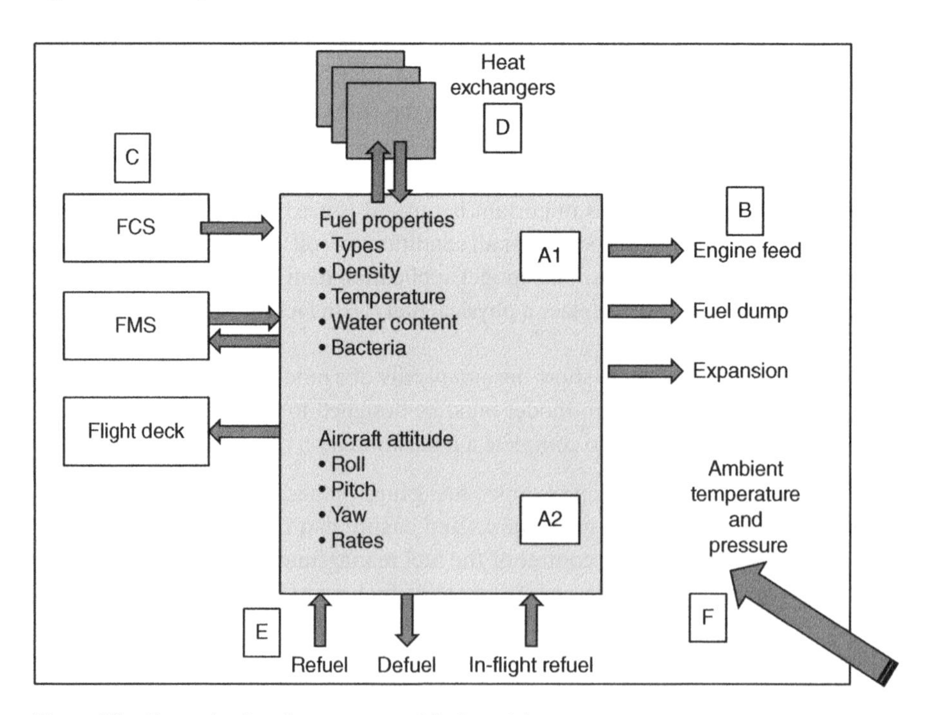

Figure 7.9 Example of an interconnected fuel model.

gauging probe position to maintain accuracy, and also causing the fuel to slosh in large tanks. A model of this can be used to select the optimum position of baffles or expansion tanks.

- B represents the demand of the engine in terms of fuel flow throughout the flight envelope, and the fuel dump and expansion requirements.
- C represents the avionics integration. The flight control system (FCS) requires the fuel centre of gravity to be maintained within a predetermined range for either manoeuvre or fuel economy. The flight management system (FMS) needs to know the fuel quantity on board, whilst the flight deck displays will provide the crew with information about the fuel quantity in each tank, the status of the fuel system components, and the fuel remaining.
- D represents the heat exchangers provided to cool engine oil and hydraulic fluid in which large quantities of heat may be dumped into the fuel.
- E represents the mechanisms for uploading fuel on the ground or in the air, and for off-loading fuel in a controlled manner.
- F represents external factors such as ambient pressure and temperature. Temperature is an important factor since high ground ambient can cause expansion and spillage of fuel, whereas very low temperatures can lead to freezing of fuel in long duration polar routes – a factor in at least one serious accident.

This demonstrates that a model of the fuel system is quite complicated and demands close collaboration between the owners of different sub-systems. There is a choice to be made between a physical fuel system rig or a combination of models and simulation, or even a combination of all three to obtain a clear understanding of the behaviour of a modern fuel system. The trade-off is often the cost of a physical fuel rig, the facility to house it and the costs of running it compared with the cost of a model and simulation. To many people the physical rig has real benefits – it can be observed and measured doing its job and it uses real aircraft components, against the virtual world of the model. A decision that has exercised many minds in recent projects is the subject of the trade-off process. The answer would seem to lie in the degree of fidelity required – how representative is the model of the real world? The physical rig can be built to be very close to the aircraft in terms of components, spatial layout, realistic tank shapes etc. However, the rig can only be moved within a limited range of attitudes in pitch and roll, and will certainly not replicate rates of change. The simulation can only approximate to real components but can change attitude in real time.

7.3.4.1 Modelling Techniques

The modelling techniques introduced above are described in more detail in the following sections.

7.3.4.1.1 Simple Diagrammatic Models Simple models captured in a pictorial representation are part of an intellectual process of envisaging the very early stages of design. They are mainly contained in the head of the designer and are used to envisage and explain concepts mentally. This may be a mental visualisation of a shape, a process or a mathematical expression. In order to achieve a common basis for discussion, diagrammatic models are often turned into sketches and rough notes. These are quick to produce, can be easily amended during discussion, and can be converted into 'harder' drawings or slides for inclusion in reports or presentations. This enables other people to share the cognitive model. This is the stuff of legend – ideas are quickly transferred to paper – a napkin in a restaurant, a roll of toilet paper, the typical 'back of an envelope' sketch. Some examples of diagrammatic models committed to paper are shown in Figure 7.10. These include a sketch of the retraction of an undercarriage leg on graph paper with calculations to illustrate the interactions between the leg and the landing gear bay doors, a model visualising the mathematics of ejection from an aircraft, a mission profile to illustrate the characteristics of the aircraft with an integrated weapon system to meet a specific set of mission requirements. All of these are sufficient to declare that there is confidence to move to the next phase of design.

7.3.4.1.2 Simple Scale Models Simple models are a physical interpretation of a system or individual systems components, usually in small scale. Typical materials include cardboard, foam, modelling clay, plastic, balsa wood, metal or acrylic. Such models are good for developing the information provided by cognitive models and they add value by being three-dimensional and tactile.

Very precise models can be made by a process known as laser stereo-lithography, in which a 3D CAD tool model is used to produce a small-scale model in an acrylic substance.

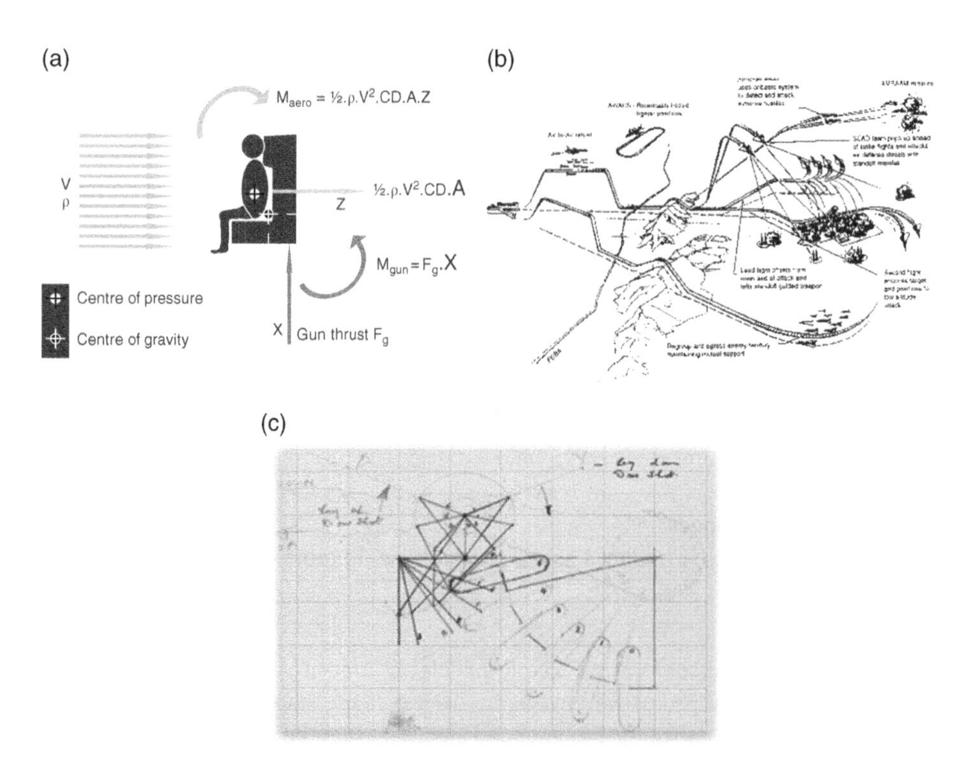

Figure 7.10 Examples of simple diagrammatic models: (a) model of ejection forces and moment, (b) sketch to understand a mission profile, and (c) an undercarriage retraction sequence.

Because the process is driven by the output of a CAD tool the dimensions are an exact scale of the original, and the final product has a high degree of fidelity. The resulting models are often used for marketing displays or in support of bid presentations to customers. This facility is now readily available using commercially available 3D printers, producing results in plastic or metal which can be suitable for use in a final product.

Another use of a simple scale model is a wind tunnel model, which is used to obtain real data pertaining to a full-scale product. In this case the overall shape is more important than detailed fidelity. This has the benefit of subjecting scale models to a range of speeds to emulate almost the entire flight envelope.

An extreme form of a scale model is the full-scale mock-up produced in wood or metal. Whole or partial mock-ups of aircraft were often produced to trial installation concepts, to validate human–machine interfaces or to provide models for marketing purposes. For design purposes a mock-up can be made to the aircraft drawings in metal, often referred to as a Class 1 mock-up. They formed a good familiarisation tool for air crew and for engineers testing their concepts and can be used as a template or former for pipe and wiring harness shapes. Glass fibre models are often used at air displays – they are easier to transport and maintain than real aircraft.

Figure 7.11 shows some examples of models constructed for a specific purpose. The Merlin Flight Simulation Group MP521 simulator can be used for all aircraft design-related principles from stability and control evaluation through to cockpit ergonomics, systems engineering, avionics, and psychology, and is shown here in the simulation laboratory at the University of Dayton, Ohio. The aircraft models were made by 3D printing to illustrate the concept shapes developed by graduate students on the air vehicle design course group project at Cranfield University. The F-35 model was a full-scale replica of the aircraft for use as a public relations tool; this technique is now used to make weather-resistant gate guardians. An alternative is the full-scale mock-up, which is a realistic replica of the actual aircraft used as a design and installation tool. This type of mock-up has largely been superseded by 3D CAD images, which can be made relatively cheaply and can be rotated for different views. However, there is a still a lot to be said for a real-life model that, with hands-on application, can be put to good use in solving real-world problems of installation, access, and human factors.

7.3.4.1.3 Mathematical Models and Simulations There are tools available that allow designers to model or simulate some aspects of their systems on a computer or desktop PC. Commercially available spreadsheets can be used to build simple models, and commercially available tools such as Matlab/Simulink (http://www.mathworks.com/support/learn-with-matlab-tutorials) and FloMaster (http://www.mentor.com/floMASTER/products/mechanical/floMASTER application example) can be used to set up more domain-specific models and simulations. This allows functions to be tested and visualised in animated display sequences supported by data in real time or slow/fast time. This has some disadvantages because the quality of some animations and the amount of data produced can give people a false impression of the reality of the simulation. Nevertheless, modelling is a powerful tool in providing the facility to test systems under controlled experimental conditions. Example systems that benefit from such modelling include:

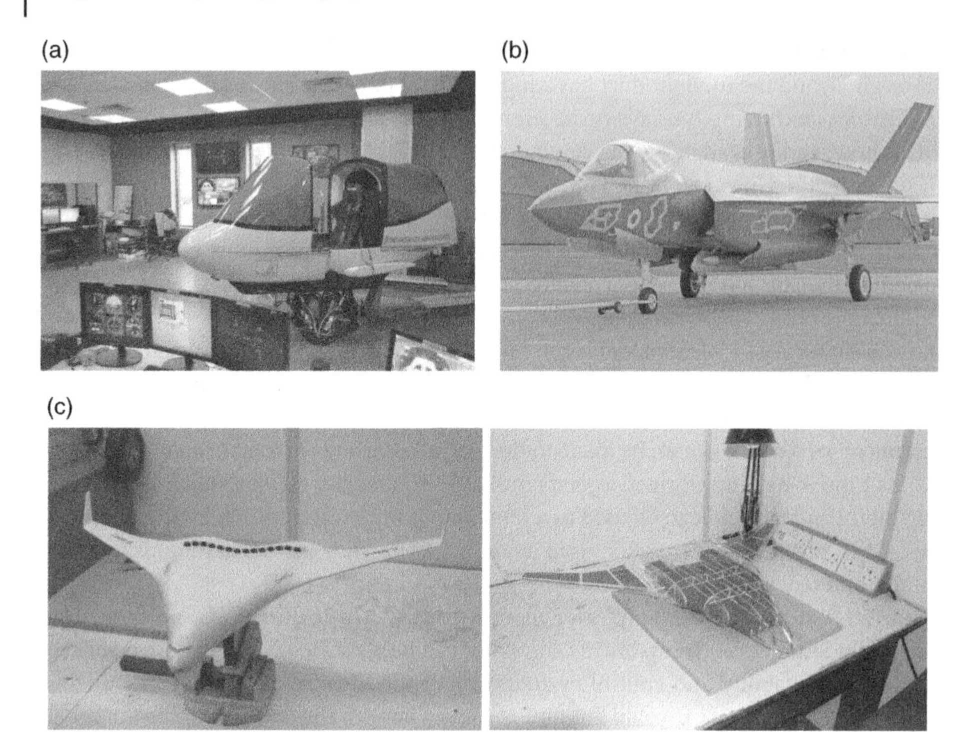

Figure 7.11 Examples of simple scale models: (a) the MP 521 simulator in the simulator laboratory at University of Dayton, Ohio (photo: Merlin Flight Simulation Group), (b) F-35 full-scale replica (photo: Allan Seabridge), and (c) models from Cranfield University (photo: Allan Seabridge).

- *Thermal/cooling systems*: to model the performance of closed loop air or vapour cycle machines, and to model air flow distribution to passengers or equipment.
- *Fluid flow systems:* to model fuel system tank height/volume characteristics for gauging system design, to model transfer sequences, to model fuel behaviour in manoeuvre conditions, and to model flows and pressures in pipes for fuel and hydraulic systems.
- *Electrical power systems*: to perform load analysis for various phases of the mission, to calculate the impact or resistive and reactive loads on phase balancing, and to perform sneak circuit analysis to look for incorrect earth connections or short circuits in circuit design.
- *Control systems*: to simulate the dynamic performance of closed-loop systems under differing phasing and time domains.
- *RF antenna interoperability*: to predict the performance of transmitters and receivers in a multiple and simultaneous transmitting environment to avoid mutual interference effects, and to examine the impact of jamming techniques.
- *Route planning*: to model commercial and military routes in order to predict ideal economic or timely routes and to plan loads for the FMS.
- *Airport management*: to model the density and movements of traffic in the air and on the ground, to model passenger movements, and to predict transportation demands.
- *Vulnerability/battle damage susceptibility*: to predict the damage effects of projectiles of fragments on aircraft structure and internal equipment to assist in the physical separation of equipment to avoid common mode damage effects.

- *Data bus loading*: to examine the impact of data density and transmission rates on the loading of data buses.
- *Weapons ballistics*: to predict the separation characteristics of weapons from the aircraft under varying conditions, and to predict or confirm the accuracy of achieving the target destruction.

7.3.4.1.4 Modelling Tools and their Application A number of tools have been developed to provide a ready-made modelling capability. These tools are available commercially and are constantly being developed to improve their performance. There are advantages to industry in adopting off-the-shelf tools:

- costly tool development is avoided
- tool development costs are borne by the tool industry
- the experience of many users is built into tool developments
- application licences can be bought to suit the number of users and can be renewed as required
- tool providers offer a consultancy service to assist with application problems
- user communities form to pool experience for the common good.

7.3.4.1.5 3D Modelling A number of CAD graphics packages are available which provide a 3D representation of a structural design. The data can be manipulated to provide images which can be rotated. CAD tools allow the entire product to be designed and stored as a database of models that can be used by many users simultaneously in 2D or 3D format. The forms of image presentation allow human–machine interfaces and installation interfaces to be viewed and tested without the need to build a physical representation. CAD tools can be combined with other modelling and analysis tools to allow the system to be analysed in many different ways. Such techniques enable engineers to look closely at installation clearances and confirm that there are no fouls between moving mechanism and structure. Figure 7.12 shows an active CAD model of a fuel system. The model is active in the sense that it can be animated to show fuel transfer from tank to tank, and tank to engine feed. The CAD tool can be combined with a computational fluid dynamic model to examine the effects of manoeuvres on fuel 'slosh' in tanks (Tookey et al. 2002).

Also shown is a simple CAD image of an undercarriage leg that can be animated to verify the simple sketch model shown in Figure 7.10. An example is also given of a CAD model transformed into a solid model by 3D printing. Two alternative constructions of a metal camera bracket are shown.

7.3.4.1.6 FloMaster FloMaster is a 1D network flow solver that allows engineers to analyse a comprehensive range of problems associated with fluid flow. It allows piping networks of virtually any size and complexity to be analysed rapidly and accurately in order to establish design integrity.

The flow distribution and pressure losses in complex, multi-branched, and looped fluid systems (such as fuel and environmental control systems) can be assessed in steady-state

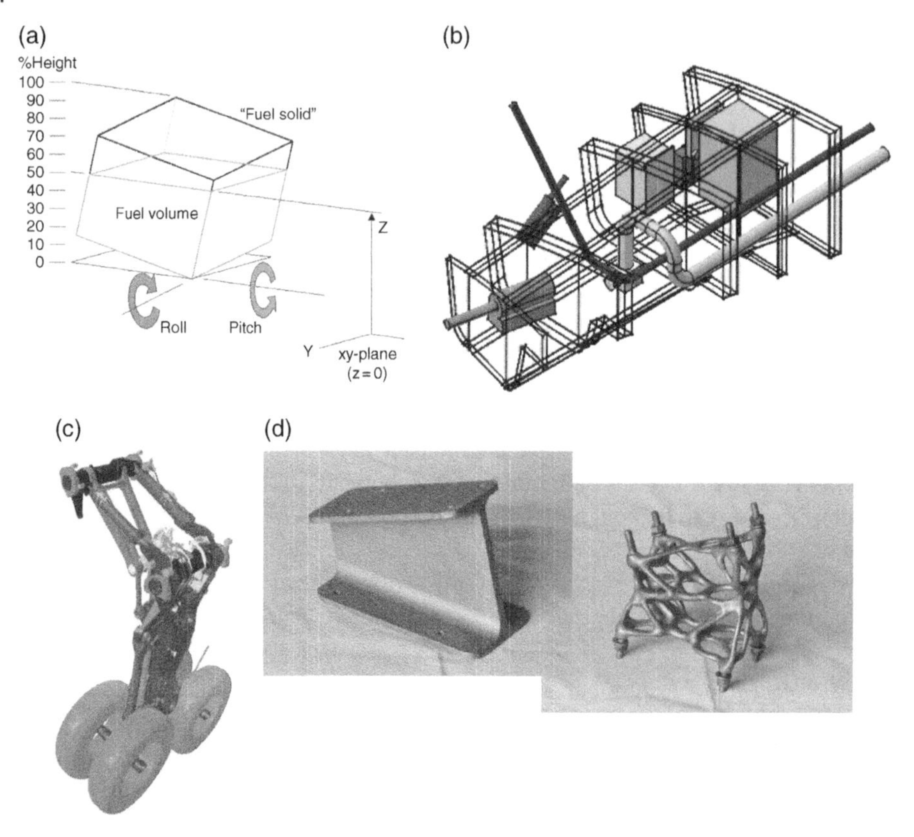

Figure 7.12 Computer-aided design (CAD) models: (a) and (b) computer aided, animated images of part of a fuel system, and (c) an undercarriage retraction sequence animated model, and (d) 3D printer models of a camera bracket (photo: BAE Systems).

conditions. With transient simulations fuel tank levels can be predicted and the sequencing of valve and pump operations to transfer fuel between tanks on the aircraft can be defined for different flight profiles. Additionally, refuelling scenarios whilst on the ground or flying can be analysed to predict pressure transients as a result of rapid operation of level control valves.

The modelling of hydraulic systems such as landing gear, nose wheel steering, and flight controls can be simulated with the FloMaster fluid power package. The transient effects of the variable loads applied to the cylinders (mechanically linked or independent), and the operation of the directional control valves allows for assessment of the systems' operational behaviour and interactions as a whole. This can reduce the need for significant 'iron-bird' testing and ensuring better 'right first time' prototype designs.

7.3.4.1.7 Tool Capability The FloMaster tool set contains a number of components that enable an engineer to build a range of models applicable to aerospace domains, including hydraulics, thermal management, environmental controls systems, and de-icing systems. Specific components that are applicable to the environmental control system include:

- heat exchangers
- ducts
- compressors
- orifice plates
- valves
- honeycomb flow straighteners.

These components can be combined to represent a particular aircraft cabin conditioning system which can then be modelled to examine aspects of performance under various operating conditions such as ground operations, cruise, climb, descent etc. System performance can then be accurately predicted in steady-state or transient conditions in terms of:

- air-flow rates
- air-flow velocities
- air-flow distribution
- air pressure
- air temperature
- humidity
- air mixing strategies, e.g. re-circulated or mixed with bleed or fresh air.

7.3.4.1.8 Human–Machine Interface Prototyping Using VAPS VAPS is a commercially available tool for developing real-time interactive human-machine interfaces (www.presagis.com). It is a product of virtual prototypes is a tool for building data driven, interactive, visual human–machine interfaces. These interfaces display application data as graphics, which are re-drawn to reflect changes in the data. In real-time applications, a rapid refresh rate results in the perception of smooth animation. At the same time user interaction can be provided by directly manipulating the graphics with a mouse or touch screen (VAPS).

VAPS allows the systems engineer to simulate human–machine interfaces such as cockpit displays and their control mechanisms on a desktop PC. This enables graphic dynamics, fonts, symbols, and colours to be tested and modified until an acceptable solution is achieved. The developed layout can be transported easily to a test facility to allow tests to be repeated under differing ambient lighting conditions and with a variety of users to achieve an optimum solution.

VAPS is widely used in the defence, aerospace, medical, and automotive industries to achieve common agreed human–machine interfaces.

Figure 7.13 shows a primary flight display which has been constructed on a desktop workstation or laptop using VAPS. The model can be animated to simulate real-time dynamic display, and all fonts and colours can be changed.

7.3.4.1.9 Bond Graphs Bond graphs are a combination of a notation and a method that form a good way of modelling systems. Bond graphs are essentially networks of physical objects bound together by energy. The method represents power flow around a network and the philosophy is object-oriented. The mapping between system and model is one-to-one and the method is graphical with an underlying equation-based model and

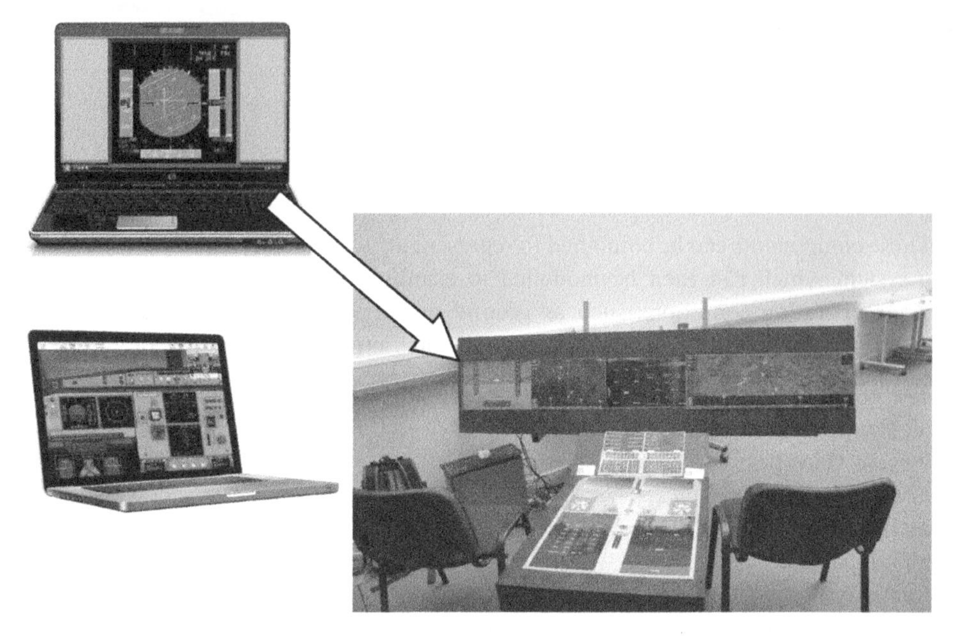

Figure 7.13 A VAPS model of cockpit displays and controls. VAPS was used to develop the displays for this model built by the Air Vehicle Design group project at Cranfield University (photo: Allan Seabridge).

explicit relationships. The method is particularly suited to the modelling of mechanical, hydro-mechanical, and electro-mechanical systems.

7.3.4.1.10 A Fuel System Model Using Simulink The Global Express™ is a long-range business jet developed by Bombardier Aerospace of Canada in the mid-1990s. Parker Aerospace is a US company that was contracted to supply the fuel system and as part of that effort developed a complete model of the fuel system to support the fuel system design and verification aspects of the programme.

The fuel control panel on the flight deck indicates the state of key pumps and control valves to the crew, and provides the ability to control the system manually during fault conditions. The engine indication and crew alerting system (EICAS) multi-function display contains a 'fuel page' that includes a system synoptic display showing fuel quantities in each tank and total fuel on board. The EICAS display also indicates system status, advisories, and warnings in the event of equipment failure.

The heart of the system is the fuel management and quantity gauging computer (FMQGC), which controls the refuel (and defuel) process, and measures fuel quantity and fuel temperature in each tank. The FMQCG also controls the transfer of fuel between tanks to co-ordinate the fuel burn sequencing and to maintain the lateral balance of the aircraft.

The pumps and valves are the effectors that result in correct fuel movement and ensure that the engines and auxiliary power unit have a sufficiently high source of fuel pressure. This is particularly important at high altitudes, where vapour can come out of solution and excessive vapour/liquid ratios can result in an engine shutdown.

In order to have a complete understanding of the system performance throughout all combinations of normal operation and in the presence of failures, a model was established using SIMULINK, which is a subset of the MATLAB product. This general-purpose simulation tool provides a quick method for assembling dynamic models of subsystems into a fully integrated model to allow the systems engineer to visualise how the system behaves. In this case the building blocks included:

- a model of the atmosphere (to take into account the operating conditions that affect engine fuel consumption, for example)
- an engine model to determine the fuel consumption as a function of flight condition and throttle setting, and, in turn, to determine engine low-pressure shaft speed and hence the engine electrical generator frequency
- a fuel network model comprising piping, pumps, and control valves
- a fuel tank model to determine the fuel quantity in each tank
- a computer model comprising the control algorithms associated with the fuel management task, including the generation and transmission of system status messages to the flight crew via the EICAS display.

Whilst a SIMULINK model can be developed very quickly, it was determined that a special purpose graphical user interface (GUI) would provide the users (the systems engineering team) with a more in-depth insight into the functional behaviour of the aircraft from a fuel system perspective. This GUI comprises:

- A panel drop-down menu for mission, engine, fuel (cockpit panel), pump, and valve faults. The selected panel fills the lower left half of the screen.
- A display drop-down menu, the main option being a system schematic diagram. This fills the right-hand side of the screen.
- A simulation menu for starting and stopping the simulation.
- Number crunching and plotting tools provided under an ANALYSIS menu.

The model allows the user to load system design data, which for the most part does not change. The user can then decide to use the model in a fully interactive mode using the mouse to change throttle settings, altitude, and Mach number, and to introduce faults and observe the resulting system behaviour. It is also possible to choose to simulate a predetermined mission profile by loading a mission file with events versus time.

7.3.4.1.11 Model Considerations Level of fidelity is the first consideration. If a model is very complex it will be expensive to develop and maybe unavailable early in the programme, when the most benefit can be had (remember Figure 3.3 showing the cost of correcting errors versus the phase of the programme). In the words of Einstein, *One should simplify the problem as much as possible – but not too much.* This is where engineering judgement comes in.

The speed of execution of the model may also become excessively long. For example, in order to evaluate a large number of cases, it may be desirable to have a model capable of running several times faster than real time.

In this case it was decided to use a very simple engine model since the steady-state behaviour was all that the fuel system sees. The fuel handling network was simplified by eliminating the refuel distribution system. Thus the model was used to study operational mission scenarios. A separate model was developed to evaluate the performance of the refuel/distribution (pre-flight) system.

Also, whilst the effect of pitch angle is important, because in a rear-engined aircraft the head pressure on the boost pumps can vary substantially, the effect of roll angle is negligible and was assumed to stay nominally zero.

When operated initially this model was capable of running about three times faster than real time and as such provided valuable insight to the design team. On today's PCs this model will run much faster.

7.3.5 Test Rigs

A test rig is used to build a full-scale representation of a system which requires a degree of high fidelity testing, that is to say that the behaviour of the system to be tested must be as close to the behaviour of the real system as possible. Although test rigs are costly to design, build, and maintain they do have the advantage that conditions are controllable and test rig operating time is less expensive that product operating costs, especially for ships and aircraft. Typical factors that determine the need for a test rig include the following:

- *Safety*: where there is a need to explore the behaviour of a system in an environment where failure will not lead to a hazardous situation. An example of this is ground testing of a complete aircraft flight control system, including full pressure hydraulics, on an 'iron bird' test rig.
- *Endurance*: where a system needs to be tested under controlled conditions for a time equivalent to its life in service, and sometimes to destruction. Examples of this are a fatigue test specimen undercarriage operation to simulate a representative number of raise and lower cycles or flap actuation.
- *Human factors*: where tests need to be conducted to demonstrate ergonomic design or where difficult external conditions must be simulated, such as high-altitude sunlight or extreme night conditions to test the performance of cockpit display legibility.
- *Integration*: where complete systems need to be progressively assembled and tested to explore all functional and physical interfaces.

Some examples of test rigs used to develop systems are as follows:

- *Fuel systems*: a full-scale partial or fully populated representation of the aircraft fuel tanks and interconnections on a moving platform allows transfer of fuel under realistic pitch and roll attitudes, but with limited rates of change. This will test transfer sequences, control system logic or software and quantity measurement system accuracy.
- *Cooling systems*: a rig to allow the air/vapour cycle cooling system to be operated under a range of conditions and cooling loads.
- *Electrical power generation*: to operate the generators at the full range of engine speed conditions using load banks to simulate a range of resistive and reactive loads.
- *Hydraulics power generation*: operation of hydraulic pumps with a range of representative loads and differing demand rates.

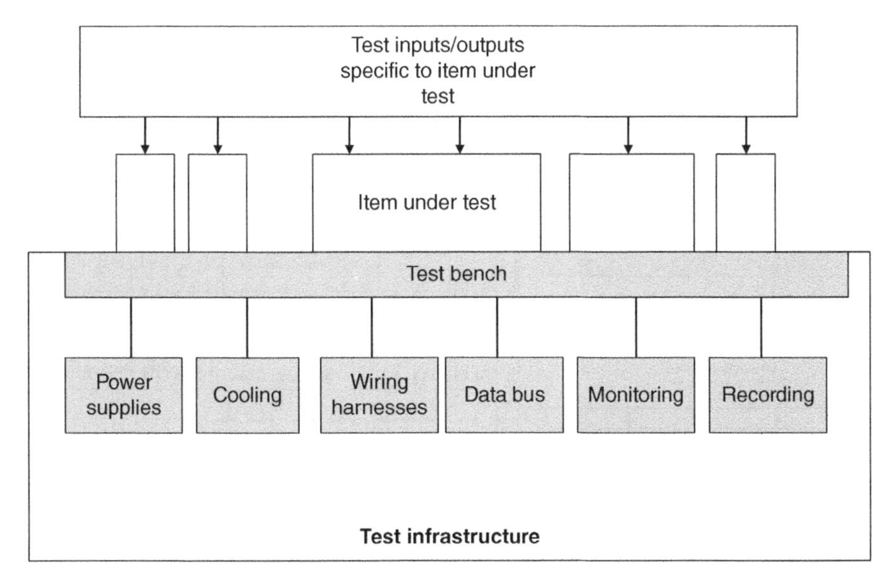

Figure 7.14 Example of a single system or component test rig.

Figure 7.14 shows a block diagram of a rig designed to test a single component or a single sub-system. In this case the test bench is designed to provide a housing or mounting tray for the equipment to be tested and is equipped with wiring and data bus types applicable to the project. The test bench also provides lab instrumentation to allow the system to be monitored during testing and the data results to be recorded. Standard electrical power supplies and cooling are provided.

A carefully designed test bench can be used to test many different systems with minimal changes to the test facility and goes a long way to providing standard conditions for testing that can be replicated for re-test. Signals for demand and for monitoring specific to the system under test can be connected to the components as required.

7.3.6 Environmental Testing

Environmental test rigs are used by suppliers to demonstrate that equipment meets the specification requirements for environmental conditions such as temperature, humidity, vibration, and fungus resistance, as defined in Chapter 4.

7.3.7 Integration Test Rigs

When testing has been completed on individual system test rigs there is often a need for some systems to be combined to understand the behaviour of an integrated system. This can range from purpose-built rigs to bring together a functional system, rigs to examine integration effects, and even ground testing the whole aircraft to lead to flight clearance. An example of an integrated test facility is shown in Figure 7.15, where vehicle systems,

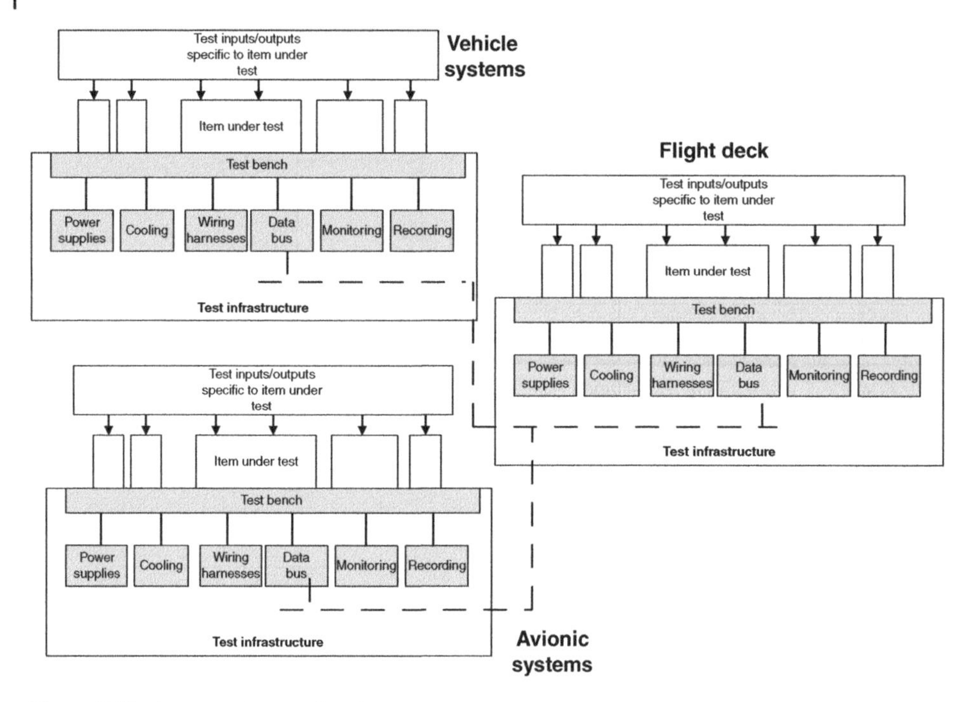

Figure 7.15 A system integration rig.

avionics, and flight deck rigs are integrated to form a whole-aircraft system test rig. Some examples of integration rigs are described below:

- *Displays and controls*: to simulate the cockpit, flight deck or mission crew working environment to test displays and controls acceptability, to confirm ergonomic and workload analysis, and to allow the crew to gain early familiarity with the aircraft.
- *Avionics integration*: to allow the complete avionic system to be progressively built up and tested.
- *Ambient lighting testing facility* (Vassey 1998): to expose the cockpit displays and the pilot to simulated conditions of sunlight from all directions and a range of day, night, and altitude conditions, or to degrees of night-time ambient light.
- *Lightning strike test*: to subject individual items of equipment or the whole aircraft to high field strengths to simulate lightning or electro-magnetic pulse effects.
- *Altitude test facility*: used by engine manufacturers to operate the engine under realistic air density and temperature conditions to simulate world-wide operation.
- *Electronic warfare*: to subject individual items of equipment or whole aircraft to a variety of radio frequency transmissions to examine their susceptibility, or to measure the radio frequency emissions of the aircraft.
- *Evacuation and escape test*: to demonstrate that facilities for safe evacuation of occupants are available. For passenger-carrying commercial aircraft a demonstration rig representing a full-scale fully furnished passenger cabin is used in a realistic test in which passengers can release seat belts, find their way to the exits, and use the escape slides. This must be timed and supported by cabin crew (the A380 evacuation process can be viewed

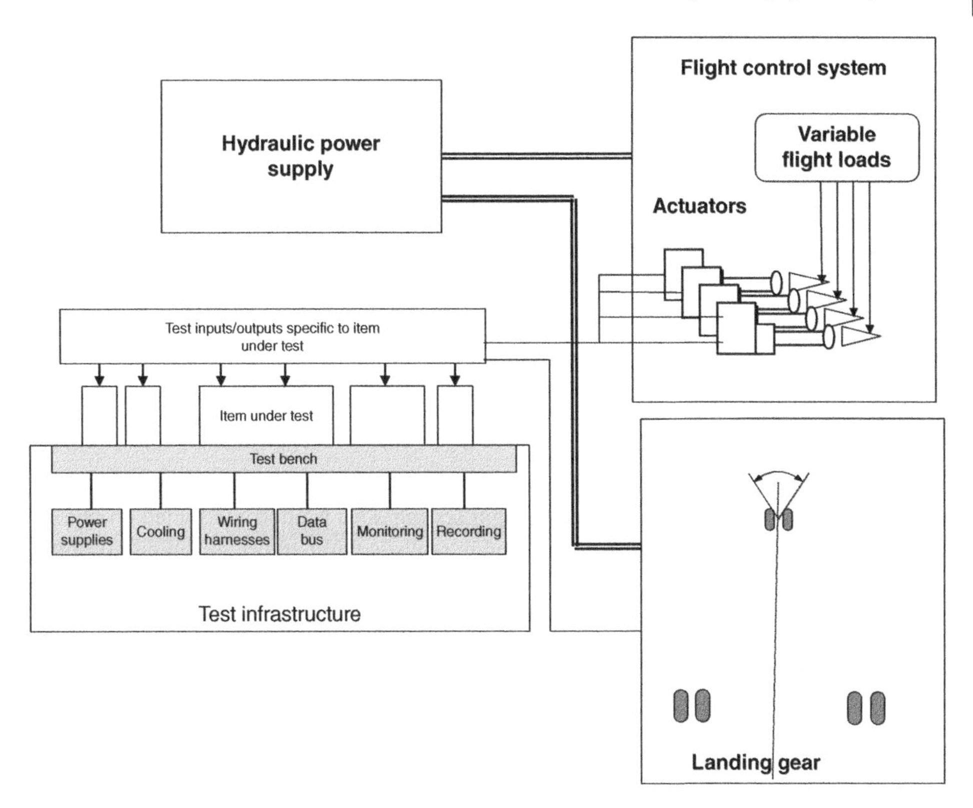

Figure 7.16 An 'iron bird' rig.

on YouTube). For fast jet military aircraft a ground rig will use a rocket-powered slide to simulate high-speed conditions and an ejection seat equipped with a realistic size and weight dummy will be used with high-speed video to demonstrate safe ejection.

- *Iron bird*: a combination of hydraulics power, flying controls, and landing gear that allows full movement of all surfaces and components to check range, rate, and freedom of movement. All the equipment required to provide control and demand signals, e.g. flight control computers and landing gear computers, will be mounted on the test bench and wired up to sub-systems such as flight control surfaces with actuators providing realistic flight loads. A source of hydraulic power can be provided from a ground truck or from a test rig operating an aircraft standard hydraulic system. A block diagram of an iron bird rig is shown in Figure 7.16.

7.3.8 Aircraft Ground Testing

There will come a time when all the individual systems have been tested and the aircraft prototype will have been assembled and all wiring installed. The wiring is tested for end-to-end conductivity and resistance, and an insulation test is conducted (Google 'DITMCO for products'). It is usual to install a ground electrical system and power this up to provide suitable aircraft power supplies before installing and connecting other equipment.

This enables the aircraft connections to be tested to ensure the correct electrical power and ground connections have been wired to the equipment connectors. Voltage, polarity, insulation resistance, and earth conductance will be measured. When this is complete it is safe to install the aircraft equipment and start to test the systems individually to agreed test procedures. Any defects or failures to meet test procedures will be signalled to design for appropriate rectification action.

This continues until all tests are complete and the aircraft is cleared to move onto ground runs with engines providing power for the electrical and hydraulic generation systems. Once these tests are complete the aircraft is cleared for engine runs, taxi trials, and prototype flying.

7.3.9 Flight Test

Following satisfactory completion of ground testing all the evidence from testing is reviewed to obtain clearance to proceed to flight testing. This is often conducted on one or more prototypes to enable tests to be conducted in realistic flight and environmental conditions. For more information read the series of books by Longworth (2012, 2013, 2014).

Some examples of test flying of prototypes and production aircraft are described below.

- *Prototype flight testing*: In order to eliminate risk from the series production programme many product manufacturers produce one or more prototypes which are subjected to a rigorous test programme to demonstrate correct performance. Full-scale prototypes are an extremely costly form of model, but the penalty of proceeding to series production without fully understanding the performance issues may be even more costly. The aerospace industry has long made use of prototypes to explore the full flight envelope and to demonstrate the performance of avionic systems. The results gained during such testing are fed back into the early models to improve their fidelity and to validate the model results. A model validated by flight test results establishes confidence in the model so that it can be used to support future changes and developments. The number of prototypes in any one programme is determined by the amount of testing to be done; the load of general and specialist systems tests is usually spread across the available aircraft to enable concurrent testing to take place.

Figure 7.17 shows four different prototypes from a period of 50 years of history of the European aerospace industry. The P1A prototype was the first of many that were used to develop the handling characteristics and speed of the UK's first supersonic production interceptor, the English Electric Lightning. The Lightning prototypes also demonstrated the capability of the aircraft as an integrated weapon system integrating the capabilities of the pilot, the AI 23 radar, a missile, and the aircraft avionics.

Tornado was designed as a multi-role capability aircraft (MRCA) by a grouping of three European aircraft companies, known as Panavia. Nine prototypes developed this swing-wing supersonic aircraft that led a successful service life. Tornado P01 is seen on its first flight from Manching in Germany on 14 August 1974. The type retired from RAF service in 2019 in its GR4 configuration but still remains in service with other users.

The experimental aircraft programme (EAP) was a single demonstrator aircraft to explore the handling of an aircraft designed to be aerodynamically unstable with a quadruple

Figure 7.17 Examples of prototypes: (a) English Electric P1B, (b) the Experimental Aircraft Programme (EAP), (c) Panavia Tornado prototype P01, and (d) Typhoon UK prototypes (photos: BAE Systems).

fly-by-wire system. Other technologies demonstrated included composite materials, a full-colour 'glass' cockpit, a computer-controlled set of vehicle systems, and a real-time processing system using a high-order language and the first flying MIL-STD-1553B data bus. After 259 flights of testing and demonstration flying this aircraft was retired as a teaching aid at Loughborough University. It was donated to the RAF Museum at Cosford in the UK in 2014.

Typhoon used much of the technology that was demonstrated on EAP and developed it further. Flight testing demonstrated the aircraft's handling and its comprehensive systems and armaments capability, performed by pilots of all the participating nations. The single-seat and trainer versions are shown in Figure 7.17. Some examples are described below.

- *Production acceptance flight testing*: A short production acceptance test is performed on each aircraft leaving the production line before it is handed over to the customer.

7.3.10 Trials

Trials are conducted by the customer's specialist agencies to evaluate specific requirements and the ability of the product to meet those requirements. This may mean that the aircraft is tested on specific range assets such as radar range, weapon delivery range or in theatres of operation such as the tropics, the Arctic, and deserts. For commercial aircraft it may be necessary to evaluate the aircraft at new air terminals to demonstrate compatibility and the ability to board and disembark passengers,

7.3.11 Operational Test

Operational testing will be conducted when the aircraft is in service to meet new situations as and when they arise. This will allow the product to be progressively adopted in new scenarios.

7.3.12 Demonstrations

There are occasions when a demonstration is needed before final acceptance of some requirements is completed. Often this is for requirements which have statistical outcomes. Examples are reliability or availability during which operational information is collected over a period of time. Another example which requires the involvement of operational personnel is the demonstration of supportability.

7.4 An Example Using a Radar System

This section summarises the examples described above for a simple radar system to illustrate what the resulting documentation will look like. Figure 7.18 shows the block diagram of a radar system.

There are a number of requirements that flow down from the top-level aircraft requirement specification that determine the primary performance characteristics, for example

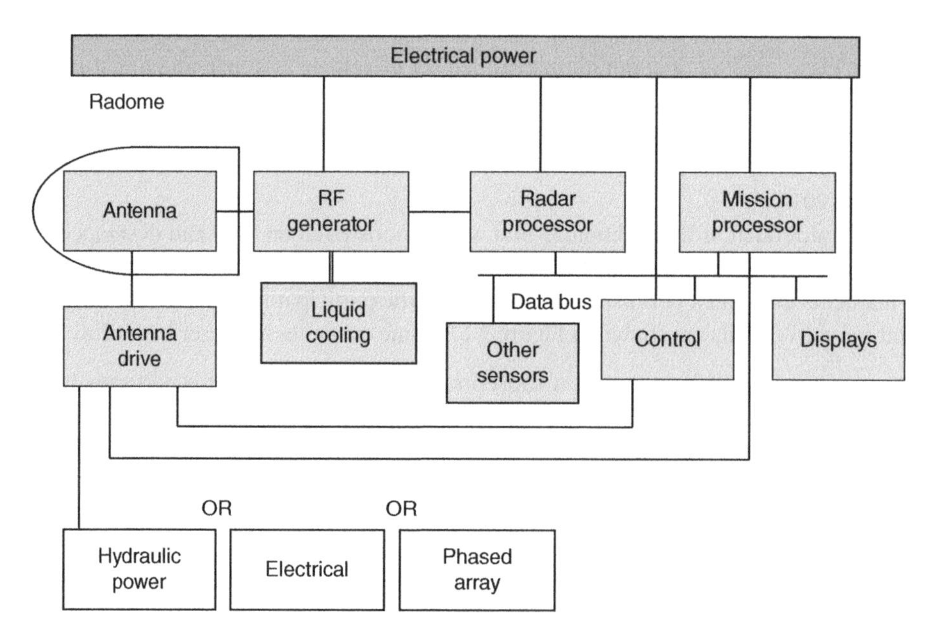

Figure 7.18 Simplified block diagram of a radar system.

detection range, target size capability, clutter discrimination, classification and identification of targets, and ability to track and lock on. There also a number of derived requirements that arise from incorporating the radar into an aircraft. These are illustrated in Figure 7.19.

These derived requirements arise from a need to design a radome that will include the swept volume of the dish, that is aerodynamically sound as well as strong (to withstand bird strike and driving rain), has low weight, can be securely attached yet easy to open for maintenance, and has the appropriate transmission characteristics so as not to attenuate the radar pulse and its return signal. There is a health concern to reduce the occurrence of any non-ionising radiation being transmitted into the cockpit from any side lobes. A plan for obtaining qualification information can be summarised in a verification cross-reference matrix, as shown in Figure 7.20. The matrix identifies the requirements that need to be tested and by what test mechanism, shown as shaded blocks. In a firm plan these blocks

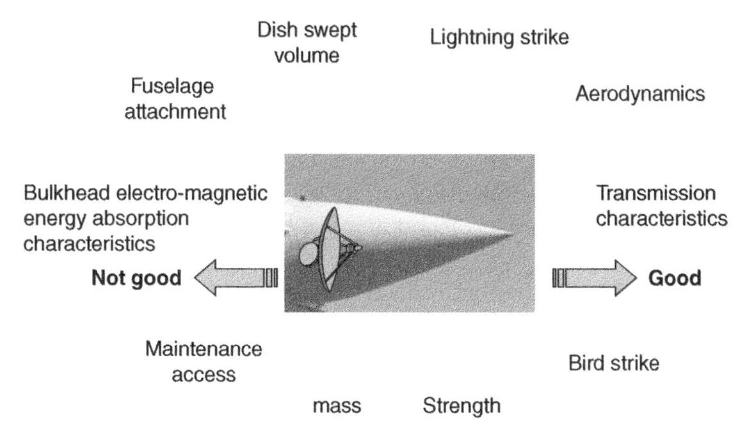

Figure 7.19 Derived requirements for a radar system.

	Design	Calc	Analogy	Model	Test rig	Env test	Int test	Trials	Operational	Demo
Detection range		■		■				■	■	
Target CSA				■				■	■	
Clutter discrimination		■		■				■	■	
Classification				■				■	■	
Identification				■				■	■	
Tracking				■				■	■	
Lock-on				■				■	■	
Mass		■								
Power		■								
Volume	■	■								
Radome swept area	■									
Transmissivity		■	■							
Aerodynamics	■	■		■						
Bird strike					■					
Driving rain			■			■				
Attachment strength	■	■			■					
lightning strike		■	■			■				
RF energy SHE	■	■								
Maintenance access	■			■						■
Integration							■	■	■	■

Figure 7.20 Verification cross-reference matrix for a radar system.

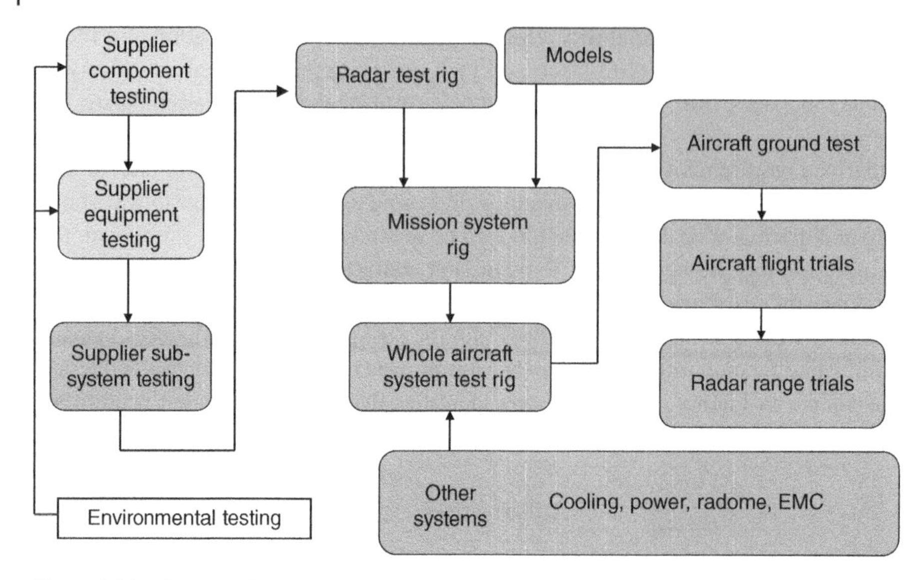

Figure 7.21 Sources of test information for a radar system. EMC, electro-magnetic compatibility.

will contain dates to allow suitable project management to be conducted. As the testing progresses the spreadsheet is used to record test report numbers and file locations leading to a complete record of testing.

A brief indication of the testing required to evaluate these derived requirements is illustrated in Figure 7.21.

7.5 Summary

Despite best intentions it is a fact of life that some systems fail when they are in service. Sometimes the customer discovers the fault, sometimes a suspicious trend of incidents highlights the fault, sometimes the customer agrees to carry out trials. The worst-case scenario is that items must be recalled and the fault rectified. This is not only costly but it affects the reputation of the supplier. This sometimes leads to controversy, especially if led by the media.

Within the contractor organisation there will be a question: Who is at fault? The designer for not checking the validation results? Quality assurance for not checking the results and the product quality? A lack of robustness in the testing regime? The standards authorities for not auditing the supplier process? Or a failure to apply a rigorous review process? These are not questions that a contractor needs, especially if they throw the quality of the product into question and affect sales.

Unfortunately, there have been numerous examples of failures occurring in service in the consumer market which have led to recalls in the motor vehicle and white goods industries, leading to significant costs and a lowering of customer perception of the company reputation. Failures in the aircraft industry may be manifested in reduced performance and in the worst case by serious accidents.

A sound design process, which includes a robust testing and qualification process, is essential. It must be understood that:

- qualification is a logical completion of the design process
- attention to design in the earliest stages is essential to reduce errors and poor design
- testing is to verify correct operation *not* to find faults
- acceptance by the customer depends on sound design and sound evidence of excellence of design, which is achieved using people, process, and tools in the most appropriate manner to delight the customer.

Exercises

1. Describe a physical model for a system of your choice. Estimate how true to life it is – what is its fidelity (how true to reality is it)? Produce a rough estimate of cost to produce the model. Then propose a simulation or mathematical model of the same system and estimate its cost. Compare and contrast the two models and decide which will give the most cost-effective evidence of correct operation.

2. Look for examples of product recalls in the consumer industries. Try to determine if they are the result of poor design or poor qualification. What was the impact in terms of cost and customer perception?

3. Can you find any instances of product recall in the aerospace industry and what were the implications?

References

Law, A.M. and Kelton, W.D. (1991). *Simulation Modelling and Analysis*. McGraw-Hill.

Longworth, J.H. (2012). *Test Flying in Lancashire, Vol 1. WW2 to the 1960s*. BAE Systems Heritage Department.

Longworth, J.H. (2013). *Test Flying in Lancashire, Vol 2. From the 1960s to the 1980s*. BAE Systems Heritage Department.

Longworth, J.H. (2014). *Test Flying in Lancashire, Vol 3. From the 1980s into the new millennium*. BAE Systems Heritage Department.

Stewart, I. (2012). *Mathematics of Life: Unlocking the Secrets of Existence*, 317–318. Profile Books.

Tookey, Roger; Spicer, Marcus and Diston, Dominic. (2002). Integrated Design and Analysis of an Aircraft Fuel System. NATO AVT Symposium on the Reduction of Time and Cost through Advanced Modelling and Virtual Simulation.

Vassey, K. (1998). *Specification and assessment of the visual aspects of cockpit displays*. Society for Information Displays International Symposium.

Further Reading

AGARD (1996). *Flight Vehicle Integration Panel Working Group 21 on Glass Cockpit Operational Effectiveness*. Advisory Group Report 349.

Belobaba, P., Odoni, A., and Barnhart, C. (2016). *The Global Airline Industry*. Wiley.

De Neufville, R. and Stafford, J.H. (1971). *Systems Analysis for Engineers and Managers*. McGraw-Hill.

Diston, D. (2009). *Computational Modelling and Simulation of Aircraft and the Environment*. Vol. 1 Platform Kinematics and Synthetic Environment. Wiley.

Garrett, D.G., Wolff, J and Johnson, T.F. (2000). *System Design and Validation through Modelling and Simulation*. INCOSE 10th International Symposium.

Gawthrop, P.J. and Smith, L. (1996). *Metamodelling: Bond Graphs and Dynamic Systems*. Prentice Hall.

Karnopp, D.C., Margolis, D.L., and Rosenberg, R.C. (1990). *System Dynamics: A Unified Approach*, 2e. Wiley.

Middleton, D.H. (1985). *Test Pilots*. Collins Willow.

Thoma, J. (1975). *Introduction to Bond Graphs and their Applications*. Pergammon Press.

8

Practical Considerations

8.1 Introduction

The design and development of aircraft systems is something that takes place in collections of organisations including customers, prime contractors, and suppliers. In order for the process to work correctly there must be some disciplines imposed in the organisations. This chapter examines some processes which, whilst not technical, are absolutely required to result in the right technical product. First, the chapter looks at the business process and in the later sections some more technical aspects are examined.

The good systems engineer is always prepared to learn from other people's experience, and this chapter is intended to provide an insight into the practical world of systems engineering. Learning from experience, from one's peers, from one's competitors, and from the good and bad experiences of others is the pragmatic approach to learning, and some research has shown that formal methods of introducing learning into teams can be applied to good effect (Meakin and Wilkinson 2002). Figure 8.1 illustrates the learning from experience model. This model shows that learning from experience is different to knowledge management. Whilst explicit knowledge can be collected, stored, and manipulated, wisdom is the result of experience and other factors such as luck, insight, judgement etc., and creation of new wisdom is the result of shared experience rather than extracting information from a database. This seemingly trivial statement is important because the widespread use of databases and the publication of weak statistical 'evidence' in the media is leading people to put too much faith in computer-generated data. All data, information, and 'evidence' must be validated before it is used.

Learning from experience is a powerful tool that enables knowledge and experiences to be shared, often to the mutual benefit of the parties involved. With that in mind, this chapter can only be an appetiser, the reader must continue to learn and improve, to seek out others with wisdom and, if possible, create learning from experience communities along the way. The phrase 'lessons learned' has become popular but it remains to be seen how much has been really learned, or whether much is discarded soon after a project ends.

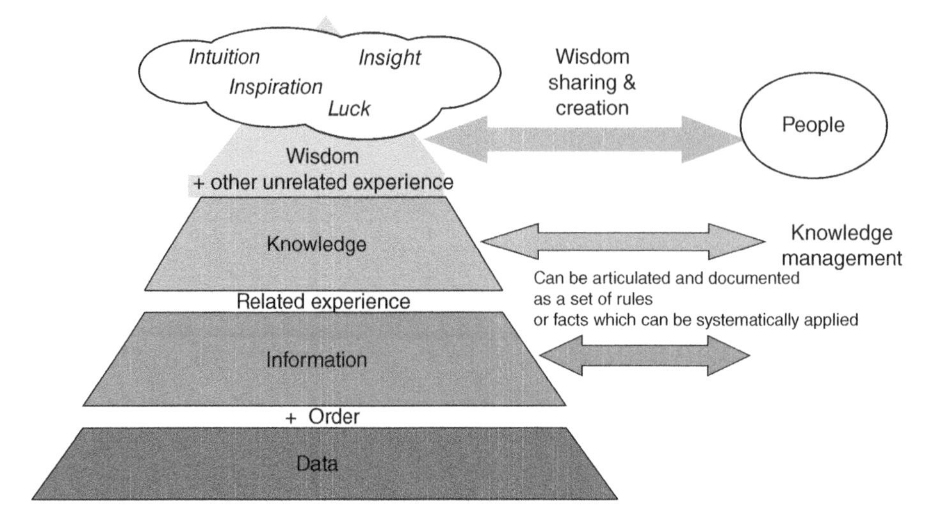

Figure 8.1 The learning from experience model.

8.2 Stakeholders

8.2.1 Identification of Stakeholders

The identification of all interested parties or stakeholders in a system is of vital importance in ensuring that all parties involved can be informed about the progress of system development. A stakeholder is a party or organisation that can affect a project or that can be affected by being part of that project. Figure 8.2 shows an example of internal and external stakeholders in a project with typical information paths. A stakeholder should be informed that they have been identified; this ensures that their role and their need to communicate is acknowledged. It also allows the stakeholder to understand their own responsibility in ensuring that the project is successful. Correct management of the stakeholder community builds up an atmosphere of trust that is beneficial to the project and all parties involved. The stakeholder community should not be regarded as a fixed entity: new stakeholders can join the community and stakeholders are allowed to leave depending on the product lifecycle phase.

As well as identifying the stakeholders, it is important to understand the nature of communication – in most cases it will be two-way and direct. However, there are cases where it will be appropriate that communication is not direct. In Figure 8.2 communication between a project's prime supplier and his own suppliers should always be conducted by the prime supplier. In circumstances where tension is observed between the prime supplier and his supplier the temptation to intervene directly must be resisted to avoid contractual issues arising.

Stakeholders can be managed by regular meetings or by direct communication to ensure that they all feel involved and consulted about the progress of the project and about key decisions that affect them all. Good stakeholder management goes a long way to ensuring a smooth programme.

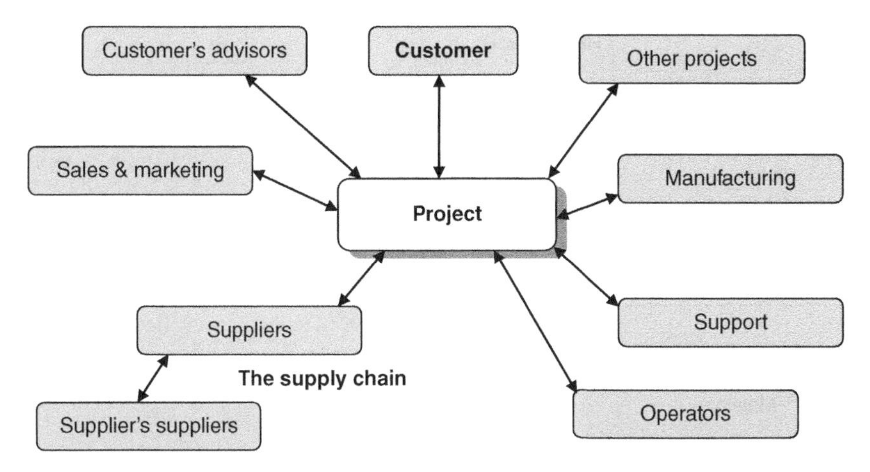

Figure 8.2 Examples of stakeholders.

There are generally available business textbooks on the subject of stakeholder management. The Association of Project Management is also a good source of information (apm.org.uk).

It is almost inevitable that some future projects are going to involve international partnerships. As well as imposing some difficulties with language there may also be cultural differences that need consideration. This is particularly true with showing respect, consideration of diet, hand and body gestures, and tolerance of bad language. Disregarding these issues can lead to poor relationships and intolerance in the project teams. This is an issue that sales and marketing teams adopt with enthusiasm but it must be passed on through all project teams. It is wise to examine the culture of potential partners and if necessary to issue a project guidebook. There are generally available business books that offer guidance on this issue.

8.2.2 Classification of Stakeholders

It is often useful to classify stakeholders in a way that ensures that they are correctly managed. Figure 8.3 illustrates how stakeholders can be classified in a simple four-box model. In this model there are four boxes with axes that indicate the relative importance of the stakeholders in each box. The ovals here represent individual stakeholders.

In Box A are important stakeholders who have real power to make decisions and act upon them. These stakeholders include the customer and their advisors, the certification authorities, airworthiness departments, and suppliers, who all need to be managed closely to ensure that they are given the most appropriate view of the project and its progress. Box B contains those stakeholders who want to keep a close eye on the project but have less authority over technical and project decisions. Examples are marketing, operators, maintainers etc.

Box C contains those with low interest and low power, including suppliers of generic materials, stationery, IT providers etc. These include those stakeholders who have power to influence the project but little interest as the project may be one of many with which they

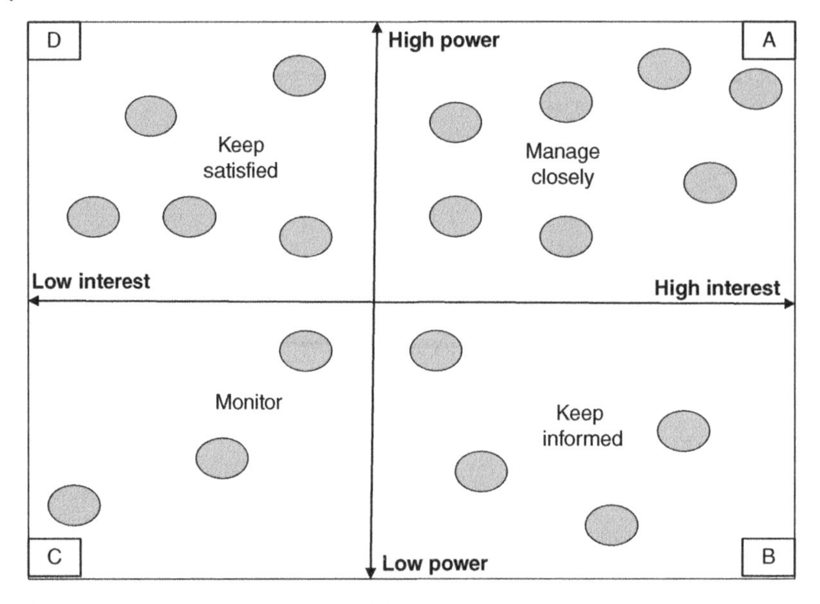

Figure 8.3 Classification of stakeholders.

are involved. This includes public relations, press, the general public, local politicians etc. Box D contains stakeholders who are peripheral to the project but may be relied on to provide support when required.

This model is a good starting point for managing stakeholders but it should not be considered as an invariant model. At different phases of a project lifecycle the stakeholder roles may well change and the model should be re-evaluated at each phase. For discussion, an example of the aviation system, based on Figure 1.1 of this book, is shown in Figure 8.4.

8.3 Communications

In order to operate smoothly any organisation needs to communicate, if it does not then it may as well not exist. Organisations are like communities – they need to maintain effective communication to establish needs, to define boundaries, to establish a basic understanding, to make contacts with other organisations, and to do business. Such communication takes place inside the organisation at and between all levels, and there must be two-way communication between the organisation and the outside world. Ideally, in all cases the communication should be:

- clear
- unambiguous
- concise
- accurate
- authorised
- traceable.

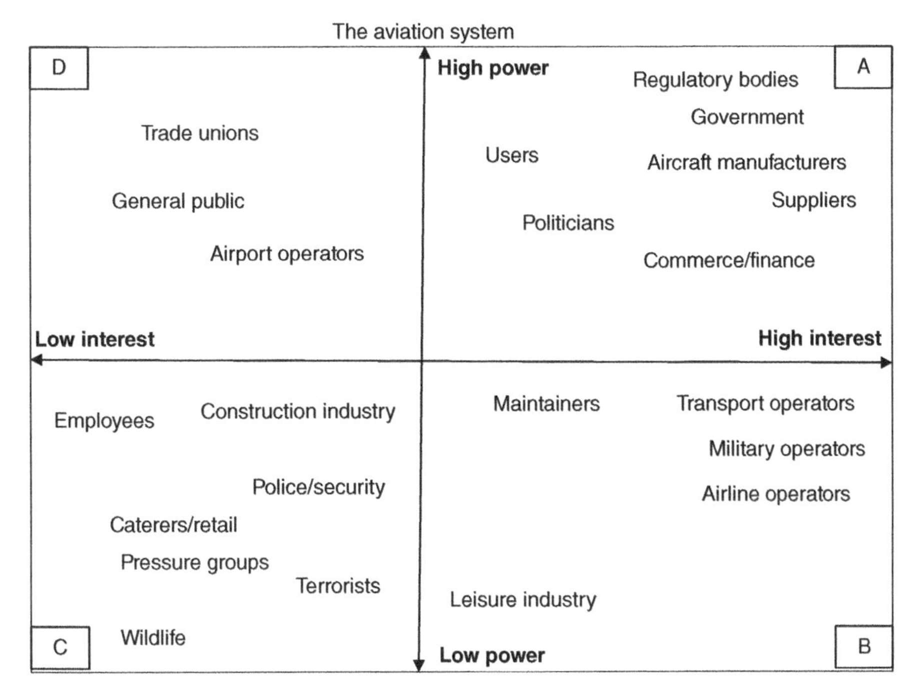

Figure 8.4 Classification of stakeholders for the aviation system in Chapter 1.

Organisations are made up of people, and people communicate using their natural language skills. The way that people use language is important: *Getting the language right is a major issue in almost every corner of society. No one wants to be accused of ambiguity and obscurity, or to find themselves talking or writing at cross-purposes. The more we know about the language the more chance we shall have of success, whether we are advertisers, politicians, priests, journalists, doctors, lawyers – or just ordinary people at home, trying to understand and be understood* (Crystal 1995).

There is a need for all people to understand their language. The point made about ambiguity and obscurity is well worth noting, industry cannot afford to tolerate misunderstandings that may have an impact on cost or poor customer perception. Paradoxically, the use of jargon often provides clear, concise, and accurate communication for the 'insiders', if not for 'outsiders'. Modern organisations tend to speak in jargon; they often do this to be clear, concise and accurate within the organisation, although this may not be immediately clear to external parties. An increasing use of acronyms adds to the obscurity of language.

All departments that make up an organisation communicate with each other. Communication is two-way, conducted by a variety of means, and essential to ensure mutual understanding. Communication is complemented by listening, that is by hearing, assimilating, and understanding what is being communicated. The skill of active listening is of paramount importance to the systems engineer. Encouraging communication, responding actively, and confirming understanding by summarising is a key skill.

No organisation can exist without communication with the outside world. This outside world includes customers and suppliers, as well as those parties that merely have to coexist with the organisation, e.g. its neighbours or the local community. Communication with external agencies tends to be more formalised than internal communication, taking the form of letters, requests for information, or contractual documents. Today, however, email is replacing the strict formality of a headed letter.

8.3.1 The Nature of Communication

Organisational communication can be said to have two facets, one of which is a formal and permanent record of proceedings, the other is rather less formal and may not need a record that enters the design record. Each method of communication has a different impact on the recipient of the message and creates an impression that lasts for different periods of time.

A permanent and lasting impression is left by the mechanisms listed below, which should be filed carefully:

- newspaper articles
- letters, memos, and faxes
- emails and attachments
- text messages
- reports
- catalogues
- recorded speech
- books
- professional journals
- contracts
- meeting minutes.

On the other hand, a transient or fleeting image is left by the following list, although a record of these can be kept by entry into a diary or a follow-up memo:

- telephone and conference calls
- speech – face to face
- brochures
- posters and flyers
- meetings
- presentations (PowerPoint)
- text messages if not saved
- social media
- TV and radio broadcasts.

A permanent or lasting impression is conveyed by written communication, a medium that can be referred to at leisure, copied, shown to other people, and used many times. It can also be stored for future reference. The transient image is exploited by advertising and by influencers. The ultimate expression of this is the subliminal impact of TV advertising and political sound bites.

Each of these methods has a different impact and is used differently by communicators. Transient communication tends to be informal, chatty, colloquial; meetings and presentations reinforce the transient image, but sometimes convey vital information.

Permanent communication tends to be more formal, restrained, and even convoluted, as in legal documents.

In everyday domestic life communication tends to convey information, gossip, chat, family news – information that is of interest. A specific statement or formal manner indicates when a communication implies a 'contract', this is reflected in a change in behaviour. The following apply to everyday life communications:

- much of everyday domestic communication conveys information
- much of that information is of interest to both parties in a dialogue
- information is rejected, stored in memory, or used; it is not usually recorded
- unless there is a specific statement made, there is no contractual arrangement implied; if a contract is intended, there is usually an exchange of letters or a handshake.

Remember, however, legal obligations apply to advertising: Advertising Standards Commission, Broadcasting Standards, Sale of Goods Act, Trades Descriptions Act. Laws of verbal and written contract differ throughout the world, e.g. between England and Scotland.

In newspapers and advertisements there is a 'contract' that can be legally enforced through Acts of Parliaments or Trade or best practice obligations, particularly with regard to the provision of misleading information.

At work, however, there is a subtle difference. Much of the information exchanged is no longer merely of interest; it will be accepted in good faith and used. In other words, what you say to other people may be used in their work. What is more, what other people say to *you*, they will expect you to use in your work. The following apply to working life communications:

- much of everyday communication at work conveys information
- much of that information is of interest to both parties in a dialogue
- information is often used and acted upon immediately in good faith
- what is said and what is written is usually accepted as a contract between the parties
- minutes, memos, letter, e-mail *will be used as evidence of an intention to do something*
- all written communications should be checked for accuracy and to obtain approval to issue, for the protection of all parties involved
- all documents should be numbered, dated, controlled, signed, and filed.

8.3.2 Examples of Organisation Communication Media

Examples of the use of transient or non-permanent methods of communication include telephone, meetings, and presentations. However, each of these implies a contract in the same way as direct conversation. Remember that meetings are often minuted, and you will be expected to complete actions placed, and to abide by agreements and decisions made at meetings. More permanent methods of communication include paper records and, increasingly, electronically transmitted messages.

The following lists some examples of document types that engineers use in the course of their work. These documents need to be maintained so that they are well ordered and easily retrievable by all stakeholders in the project. It should also be noted that document records are part of the customer requirement and must be maintained for a period of time after the product has been withdrawn from service. Documents like this usually form the content of a data requirements list (DRL) and may be part of a payment plan:

- brochures
- leaflets
- plans
- specifications
- contracts
- drawings
- statements of work
- technical reports
- financial statements
- confidentiality agreements
- teaming agreements.

All of these documents will be prefaced, usually on the first sheet or cover sheet, with information that includes:

- title
- number
- date of issue
- issue/revision number
- record of changes
- originator signature
- approval signature
- authorisation signature.

Formal documents will be subject to configuration control so that the document can be positively identified and changes in the document can be traced. This is vital to establish a common agreement to the content of the document, especially if it is to be used as the basis for work to be conducted – in other words, an authorisation for money to be spent. This discipline is worthwhile even for informal exchanges of information, especially for attachments, to ensure that there are no misunderstandings. This should be done by adding the issue or revision to the file title and to the header or footer of the document.

As well as the administration detail of titles, dates, numbers, and issue record, it is essential to obtain signatures to give the document and its content some veracity. This can act for the benefit of all parties in the transaction. The status of the person allowed to sign a document and the giving of authority to sign is a vital part of the process of authorisation of design. The use of electronic signatures or a code word is an acceptable and more usual method of signing.

It should be noted that the use of the word 'document' in this section includes the electronic image of a record – a text document or a drawing will most often today exist as an electronic file created using a word-processing or drawing package on a computer, either a central computer or a collection of desktop or laptop machines. Although formal documents will be entered on the record, there is a significant risk that emails will be stored in personal files on many laptops. A project process for producing, handling, and storing project information must address this, especially if this form of correspondence with technical attachments leads to a design decision.

The use of electronic means of communication is widespread in both domestic and commercial worlds. This has benefits in improving the speed and responsiveness of communications and in moving towards the 'paperless office'. Although generally seen as beneficial there are some aspects that need careful consideration to ensure that information is correct and is protected from inadvertent dissemination to parties that have no right to see it, and to ensure that security is maintained in terms of commercial, proprietary, company, and national interests. Some of these concerns will be described below.

8.3.2.1 Mechanisms for Generating Information

An organisation will provide its employees with a mechanism for generating the information that is required to develop and define a product. This will often be based on a computer network of desktop machines with the appropriate tools to allow work to be generated, stored, and merged in a standard form. This is usually supplemented by individuals with other portable tools, sometimes supplied by the company, but often provided by the individuals. These include:

- laptop, own or company provided
- hand-held tablet device with email access
- telephone (smartphone) with speech, text, and email access
- notebooks and pen/pencil.

This collection of mechanisms allows information in the form of emails, texts, voice, and handwritten notes to be compiled and exchanged. At intervals this information should be consolidated and entered into the company network. Unless each individual is sufficiently disciplined to delete the information, it may still exist in these other portable devices. A suitable secure interface is essential to prevent connection to the company network that will open the door to the introduction of viruses or malware that could lead to hacking.

Figure 8.5 shows an example of an integrated digital data management system that interconnects all users and imposes a configuration management structure on all information registered in the system. It should be noted that emails can be sent without entering this system, hence by-passing the configuration management tool.

Each of these devices is a risk if it is left unattended, lost, stolen or otherwise mislaid since the data it contains is now available to unauthorised people. There is an added danger that these devices can be viewed or heard by other people when used in public places, such as the train, bus, café, or hotel. There is also a major risk of virus contamination or 'hacking' and the company systems must have sufficient security to prevent this.

8.3.2.2 Unauthorised Access

Unauthorised access to data is difficult to prevent with such a proliferation of different types of storage device and with many people using their personal portable computing devices and undertaking different forms of travel that expose them to risk of being observed

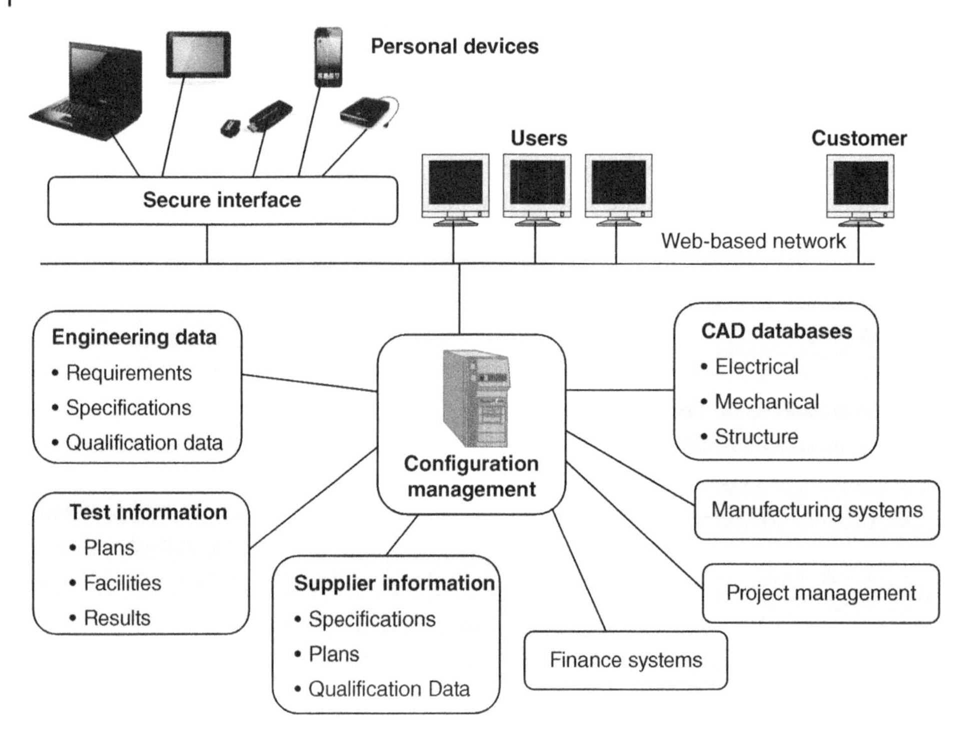

Figure 8.5 An example of a digital data management system.

accidentally, of deliberate industrial espionage or of theft. A typical journey may have the following risk areas:

1) Taxi to airport – risk of being overheard, leaving device in taxi.
2) Departure lounge – risk of being observed and overheard, theft, leaving device in lounge.
3) Flight – risk of being observed and overheard, leaving device on aircraft.
4) Taxi to hotel – risk of being overheard, leaving device in taxi.
5) Hotel – risk of being observed and overheard, theft from hotel.
6) Taxi to office– risk of being overheard, leaving device in taxi.
7) Meeting – Risk of open access meeting room, risk of unauthorised participants/ interlopers.
8) Return journey risks similar to outward journey.

This sounds a bit extreme, but for frequent travellers it will be difficult to remain vigilant on every journey.

8.3.2.3 Data Storage and Access

Apart from permanent records such as notebooks or printed paper versions of information, data is stored on a number of portable devices for use with laptops and for transportation between different sites. Some may be issued by the company or, more usually, purchased by employees to suit their personal computer. There needs to be strict control over connection of such devices as they may contain viruses. They should certainly not be used without some form of encryption and care must be taken to avoid unauthorised download of company information. Examples of data storage devices and associated risks include:

- optical discs such as CDs or DVDs – these are not durable and may have a limited read life
- flash drives or memory sticks – these are very commonly available, but they are small and easily mislaid or stolen
- external hard drives – these are also readily available and also easily misplaced
- cloud storage – this is available but is a slight risk because access is in the control of a third party.

It should be noted that previous forms of data storage, such as floppy discs and diskette drives, may also need to be accessed to obtain legacy data. If the storage devices are available data can only be extracted if a suitable read device is available. It is vital for any major organisation to keep a sample of such devices in an operational condition or to progressively move data onto more modern readable storage mechanisms as each medium becomes obsolete.

8.3.2.4 Data Discipline

With the proliferation of portable devices and office email accounts discipline is needed to ensure that data is available to the project and that it is correctly stored. A great danger, especially with emails, is that the main recipient will tend to store the message in their own personal file structure on, say, Outlook. If this is on their personal device then it may not be uploaded onto the desktop device, and even if it is the message will probably go to the individual file structure. This means that it is not stored centrally. For information that is a part of the design decision making, this is a serious omission in the traceability of decisions. Even worse is the fact that if an individual leaves the project or the company their data may be deleted as part of the company security policy.

Information received or disseminated by text or telephone is similarly unavailable to the project.

8.3.3 The Cost of Poor Communication

Poor communication costs money. The causes are many, but they usually result in a loss of productive time or scrap product. However, the impact on the recipient of a document or a statement must not be underestimated. As well as the potential for creating a poor impression of the company, this poor impression may also be prejudicial to the author.

The cost of putting things right is always high, but increases considerably as a project migrates from concept (correction to paper design) to use by the customer (recall and modification to hardware), as described in Chapter 3 and illustrated here in Figure 8.6.

Poor communication can lead to internal strife, discord, poor personal relationships, inefficiencies, and poor morale. Many organisational failures can be traced to failures of communication.

Poor communication is not helped by poor spelling, and there are circumstances in which this can lead to serious issues. In international programmes English may well be the main contender for the approved language for the project, but even if this is so there will be

- Time is spent correcting documents or responding to requests for clarification.
- More time is required to understand a document, e.g checking, asking for clarification.
- Instructions may not be carried out right first time.
- Lack of understanding, leading to poor quality of product results in costly re-work or modification.

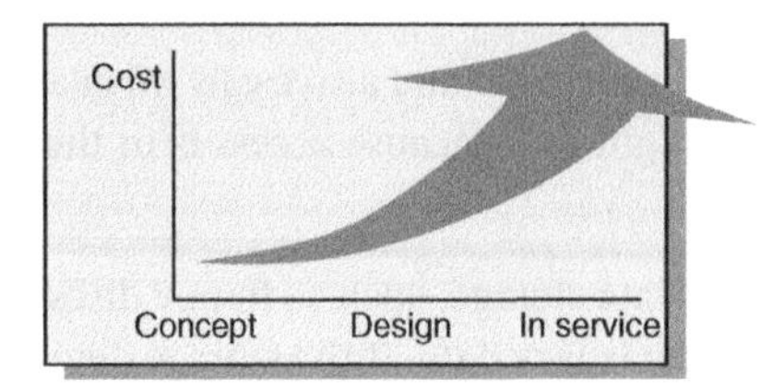

Figure 8.6 The cost of poor communication.

many participants who need to read and communicate in their own language. Traditionally translation from one language to another has been done by human interpreters, but increasingly automatic translation is being performed by speech recognition machines and character recognition for written material. English contains a number of homophones that may not be translated correctly into another language. Some commonly mis-used/mis-spelt words are:

- break and brake
- pain and pane
- great and grate
- sale and sail
- red and read
- dependent and dependant
- its and it's
- there and their.

Although some people will complain that as long as the meaning of a sentence is understood in English, regardless of the spelling, this will not be the case if the translation is performed exactly on what is written.

8.3.4 A Lesson Learned

A classic example of poor communication that has entered the language (at least in the UK) is illustrated in Figures 8.7 and 8.8 to illustrate the weaknesses of verbal communication and the strengths of written communication. This is in all likelihood an apocryphal tale about messages sent from the General Staff to the frontline during World War I and is one of several such tales. It does, however, serve to provide an example of different types of communication, and can be heard in conversations today to indicate a lack of organisation or planning.

Passing the message on verbally through a number of stages requires great skill in:

- clearly explaining the message to be given
- clearly understanding the message received

Figure 8.7 A weakness of verbal communication.

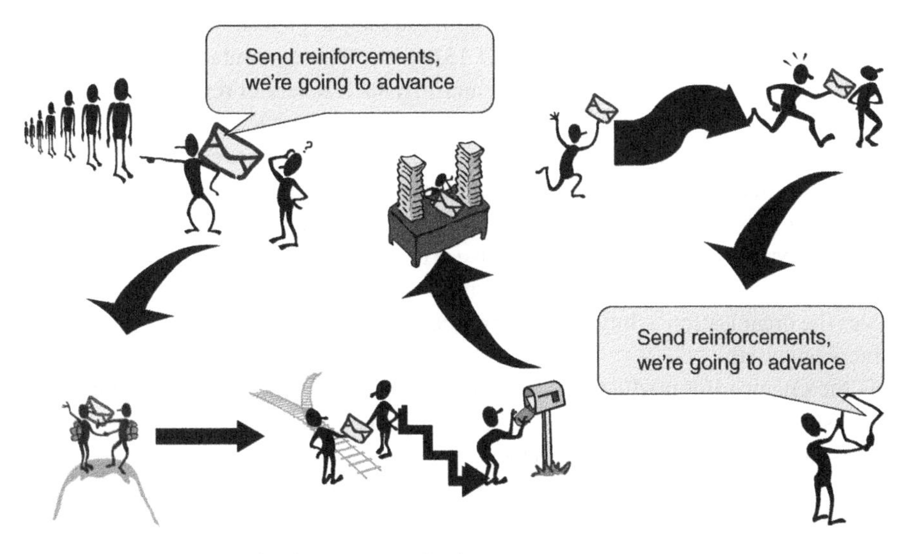

Figure 8.8 A strength of written communication.

- memorising the message
- avoiding distractions.

There is no immediate means of checking the message at each stage of handover, other than by going back to the sender.

(Note: Three and fourpence refers to the pre-decimal currency of the UK, now equivalent to about 17p in decimal currency.)

In this case the message has remained unchanged at each handover, and can be given credibility and authority by a recognisable signature. This technique is often seen in email correspondence by forwarding with attachments.

8.4 Giving and Receiving Criticism

Giving and receiving criticism is an important aspect of any creative process, whether that criticism is self-criticism or public review. Artists who display their work in public, the musician who plays before an audience, or the author who publishes their work, they all invite criticism from their peers or the wider public. Their work has been created and polished by their own self-critical appraisal, but it takes courage to expose it to others for critical review.

Criticism is also an essential aspect of systems engineering – as it is indeed of any sound engineering or design activity. Whilst self-criticism is an essential stage in the design process, there is no substitute for an external, objective review of engineering activities and products. There is no shortage of volunteers to provide criticism – everybody believes that they are ideally qualified to do so.

8.4.1 The Need for Criticism in the Design Process

Systems engineering is carried out according to defined and accepted processes. These processes may often be defined in accordance with industry or government standards, for example MIL-STD-1521B, MIL-STD-499, DO 15288 etc. These standards and industry processes include reviews at specific points of maturity such as tender review, preliminary design review, critical design review, test readiness review etc. The reviews may be performed by other workers (peer review), by senior management (management review), by specialists in a particular field or by workers from other programmes with no technical or business connection to the project being reviewed (often known as non-advocate review).

Whatever the type of review the main objective is to invite criticism with the intention of verifying that the review material is fit for purpose or to improve the quality. The review also enables the organisation to buy in to the product and to understand the feasibility, risk, and business potential.

What is the engineering product that is reviewed? It can be drawings, documents that define the design process or the product, management plans, test plans and results, hardware as in the first article verification (FAV) review, or financial and risk analysis. The review can be conducted by presentation of the material or by studying the material or product in paper or soft format. In many modern programmes the review material takes the form of a PowerPoint presentation. The audience at the review can be supplemented by remote reviewers viewing the same slides using a tool such as NetMeet. Experience has shown this to be a good way of presenting information to a lot of people simultaneously, but the overall effect is more fleeting; it is not easy to retain images over the period of the review, unlike a drawing or document that can be written on. Care must be taken with this format of review to make sure that it is sufficiently rigorous. If it is not acceptable, the reviewer must ask for the information to be provided in an alternative format.

8.4.2 The Nature of Criticism

Criticism is often perceived as a negative act, a perception which is not helped by the dictionary definitions. The dictionary contains two definitions, the first entry being:

The act or an instance of making an unfavourable or severe judgement of a work of art, literature, etc. (*Collins Dictionary and Thesaurus*). The *Thesaurus* includes terms such as animadversion, bad press, disapproval, disparagement, fault-finding etc.

It is only the second entry that has a positive aspect, and this definition is the one that most creative people would recognise: *A work that sets out to evaluate or analyse.* The Thesaurus in this instance includes analysis, appraisal, appreciation, assessment, comment, critique, and evaluation, which are all positive terms.

As well as preconceived negative perception, the way in which criticism is given often influences the way in which it is received – the medium becomes the message, to paraphrase Marshall McLuhan (McCluhan 1964). The tone of voice, the form of words, and the tone of the message can be deeply destructive. Consider four ways that criticism is often given at reviews:

- *Destructive*: short, abrupt comments without explanation such as 'Rubbish', 'Waste of time', or worse are examples of criticism that is destructive, can be crushing to the people being reviewed, and does not provide any information about how to correct the review material.
- *Belittling or demeaning*: phrases such as 'I could have done better myself', 'Do you mean we paid for this stuff?', and 'When I asked for ideas, I meant good ideas' are simply not relevant and serve to undermine the confidence of the people being reviewed. They also give no help for correction.
- *Non-committal*: phrases such as 'OK, I suppose' or 'so-so', or no comments at all are indicative of the reviewer that has no opinion. This could mean that the review material was not considered constructively at all, or that the reviewer didn't understand it. Again, this is not helpful in improving the quality of the review material. It is always better to give some comment or to own up to not understanding, rather than to leave the review team helpless. It is far better to recommend an alternative reviewer than to pretend to have the competence to proceed oneself.
- *Shallow*: there is a great danger at major PowerPoint-type reviews that the information presented is difficult to read and to hear. The size of the audience and the atmosphere of keeping to time is often a barrier to offering a comment or asking a question. This can result in no questions being asked or not persevering to obtain the right answer. A review is important – reviewers must avoid getting into this situation.
- *Constructive*: opinions expressed as 'That was good, but I feel that if you gave more of an explanation in a particular area with examples, then I would understand it better' or 'I believe that a diagram would make the explanation clear' or 'This is not correct, but I can give you a correct explanation or an example' provide the people being reviewed with information that can be used as improvement, and opens the door to continuing discussion to correct the review material.

8.4.3 Behaviours Associated with Criticism

The review team may have little time to get their material ready for publication; they will always be under pressure to meet schedule demands, especially bid teams, who must meet their customer's deadline or fail.

Their morale must, therefore, not be damaged by the review, they need to be at maximum efficiency to consider the review results, make the necessary changes, and publish the material. The way in which criticism is received can have an impact on both personal and team behaviours. Typical behaviours exhibited by people receiving criticism are apprehension, uncertainty, feeling threatened, discomfort, fear of personal attack, and being defensive. These behaviours may be reinforced at the review and will damage people's confidence, thereby potentially reducing their performance post review.

On the other hand, the team being reviewed must also understand how to accept criticism. They must deflect the destructive comments, if possible by interrogating the critic to get behind his emotions. They must remain dignified, formulate plans for re-work and to move forward, using the review comments to maximum advantage. Received advice used to be to acknowledge the comments, never become defensive, answer concisely, and if an answer is not immediately obvious ask for a written review comment to be answered later.

Reviewers are also subject to behavioural changes; people giving criticism often feel the need to be aggressive, superior or negative, often adopting a 'not invented here' syndrome by trying to interpret the material in terms of their own experience and/or prejudice. People may often adopt a superior stance, feeling better than the team being reviewed.

The most constructive way to approach a review is to act as a team that is part of a business team that is aiming to produce the best possible product. The reviewer must remain objective, but must also be constructive, not finding fault, but identifying mistakes or technical inaccuracy whilst also providing recommendations for improvement. The size, complexity, and geographic dispersion of teams in modern large-scale projects makes the review process important. It is an opportunity to get people together, to discuss issues and progress, and to invite project-independent views and experience.

8.4.4 Conclusions

Systems engineering is a creative process performed by a number of individuals and teams that produces deliverables throughout a product lifecycle. It is important that these deliverables are reviewed in order to ensure that all stakeholders are aware of the progress, content, and standard of the deliverables. A well-defined and robust review process is vital to establish sound quality products. All parties concerned with this review process must act for the greater good of the business by conducting the reviews in a disciplined manner.

Constructive criticism in the form of evaluation and analysis, the application of engineering and business judgement, and a positive approach to both giving and receiving criticism is of prime importance.

8.5 Supplier Relationships

A substantial proportion of a complex aircraft product is purchased from suppliers outside the prime contractor organisation, often referred to as the supply chain. These purchases, which may be as much as 80% of the contract value, may take the form of materials, hardware or software products (equipment) or services. There is generally an industrial supplier

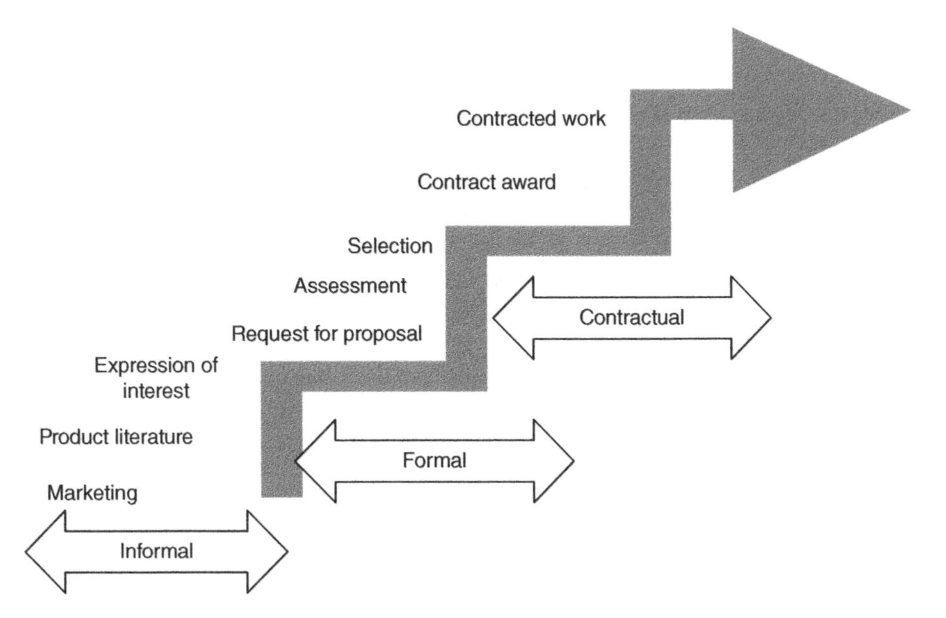

Figure 8.9 The changing nature of relationships with suppliers.

base with specialist companies able to provide support to one or more of these areas. These suppliers are key stakeholders in the project.

Understanding the supplier base and understanding how to engage with suppliers throughout the product lifecycle are key factors in establishing a sound relationship with a supplier. Figure 8.9 illustrates how the relationship with a supplier changes through the product lifecycle.

The informal relationship is based on understanding the product range available in the marketplace and at looking at emerging developments. This activity allows market awareness to be gained by studying trade literature, attending exhibitions and conferences, and participating in joint studies and trade-offs or research paper preparation. This relationship allows both parties to access information and comment constructively on product development without any contractual or intellectual property constraints.

The relationship becomes formal when the information is required to support a more formal activity such as a tender, in which a number of suppliers may be requested to provide information to a prime contractor. There is now an obligation on the part of the supplier to provide correct and pertinent information, and on the prime contractor to treat each supplier's information as confidential. The information may form the basis of a commercial tender and may be the subject of a confidentiality agreement between both parties.

The nature of a formal relationship means that possession of the information must be recorded, the information must not be divulged to other parties, and recognition that infringement of copyright can be a criminal offence. This relationship is often formalised in a non-disclosure agreement (NDA) signed by all parties concerned.

If the information requested is to be used in a competitive tender process, then technical, programme, and cost information from a number of suppliers will be formally assessed and scored to make a judgement on a winning tender. This information will be used to select which supplier is to be given the business. In these circumstances all communications with individual suppliers is severely restricted to avoid any prejudice in the selection process.

Following selection the award of a contract changes the relationship again. The supplier is obliged to meet the terms and conditions of the contract to provide the goods or services that are the subject of the contract. This means that work over and above the contract will be seen as additional work and incur additional cost.

Throughout this whole period of changing relationships it is important for the prime contractor and the supplier to respect each other's commercial position and yet work together for the good of the project. The systems engineering approach of careful management of requirements and an open relationship with stakeholders becomes important.

8.6 Engineering Judgement

Judgement can be defined as *the facility of being able to make critical distinctions and achieve a balanced viewpoint* (*Collins Dictionary and Thesaurus* 2012). Engineering judgement is the ability to do this in an engineering situation, it is an invaluable pragmatic input to many engineering decision-making processes.

It is the type of quality that is often observed in the very people who possess the wisdom so important to learning from experience. Engineering judgement cannot be taught or measured, it is a quality acquired over many years of experience on a number of projects and over many learning experiences, but it is an important quality to aspire to possess. It can be acquired by understanding how people come to conclusions, by observing how people apply knowledge, and by understanding how people use other stakeholders and their opinions to form judgements that in themselves may not be firm decisions, but help others to make decisions.

Engineering judgement helps a systems engineer to do his job and produce sound systems engineering. Unfortunately, it is not possible to quantify or to validate this property. It is inevitably personal and ephemeral, and can be subject to criticism at review; it can be a risk. If it is used as part of design it must be recorded and the qualifications and experience of the provider need to be explained.

8.7 Complexity

The systems of an aircraft are being designed to perform ever more complex and demanding tasks. The systems that are contained in the environment of the aircraft are also increasing in complexity, including the in-service support system, aircrew and ground crew training systems, airport management systems, and airport security systems. The systems engineering task demands an integrated view of all such systems to generate customer satisfaction. Some observations on complexity (Maier and Rechtin 2002) show that it can have a significant impact on engineering: *It is generally accepted that increasing complexity is at the heart of the most difficult problems facing today's architects and engineers.*

This observation is pertinent to many large-scale projects – civil, marine, aerospace, agricultural, and telecommunications projects have all exhibited examples of partial or total failure because of an inability to deal with complexity. Such failures manifest themselves as:

- cost and schedule over-runs
- performance shortfalls
- poor availability
- slow start-up
- human–machine interface issues
- maintenance issues.

Any large-scale system exhibiting one or more of these criteria will lead to poor customer perception. In the case of large public sector funded projects, there is likely to be adverse media comment.

Understanding complexity begins with understanding the requirement completely. A holistic systems approach with a careful analysis of the design drivers in different environments together with sound stakeholder communication aids that understanding. This needs to be followed by careful management of the flow down of requirements with review at all stages of the lifecycle. Maintaining a rigorous approach to matching emerging designs to the requirement, as well as clearly defining functional allocation and functional, physical and data interfaces, is good practice. Chapter 11 examines this issue more closely.

8.8 Emergent Properties

Emergent properties are those properties or characteristics of a system that are unexpected. They may be desirable because they improve the performance of the system, or they may be undesirable because they reduce its performance. The reason that they are unexpected is because they arise from a combination of functional, and sometimes physical, interactions within a system. The more complex and interdependent the functions of a system, the more difficult it is to predict the exact outcome of their combined results when they are integrated into a working whole.

Figure 8.10 shows some factors that distinguish an emergent property from a normal system characteristic. In general one expects most factors to have a linear effect on a system, one expects that the controllable variables shown in the diagram are distributed and linear, and there is usually a singular self-contained impact. For example, the addition of more mass makes a system heavier in direct proportion to the mass added, an increase in the volume of a component is directly measurable, etc. The consequences of these actions are immediately evident and are generally monitored and controlled during the early stages of the lifecycle.

Emergent properties arise when effects are non-linear and multi-variant – the impact of a combination of effects is greater than the sum of the individual effects. Examples shown in Figure 8.10 include the impact of electro-magnetic interference (EMI) between systems, which may be greatly influenced by their installation, by variations in bonding resistance or corrosion at bonding surfaces, or by changes in transmitted power and electrical noise.

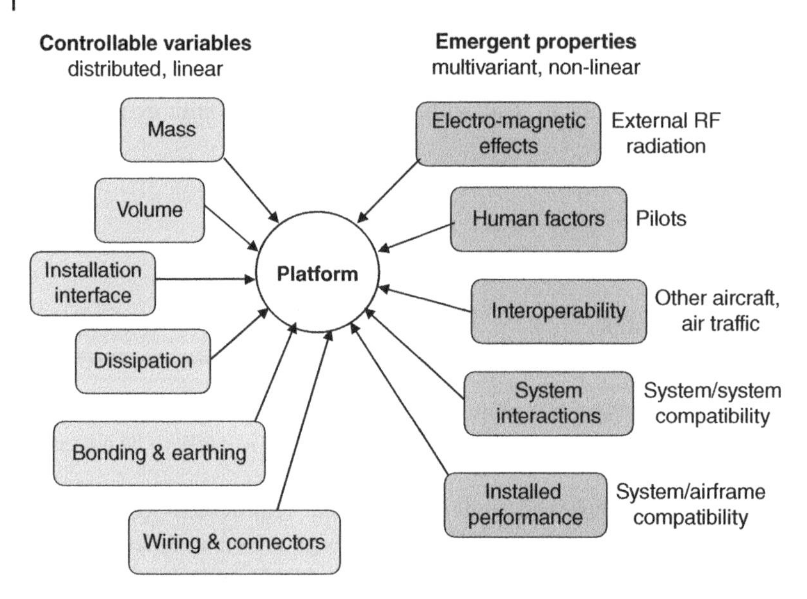

Figure 8.10 Controlled and uncontrolled interactions.

Human factors is an area in which situations can arise that can lead to a situation where the crew workload is too high under conditions of stress, thereby increasing the probability of an accident occurring. Interoperability issues may arise when aircraft from different operators or military forces are unable to operate together – different spares, fuels, radio frequencies, and protocols are factors that can lead to this situation.

Emergent properties have an impact on cost and schedule performance. Because of their unpredictable nature, their effects may not become apparent until late in the lifecycle, typically during testing, and especially in field trials conducted on the whole aircraft such as hot and cold weather trials, tropical trials, route proving, stealth etc. As described in Chapter 3, this is undesirable as it can lead to costly and time-consuming re-work and change. It is important to predict and identify the sources of emergent properties early in the lifecycle, and to constantly review the results. In large-scale complex systems this is a challenging task, but it is vital to reduce work and schedule impact at later stages of the lifecycle.

Chapter 11 provides some examples of emergent properties or unexpected behaviours taken from operational experience.

8.9 Aircraft Wiring and Connectors

8.9.1 Aircraft Wiring

Of the many invisible attributes of an aircraft the electrical wiring is not the easiest to understand and its all-pervasive extent within the aircraft is most difficult to comprehend. This section on aircraft wiring will be useful to those studying aircraft systems. Flight control runs, fuel pipes, hydraulic lines, and air-conditioning ducts are easy to visualise and identify within the airframe whereas electrical wiring is less easily identifiable. This is compounded by the variety of different connectors used for differing electrical technical reasons. A good starting point is to understand the basic aircraft structure and how aircraft wiring relates to these basic structural building blocks.

8.9.2 Aircraft Breaks

The key structural/wiring breaks within an aircraft are determined by aircraft structural breaks and in an increasingly world-wide aerospace industry there is a tendency for a level of outsourcing or subcontracting of these major areas of work. Significant structural elements are increasingly likely to be distributed to investment risk-sharing partners around the world who will accept responsibility for that element of the aircraft structure. These risk-sharing partners may also take responsibility for the installation of systems components and wiring that lie within that part of the airframe. A typical example is shown in Figure 8.11. Typical aircraft breaks include the following:

- *Forward and aft fuselage breaks*: these separating the forward, centre, and aft fuselage sections.
- *Wing/fuselage breaks*: define the boundary between the relatively benign pressurised cabin environment and the more challenging area of the wing in which flight control actuators, fuel system components such as fuel pumps, valves, and gauging, and temperature sensors reside. The aircraft wing area presents severe challenges to aircraft wiring and electronics. Typical withstanding voltages for aircraft wing wiring are twice those that would be specified for wiring residing within the fuselage compartment.

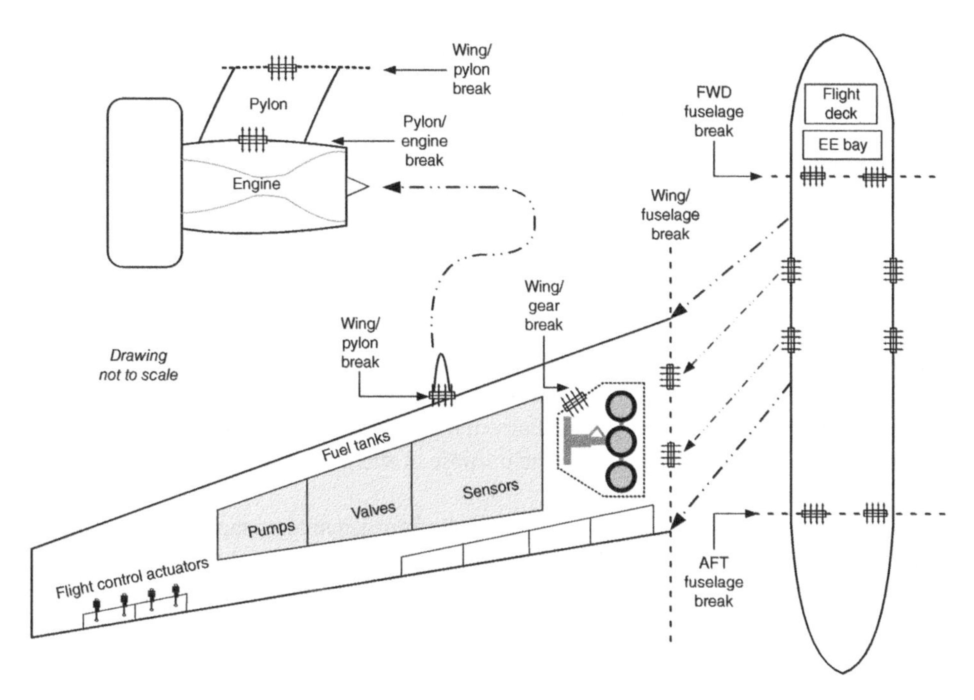

Figure 8.11 Example of aircraft wiring breaks.

- *Landing gear/wing compartment breaks*: The landing gear bays present another severe environmental area. Because of the exposure to a hostile environment during take-off, landing, and approach the wiring looms on the landing gear are usually armoured to survive within the environment as well as withstand the effect of flying foreign objects such as discarded aircraft wheel tyre treads or objects picked up from the runway.
- *Wing/pylon and pylon to engine breaks*: These breaks are important as the associated wiring carries essential information between the flight crew and the engine. The pilot transmits throttle commands and other control information to the engine.
- *Engine wiring*: The engine without doubt presents the most aggressive environment on the aircraft in terms of temperature and vibration. In a similar manner to the landing gear harnesses, wiring harnesses on the engine are typically armoured to provide protection against the severe operating conditions.

With the exception of a few key systems, most aircraft systems are constrained by the boundaries described above. Particular exclusions include:

- the routing of high-power generator feeder cables due to the possibility of high resistance/high power dissipation contacts
- the routing of high-integrity wiring such as fire warnings and hydraulic shut-off valve selection lines due to the consequence of a connector failing
- specific wiring associated with flight control.

8.9.3 Wiring Bundle Definition

Within specific zones the aircraft wiring may differ from single wires connecting two items together electrically to wiring bundles or harnesses in which a number of wires need to be routed to/from specific points within the aircraft structure. The definitions associated with individual wires/bundles/harnesses are broadly as follows:

- *Open wiring*: any wire, wire group, or wire bundle not enclosed in a covering.
- *Wire group*: two or more wires tied together to retain the identity of the group.
- *Wire bundle*: two or more wire groups tied together because they are going in the same direction at the point where the tie is located.
- *Wire harness*: a wire group or bundle tied together as a compact unit (open harness) or contained in an outer jacket (enclosed harness). Wire harnesses are usually prefabricated and installed on the aircraft as a single assembly.
- *Electrically protected wiring*: Those wires which have protection against overloading through fuses, circuit breakers or other current limiting devices. Most of the aircraft electrical wiring is protected in this way. The purpose of the protection is to protect the aircraft wiring, not the load.
- *Electrically unprotected wiring*: those wires (generally from generator to main bus distribution points) which do not have protection from fuses, circuit breakers or other current limiting devices. However, protection against electrical fault conditions will be inherently provided as part of the generator control loop, including current and voltage fault conditions.

These definitions were extracted in the main from AC21–99, Advisory Circular from the Australian Civil Certification Authority and are illustrated in Figure 8.12 (CAA 1999).

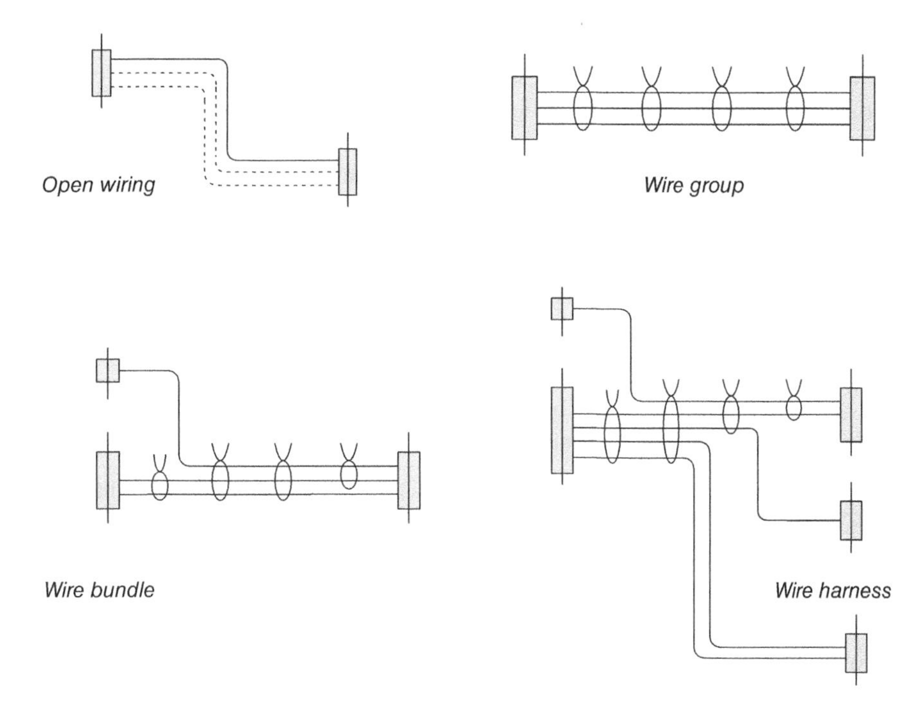

Figure 8.12 Examples of wire groups.

8.9.4 Wiring Routing

Given the foregoing constraints and the need to transit the various structural and/or electrical breaks as already described, the wiring is subject to very practical considerations during installation, namely:

- taking care not to exceed the bend radius of the wire type
- prevention of chafing between wire bundles and aircraft structure
- securing bundles through bulkheads and structure
- fastening wires in junction boxes, panels, and bundles for correct routing and grouping
- prevention of mechanical strain that may break conductors and connections
- prevention of the possibility of arcing or overheating wires causing damage to mechanical control cables
- facilitation of reassembly following repair
- prevention of interference between wires and other equipment
- permitting replacement of individual wires without removing the entire bundle
- prevention of excessive movement in areas of high vibration (armoured cables in landing gear and engine zones).

8.9.5 Wiring Sizing

Aircraft wiring is generally categorised by reference to the American wire gauge (AWG) convention. Within the AWG convention the higher the number the smaller the size of the wire. Typically, AWG 24 (Boeing) and AWG size 26 (Airbus) are the smallest wires used within the aircraft for reasons of robustness. Smaller gauge – higher AWG sizes – may exist within individual equipment as these are protected from general wear and tear within the aircraft.

The lower AWG categories are used for high power feeders, usually from the aircraft electrical power generators or for major electrical power feeders within the aircraft electrical power distribution system. A key consideration in the selection of the wire size is the voltage drop associated with the wiring run and the power dissipation associated with feeder losses. The nature and duration of the anticipated electrical faults in association with the capability and reaction time of the wiring protection devices available is an important consideration.

Tables 8.1 and 8.2 give an indication of typical aircraft wiring parameters and an indication of the amount of wire in a typical large aircraft:

- The current handling capability of aircraft wiring in a typical large civil aircraft. Table 8.1 provides indicative information for copper (Cu) wiring. Aluminium (Al) wiring is lighter but has higher resistivity so there is a scope for a selection of Cu versus Al power feeders to save installation weight at the expense of greater voltage drop/feeder losses.
- Table 8.2 gives a typical wiring weight budget for a large transport aircraft in the Boeing 747 family of around 20 years ago. This represents basic aircraft wiring and does not

Table 8.1 Typical aircraft wiring current capacity.

AWG	Diameter (in.)	Ohms/1000 ft (Cu)	Maximum current (typical)	Typical applications
0000	0.46	0.049	260	Primary
000	0.41	0.062	225	Power feeders
00	0.36	0.078	195	
0	0.32	0.098	170	
1	0.29	0.124	150	
2	0.26	0.156	130	
4	0.20	0.248	95	
6	0.16	0.395	75	
8	0.13	0.628	55	Secondary feeders and high-power loads
10	0.10	0.998	40	
12	0.08	1.588	30	
14	0.06	2.525	25	
16	0.05	4.016		Medium-sized loads
18	0.04	6.385		
20	0.03	10.150		Normal use
22	0.26	16.140		
24	0.02	84.22		

Table 8.2 Engineering snapshot of Boeing 747 wiring.

WG		Length (ft)	Weight (lb)[a]
24		162445	887.0
22		148239	594.2
20		237713	1859.4
18		82211	732.6
16		26663	276.8
14		4998	65.4
12		9872	256.2
10		4681	146.0
8		3981	231.9
6		2048	115.3
4		2622	240.9
2		1140	170.2
1		444	50.2
***1**		719	196.1
***2**		2447	418.4
***3**		55	12.5
Special		5574	219.0
	Total	**695852**	**6472.1**

[a] Includes connectors but excludes in-flight entertainment.

include the in-flight entertainment burden. Of particular interest is the extent of the wiring – almost 700000 ft of wiring weighing in the region of 6500 lb. This example is dated and it could be expected that an aircraft of more recent generation would have a much lower wiring content due to the impact of integrated modular avionics (IMA) cabinets and remote data concentrators (RDC).

8.9.6 Aircraft Electrical Signal Types

Aircraft wiring as has already been described is complex and is often installed in a hostile environment. In many cases aircraft wiring cannot be accessed following aircraft initial build. Wiring types are varied, as the following examples testify:

- RF/co-axial wiring for radios and radars; sub-miniature co-axial wiring is used in places
- power feeders for primary electrical power; conventional wiring for lower power electrical supplies
- signal wiring for aircraft sensors; often twisted/screened pairs, triads and quads
- twisted copper pairs and quads for data buses
- fibre optic wiring for data buses and in-flight entertainment (IFE) systems.

Specialised wiring is also required in the area of fuel gauging (tank wiring harnesses), landing gear, and the engine (armoured conduits). Wiring in fuel tanks has to be particularly protected to limit the amount of energy associated with a fuel content sensing probe and also limit the fault conditions associated with an electrically powered fuel pump.

Associated with the varied wiring configurations is a huge range of connector types.

8.9.7 Electrical Segregation

Due to the widely diverse nature of aircraft, electrical signalling types require segregation as certain types may interfere disproportionately with others, causing detrimental performance of vital aircraft systems.

The wide diversity of aircraft signal types may be summarised by using an Airbus example. On Airbus aircraft systems the aircraft wiring system is generally divided into two main systems and further subdivided into routes. This ensures that damage is limited and any EMI interference is reduced to a minimum.

In the Airbus system different circuits have specific identifiers (similar conventions are used by other aircraft manufacturers):

G Generation
P Power supply
M Miscellaneous
S Sensitive
R Audio
C Co-axial

8.9.8 The Nature of Aircraft Wiring and Connectors

The discussion so far has centred on the point-to-point electrical wiring as it connects components together throughout the airframe. The means by which the various controllers and sensors are connected also deserves mention.

There are three main ways in which these components are electrically connected, as illustrated in Figure 8.13:

- rack mounted
- structure mounted
- bulkhead or wiring breaks.

All of these connector types employ the separation of differing signal types as described above.

Most electronic controllers are rack mounted, usually in the electrical equipment bay or compartment in the forward section of the fuselage. The connector and mounting arrangement is defined by ARINC 404 or ARINC 600 depending upon the vintage of the equipment. ARINC 404 relates to older analogue equipment whereas ARINC 600 is associated with more modern digital equipment. Some components are mounted directly to the aircraft structure. In the case of aircraft bulkheads or aircraft wiring breaks as already described, circular connectors are usually used. Examples of the various connectors used are shown in Figure 8.14.

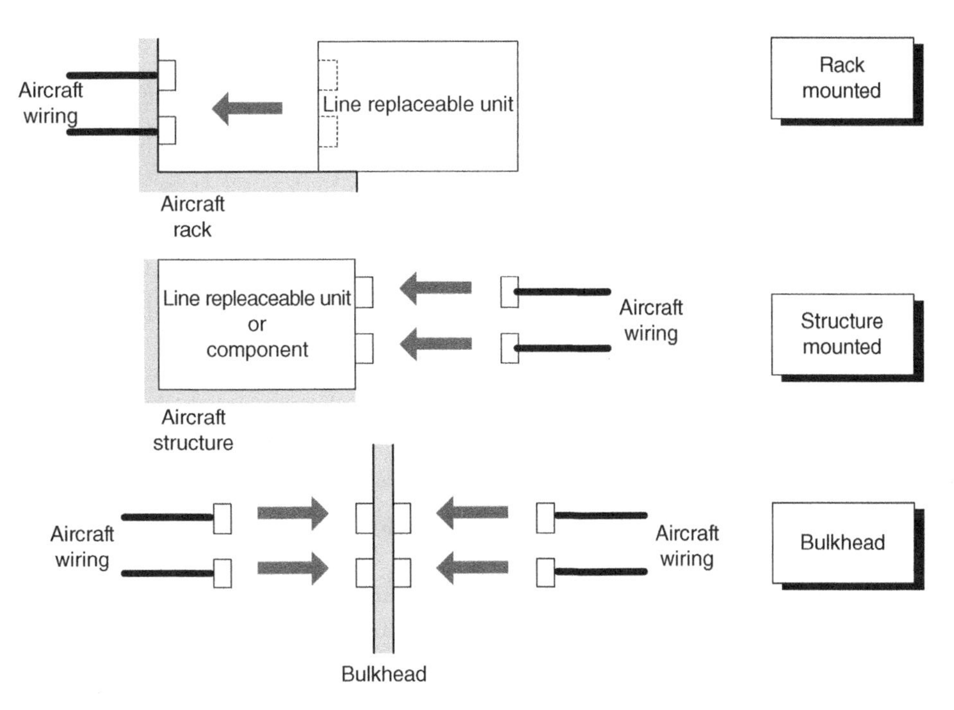

Figure 8.13 Typical equipment mounting arrangements.

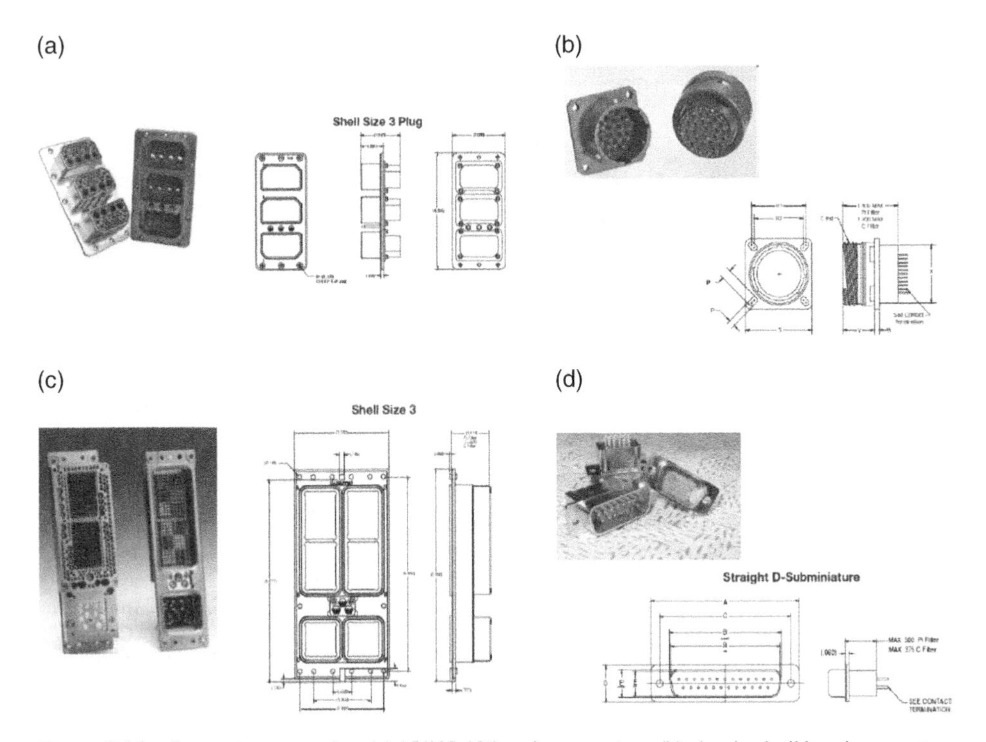

Figure 8.14 Connector examples: (a) ARINC 404 rack connectors, (b) circular bulkhead connectors, (c) ARINC 600 rack connectors, and (d) D-type connector.

8.9.9 Use of Twisted Pairs and Quads

The recourse to resisting EMI within the aircraft wiring is to resort to shielding and screening. This is usually employed for sensitive sensor signals and digital data buses (which can also be significant emitters). Practising aerospace engineers often have problems envisaging what the data buses in an aircraft actually look like so it is worth giving a brief explanation. Twisted screened wire pairs are categorised as follows:

- unshielded: the twisted pair has no metallic shroud, UTP
- shielding is where the unshielded twisted pair is contained within a metallic shroud, S/UTP, also known as FTP (foil TP)
- screening is where a twisted pair is contained within a metallic shroud, STP, also known as STP-A
- shielding and screening together for a twisted pair, S/STP, also known as S/FTP
- similarly twisted triple or quad wire arrangements may be used.

The shield or screen may be bonded or grounded depending on the installation requirements.

Examples of shielding and screening techniques are shown in Figure 8.15.

There are different ways of installing data buses – Boeing and Airbus adopt different schemes, as shown Figure 8.16, which also shows examples of single stand-alone quadrax and twinax connectors. Airbus tend to use the quadrax arrangement shown in the upper part of the diagram. A full duplex data bus – a data bus passing data in both directions simultaneously – is implemented in one self-contained cable including two twisted wire pairs.

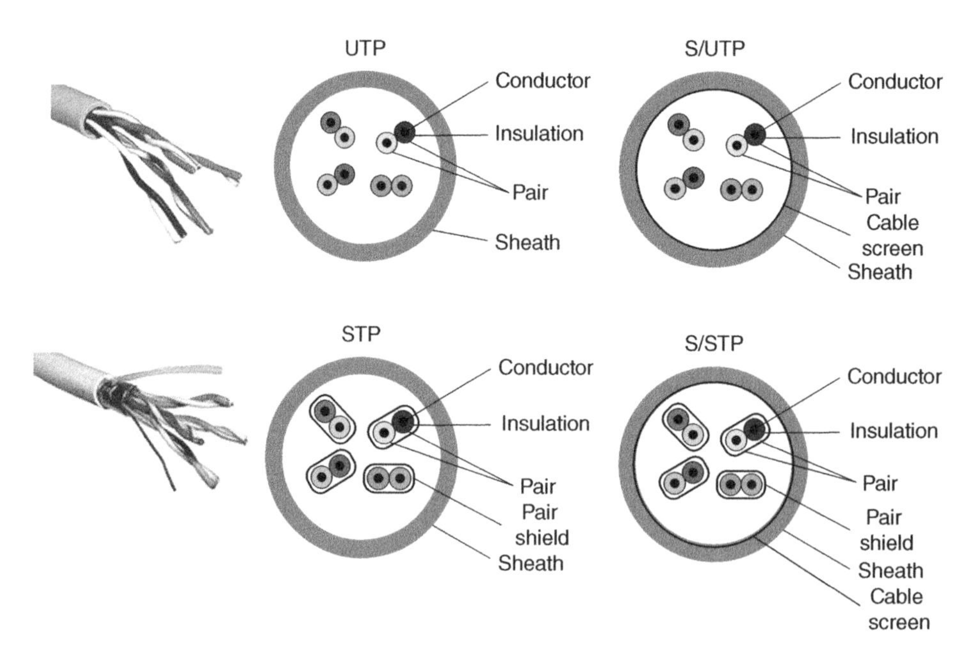

Figure 8.15 Examples of screening and shielding.

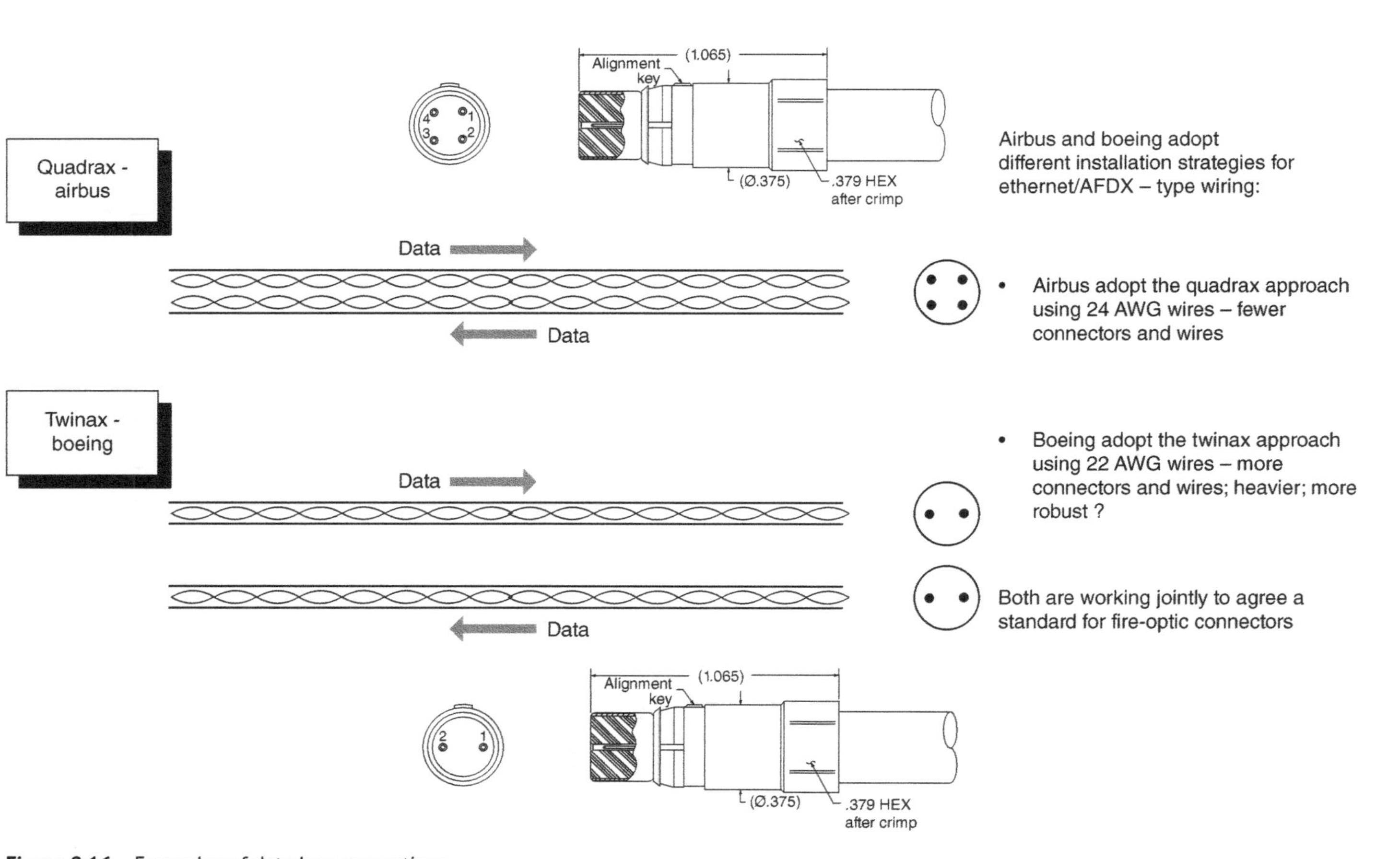

Figure 8.16 Examples of data bus connections.

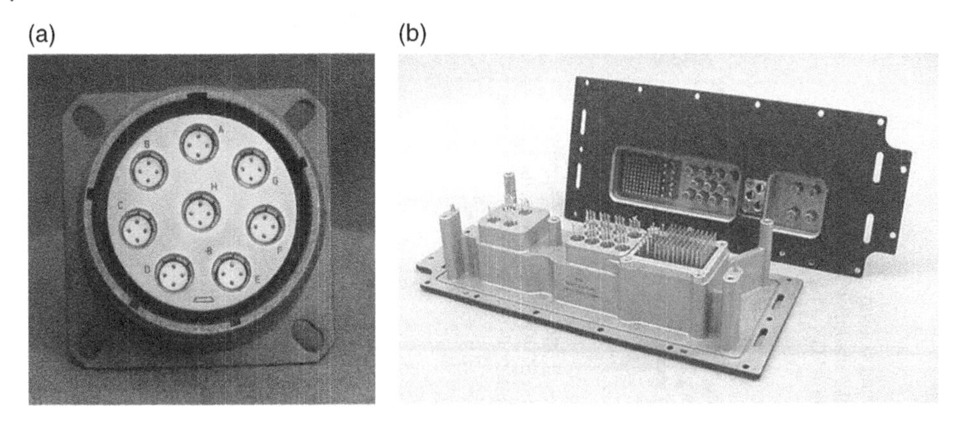

Figure 8.17 Examples of quadrax connectors: (a) military MIL-DTL-38999 and (b) civil ARINC 600.

This arrangement allows a higher packing density but arguably is less robust than a twinax arrangement. Boeing favour the twinax scheme shown in the lower part of the figure. This implements a half-duplex arrangement where each bus only passes data in one direction and is separate from the other.

In an aircraft which makes extensive use of aircraft data buses, which probably represents most of the aircraft in production today, multi-connector arrangements are used. Figure 8.17 shows a circular MIL-DTL-38999 connector and civil ARINC 600 type rack connectors.

8.10 Bonding and Grounding

There are important considerations to be borne in mind relating to the aircraft electrical wiring system: bonding, grounding, and earthing. These techniques reduce the voltage potential between adjacent items of hardware, provide a stable reference point for the aircraft electrical systems, and provide a means by which static is dissipated during ground servicing operations. Sometimes these terms are very loosely used – even by aerospace professional engineers – so it is necessary to use precise definitions. Commonly used definitions are as follows:

- *Bonding*: the electrical connecting of two or more conducting objects not otherwise adequately connected to minimise potential difference.
- *Grounding*: the electrical connecting of a conducting object to primary structure or earth electrode, for return of current.
- *Earthing*: a specific case of bonding to an earth reference to dissipate static whilst the aircraft is being serviced, particularly during refuelling and/or when an external electrical power source is connected. In this sense two further definitions may apply:
  - Static ground: an approved ground point with an impedance of less than 10 000 Ω when referenced to earth
  - Power ground: an approved ground point with an impedance of less than 10 Ω with respect to the aircraft power system neutral.

This information was extracted from AC21–99, Advisory Circular from the Australian Civil Certification Authority, for which acknowledgement is given.

Whilst the aircraft is in flight the aircraft structure represents 'ground', therefore the aircraft structure provides the return path for current flowing through an aircraft load and back to the electrical power source.

The aircraft is only 'earthed' during specific ground-servicing operations such as maintenance, refuelling, and arming. There are two ways in which an aircraft can be earthed during servicing operations:

- Using a dedicated earth lead to connect an earth stud on the aircraft structure to a specified earthing point on the airfield. This situation prevails if the aircraft is connected to a stand-alone external ground power cart. Earthing points are conveniently situated around the airfield, particularly close to where the aircraft is to be serviced (or armed). They are specifically designed and maintained to ensure the necessary high quality of earth connection required for the task in hand.
- If the aircraft is connected to a mains-generated power source (or electric/electric power source) it will automatically be connected to the external power source (including earth) via the aircraft external power connector. In this situation the aircraft is effectively connected to the National Grid earth of the country in question.

It is also necessary to make a distinction between the various types of grounding connection used for different power and signal types. This may be understood by reference to Figure 8.18, which shows a typical electronic controller as installed in the aircraft, whether a rack-mounted or stand-alone unit.

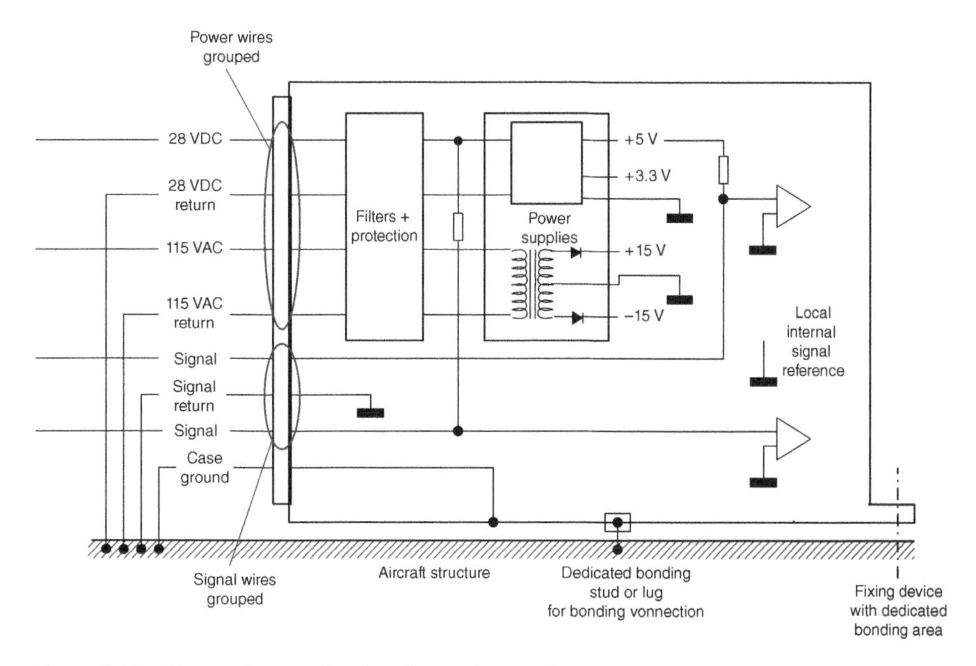

Figure 8.18 Electronic controller bonding and grounding.

The aircraft structure on which the unit is mounted represents 'ground' and provides the current return path as already described:

- The fixing or mounting device has its own dedicated bonding area.
- The unit case is directly bonded to the aircraft structure using a dedicated stud or lug.

In this example two other bonding connections are shown:

- 28VDC return connection
- 115VAC return connection.

Both these returns are from the internal unit power supply unit and are separately grouped together. These are likely to be EMI emitters:

- signal return connection
- case ground connection.

The signal connections are separately grouped together as they are more sensitive and are EMI susceptible. Within the unit there will also be local internal signal references for power and signal types.

Bonding, grounding, and earth points will all have specified low resistance values which may depend to some degree on the aircraft type and nature of the avionics fit. Aircraft and equipment-level designers need to adhere closely to these requirements if the aircraft is to have a satisfactory EMI performance.

Exercise

1 Take the examples of English homophones in section 8.3.3 and add some of your own. Then translate the whole list into French and German to understand how an automatic text translation can lose the original meaning of a message.

References

CAA (1999). AC21-99, Australian Civil Aviation Authority.

Crystal, D. (1995). *The Cambridge Encyclopaedia of the English Language*. Cambridge University Press. ISBN 0 521 40179 8.

Maier, M.W. and Rechtin, E. (2002). *The Art of Systems Architecting*. ARC Press.

McCluhan, M. (1964). *Understanding Media: The Extensions of Man*. McGraw-Hill. *"The medium is the message" because it is "medium that shapes and controls the scale and form of human action."* Frequently Asked Questions. www.http://marshallmcluhan.com.

Meakin, B. and Wilkinson, B. (2002). *The 'Learn from Experience' (LfE) Journey in Systems Engineering*. 12th International Symposium of the International Council of Systems Engineering (INCOSE).

Further Reading

Scholes, E. (1999). *Guide to Internal Communication Methods*. Gower.

9

Configuration Control

9.1 Introduction

A major consideration in the development, deployment, and support of any system is the introduction of configuration control (known colloquially as 'config control'). Configuration control serves the following aims:

- It establishes systems design baselines in a manner such that all those elements necessary for correct system function are organised to ensure full system compatibility.
- It enables changes to the baselines to be introduced in a controlled manner and all changes to be made visible with full traceability.
- It ensures that as a system evolves through different configurations or standards full compatibility is maintained at every stage.
- It maintains compatibility where desired between early and later system/product development implementations.

The key principles of configuration control are described in the following section. These aims combine to achieve baseline data sets that are common and consistent to all users, and ensure that changes or increments to the baseline are controlled. It is important to note that products have multiple design standards that co-exist through the product lifecycle – in other words sub-sets of the product may be changing at different times before they are assembled into the completed product.

9.2 Configuration Control Process

Configuration control is applied by establishing and controlling the issue of all authoritative documents and databases that define the product design. 'Documents' in a modern project covers the traditional paper documentation as well as soft products such as databases, models, and software loads. Typically the design of a product is defined by:

- requirements statements
- system and sub-system specifications
- equipment and component specifications

Design and Development of Aircraft Systems, Third Edition. Allan Seabridge and Ian Moir.

- systems architectures
- software requirements and specifications
- interface control documents (ICDs)
- wiring diagrams
- installation drawings or 3D models
- test procedures
- test results
- safety analyses
- statements of design
- equipment lists
- product build standards
- clearance standards
- plans such as engineering plans, management plans, project plans, etc.

These documents will be accompanied by concessions to record and authorise deviations from the baseline or to record shortfalls in performance. These concessions may lead to significant change or the need for re-design. This is usually authorised by a change authority. This allows the documentation set to be raised in issue to incorporate the changes into a new baseline design.

Any other systems or processes affected by the change are thereby informed and authorised to implement their own changes to ensure compatibility. This process ensures that change is understood by all stakeholders and that the product remains fit for purpose. In other words, the right part number of equipment (defining hardware and software) comes together with the right wiring and installation in the right standard of product.

9.3 A Simple Portrayal of a System

A classic closed-loop control loop is usually portrayed in its simplest form as shown in Figure 9.1. A system demand is fed into the forward path, which embodies and executes the control laws associated with the operation of the system and results in a system output. In a closed-loop system, the system output is fed back and compared with the system input

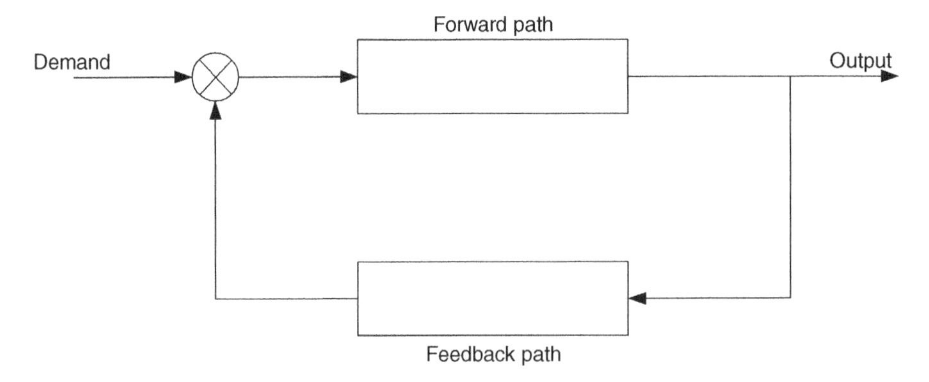

Figure 9.1 Classic closed-loop control system portrayal.

to ensure that the system is maintaining the desired performance. The feedback path may include additional control functions or compensation to achieve the required system performance or accuracy.

This is the idealised portrayal usually shown in control engineering textbooks and it is useful in examining the theoretical system issues. However, it does not address in any way the system hardware and functional boundaries, and is therefore of little use in examining the issues of configuration control.

To begin to address these issues the existence of physical boundaries needs to be acknowledged. The control functions associated with the forward and feedback paths are likely to be hosted in electronic hardware or a 'black box', therefore the existence of a hardware boundary as shown in Figure 9.2 needs to be included. This leads to an awareness of the physical manifestation of the system, including issues such as:

- the size and dimensions of the 'black box' or controller (known as the form factor)
- weight (properly mass) and centre of gravity
- how the controller is mounted or secured and the environment, temperature, vibration, etc., that it experiences
- power consumption and cooling
- wiring interface and connections with the other system components.

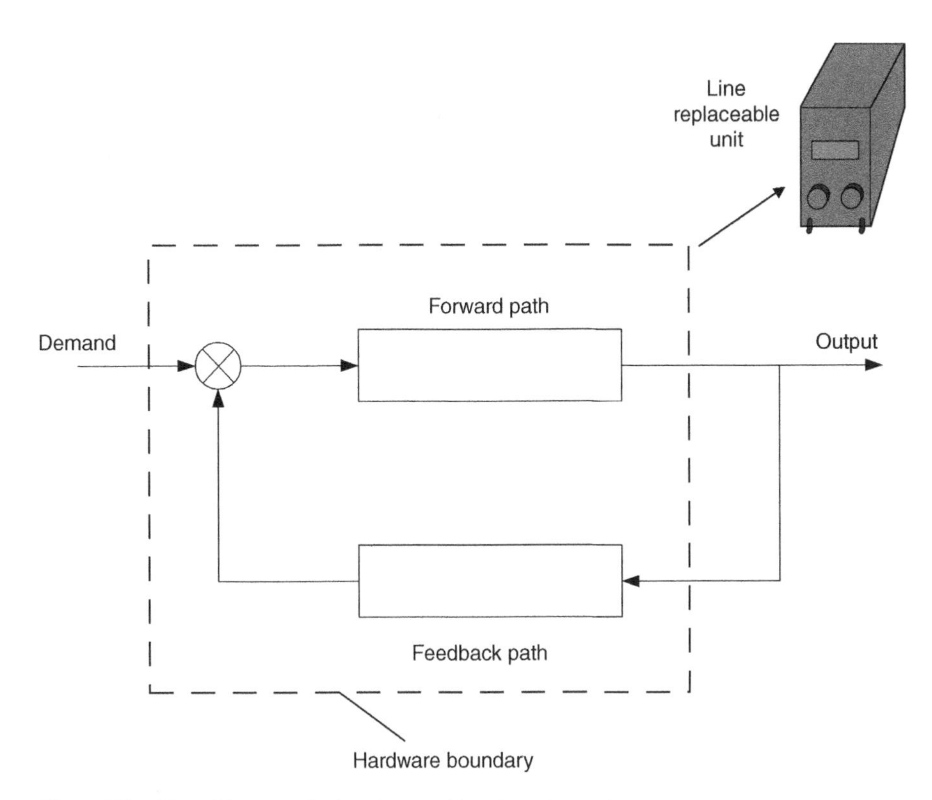

Figure 9.2 Closed-loop control system and hardware boundary.

It can therefore be seen that the consideration of the physical boundary of the system immediately raises a number of additional considerations that affect system design and its impact on other surrounding systems. The diagram in Figure 9.2 is not sufficiently detailed to examine these issues and further information needs to be considered. It is also clear that as a system is initially developed before entering service, or evolves into different forms during its useful lifetime, these physical considerations will become ever more important. Examples illustrating this theme are described in the following section.

9.4 Varying System Configurations

In this discussion the issues surrounding the system physical issues are examined as well as those that affect system modification or evolution. The example given describes a notional system as it evolves from an initial configuration, System A, to a modified system, System B, and finally to a third variant, System C. As well as examining the hardware controller or 'black box' change issues, other equally important issues are addressed as follows:

- *System wiring*: In many systems the electrical wiring which interconnects input devices (controllers), output devices (actuators or effectors), and sensors is the most difficult item to change. On an aircraft, ship or motor vehicle the wiring harnesses are installed during vehicle build and assembly, and may be inaccessible once the vehicle is completed. Consequently, the wiring can be extremely difficult to change following build and this can be a significant constraint.
- *System software*: Many control systems developed today use 'intelligence' to achieve and maintain adequate system performance. This may involve the use of computers, microprocessors or micro-controllers to host and execute the necessary sophisticated control laws. System performance alteration or modification may be effected by changing these control elements, but the manner in which these changes may be introduced or embodied in an existing system requires considerable care to be exercised.

9.4.1 System Configuration A

System configuration A is shown in Figure 9.3. This diagram has been expanded to show additional features which relate to the description of the system. The additional items identified on this drawing include the following:

- An input device or demand, in this case shown as a lever, by which an operator may introduce a demand into the system. This input device is assumed to be connected to the controller by means of two wires.
- An output device or actuator that provides the 'muscle' to move a control surface or other effector such that the system will satisfy the operator's demand. This actuator is assumed to be connected via four wires such that actuator demand and feedback signals may be exchanged.
- A controller, Unit 1, that closes the control loop for the system. This controller is assumed to have a time-variant control law $F_1(t)$ in the forward control path and a fixed gain K1 in the feedback path.

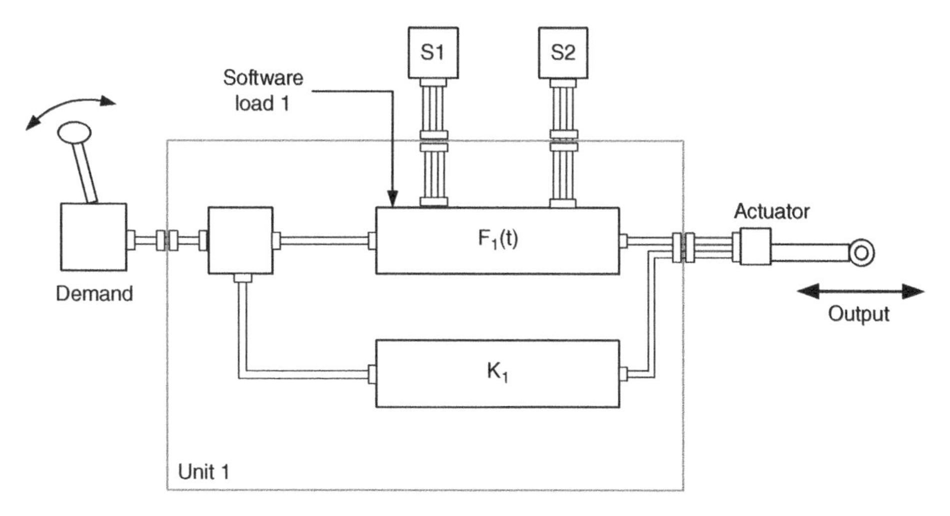

Figure 9.3 System configuration A.

- Sensors measuring parameters in the outside world or in other systems are used to modify the real-time performance of the forward path control function $F_1(t)$ and are shown as four-wire sensors S_1 and S_2.
- A system software load, termed software load 1, is assumed to be loaded into the computing device that is executing the software associated with the $F_1(t)$ function.

This diagram effectively encompasses and describes, in a shorthand fashion, all those elements necessary to achieve full and complete system operation. During the system development phase design effort will be concentrated on ensuring that all these items, hardware, software, and wiring, are developed together synchronism and tested to verify compliance with the original system specification.

9.4.2 System Configuration B

System configuration B is shown in Figure 9.4. If a system is developed or evolves from the original configuration, changes may be made to enhance performance, increase reliability, and offer other benefits and improvements. System B differs from the original configuration A as follows:

- The forward control path control laws have been modified from $F_1(t)$ to $F_2(t)$ though the sensors required, S1 and S2, are unchanged. The software load to accommodate the change in control laws is embodied in software load 2 and this software must be downloaded to implement the new control laws.
- As part of an improvement in the performance of the feedback path new time-variant control laws, $K_2(t)$ and a new sensor S3 have been added. The changes to the feedback path control laws are embodied within the software load 2.
- The effect of these modifications in hardware and wiring terms has been to add another four-wire electrical input to Unit 2 to accommodate the input from new sensor S3. This may require an additional connector, which changes the hardware configuration.

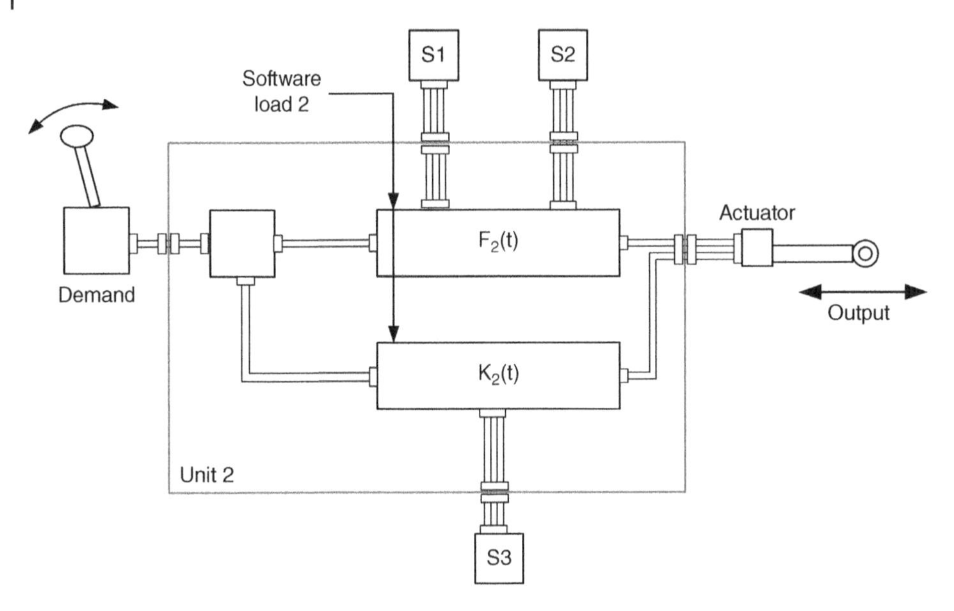

Figure 9.4 System configuration B.

- Systems A and B now have incompatible hardware as the latter needs an extra four wires to interface with sensor 3 and possibly an additional connector. In the present form Units 1 and 2, associated with systems A and B, respectively, would not be interchangeable. For a customer using both variants this would result in an additional support overhead as both variants would need to be maintained in terms of spares, technical manuals, technician training etc.

9.4.3 System Configuration C

A final system, system configuration C, shown in Figure 9.5, is considered. In this system the following is assumed:

- The control laws are modified to $F_3(t)$ in the forward path and $K_3(t)$ in the feedback path. The further modified control laws associated with this implementation are embodied in software load 3.
- The improved control laws associated with software load 3 have negated the continued use of sensor 2 in the forward path. This sensor is no longer required for correct system operation and may therefore be removed.
- Unit 3 can be seen to have the same hardware configuration as Unit 2 in system B with the exception that sensor 2 is not connected. Units 2 and 3 may be interchangeable provided the correct software load, load 2 in the case of Unit 2 and load 3 in the case of Unit 3, is also present. For an operator utilising both system configurations B and C this may result in a significant reduction in support overhead.

A summary of the three system configurations is given in Table 9.1.

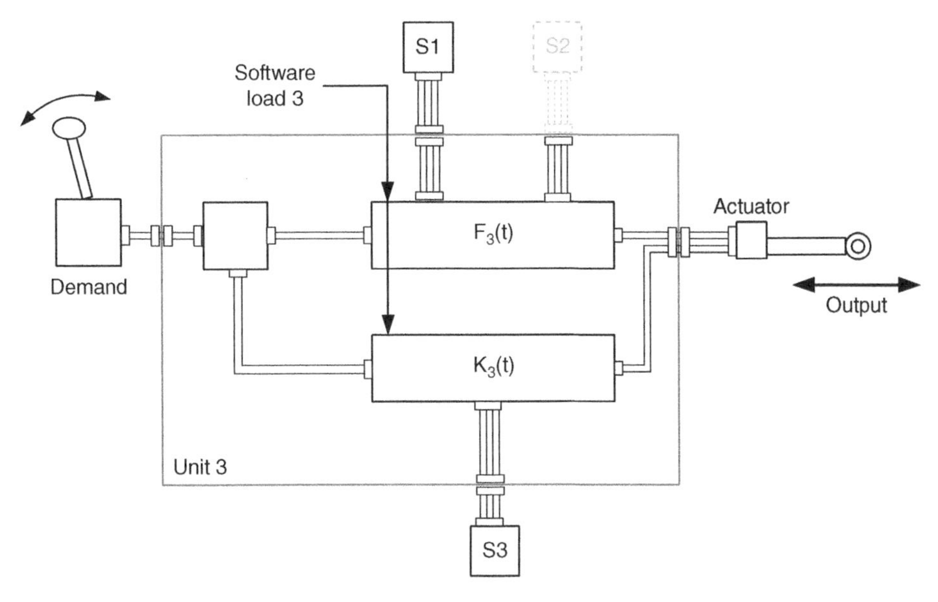

Figure 9.5 System configuration C.

Table 9.1 Comparison of system configurations.

	System configuration		
	System A	**System B**	**System C**
Sensors	2	3	2
Variables	$F_1(t)$	$F_2(t),K_2(t)$	$F_3(t),K_3(t)$
Wires	14	18	14 (18)

9.5 Forwards and Backwards Compatibility

The foregoing examples illustrate the concept of forwards and backwards compatibility, which is described in more detail below. The issue of forwards and backwards compatibility is an important consideration when a customer may be procuring similar systems that were developed in a different timescales and wishes to ensure compatibility between earlier and later versions, or vice versa.

9.5.1 Forwards Compatibility

Forwards compatibility describes the situation where an initial, perhaps early, system variant is evolved to a later system, as shown in Figure 9.6.

The customer will wish to assure himself that the later system, system Y, is compatible with the earlier system X. In this case he will wish to ensure that the form, fit, and function are compatible between the two systems.

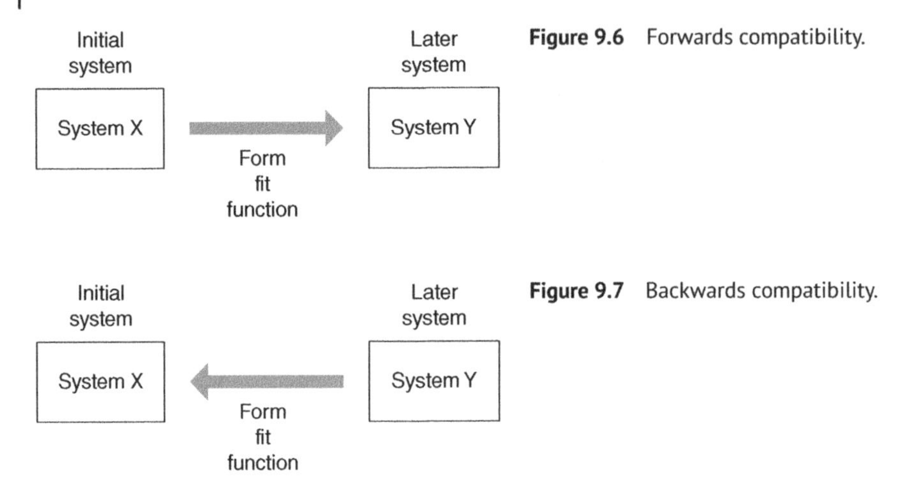

Figure 9.6 Forwards compatibility.

Figure 9.7 Backwards compatibility.

Form relates to the shape of the controller or 'black box'; clearly if Box X has to replace Box Y then it needs to be the same size and shape. Fit relates to other aspects of the physical attributes. Not only should the box have the same shape but other detailed parameters such as electrical connector types and orientation, physical mountings and tie-downs, and alignment of cooling vents and apertures should all be arranged to ensure that one box may physically replace another without modification.

Function relates to the performance characteristics of the unit. As has been described already, for many modern systems this may be affected by the software programme that is loaded and which encompasses the necessary control laws. However, some of the detailed performance characteristics of the processor or micro-processor executing the software may have a bearing on performance. Processor type, instruction set, clock speed, and memory configuration may all affect correct execution of the software as any PC or laptop owner who has transported an application from one computer to another will know.

The successful achievement of forwards compatibility must therefore ensure that the change from box X to box Y embraces all of these issues and that the change from one to the other is totally transparent to the system operator in terms of system performance.

9.5.2 Backwards Compatibility

The reverse situation, termed backwards compatibility, is shown in Figure 9.7.

Backwards compatibility relates to all of the issues associated with forward compatibility but in this case is associated with ensuring that a later system can satisfy the requirements of an earlier implementation. This often harder to achieve in practice than forwards compatibility.

9.6 Factors Affecting Compatibility

For systems of the type installed in vehicles, cars, ships, aircraft, etc., the factors that affect compatibility are shown in Figure 9.8. These may be related to three distinct areas of system design and implementation:

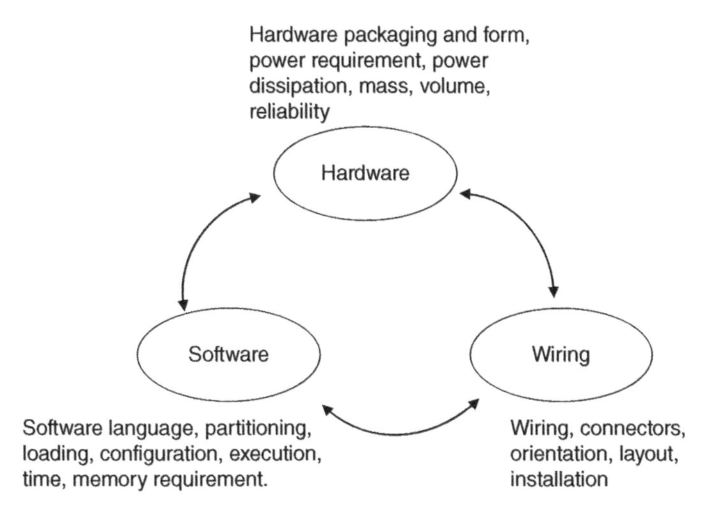

Figure 9.8 Factors affecting compatibility in a vehicle.

- hardware
- software
- wiring.

All three must be compatible to provide a viable working system that meets the performance objectives.

9.6.1 Hardware

The aspects relating to hardware have already been outlined. They are:

- physical form and fit
- physical orientation and installation
- weight and centre of gravity
- suitability for the anticipated environment: temperature, vibration, electro-magnetic interference (EMI), etc.
- power dissipation and the need for cooling
- reliability.

9.6.2 Software

The considerations relating to software execution are also important:

- type of processor
- instruction set
- software language
- clock speed
- memory configuration.

It can be seen that many of the considerations that can impact software execution are associated with the detailed design of the controller rather than some of the higher level physical form and fit issues. Due to these interactions function (performance) may be subtly affected if a processor within a controller is changed, perhaps due to component obsolescence, whilst form and fit remain unaltered.

9.6.3 Wiring

The issues of wiring are also important. Major issues that need to be addressed include:

- interconnecting wiring to control levers, sensors, actuators, and effectors
- connector types and orientation
- voltage drop
- wiring harnesses – length of cables, routing, and installation
- wiring protection against electrical faults
- harnessing heat dissipation
- screening, earthing, bonding, and susceptibility to EMI and external high-intensity radio frequency (RF) fields and lightning strikes.

The foregoing list is not exhaustive but is indicative of many of the issues that need to be harmonised to ensure the compatibility of a system and the ability to perform to specification within the intended environment.

9.7 System Evolution

So far, section 9.5 has described how the designer may need to consider forwards and backwards compatibility to ensure inter-operability between early and later implementations of the same system. Section 9.6 has illustrated how the compatibility of the triad of hardware, software, and wiring that represents the total system must be maintained for each working configuration.

In reality both of these issues must be addressed to ensure that working systems are proven for each stage of a system or product evolution, as illustrated in Figure 9.9. This shows how each of the key areas of hardware, software, and wiring evolves in time from left to right. At each stage or system configuration these elements should also be compatible to ensure that specified system performance is maintained.

For large systems the development and proving of just one system configuration may take several years. In the case of a modern fighter aircraft it is not unusual for the development programme to extend more than 10 years from project go-ahead to entry into service. The production cycle may also last anywhere between 10 and 20 years for a large programme and different production configurations will apply throughout various stages of the production phase. Finally, the in-service phase may extend over tens of years, during which time further modifications or capability upgrades may be embodied. Such a product may be regarded as a 'system of systems' and many of the systems, sub-systems, and components that contribute to the whole will have their own compatibility issues.

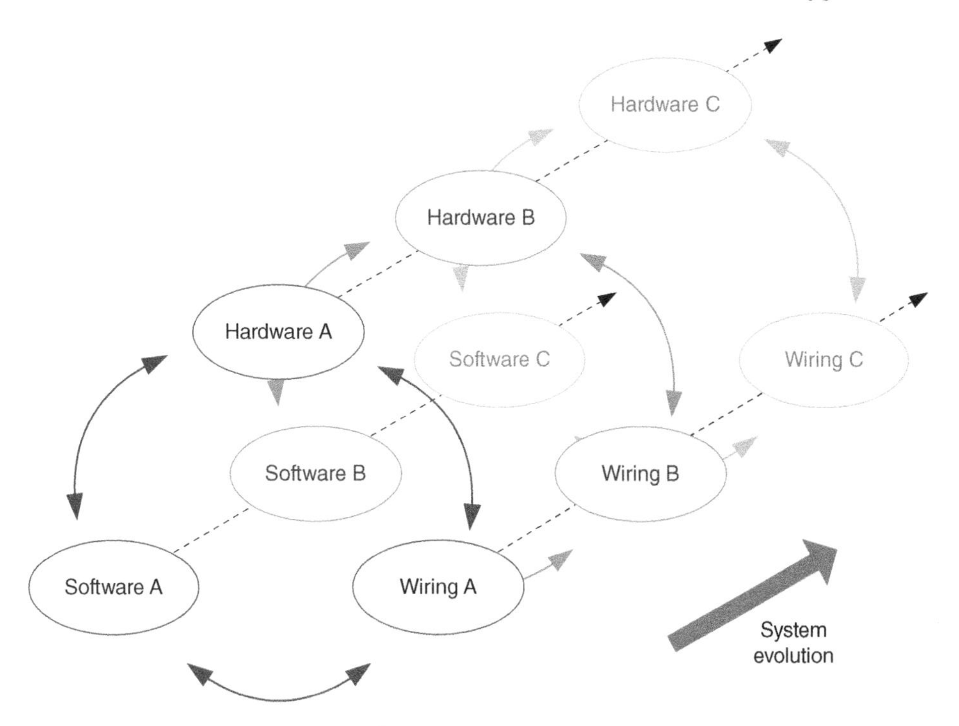

Figure 9.9 Long-term system evolution.

In this section an attempt has been made to highlight some of the practical issues that must be addressed in attaining the necessary configuration control of systems as they move through the development and in-service phases. Whilst this is a simplified overview it can be seen that many issues must be considered as a system evolves through different implementations during the course of time.

9.8 Configuration Control

The generic anatomy of any microprocessor or microcontroller used within an aircraft control system is shown in Figure 9.10.

The central processing unit (CPU) contains an arithmetic logic unit (ALU) and a control element which sequences the application software instructions. In its simplest form at least two areas of memory will be provided:

- programme memory or read-only memory (ROM), which contains the executable software
- data memory or random access memory (RAM), which contains the variable data which the unit needs to execute the programme.

In recent designs it is likely that some form of non-volatile memory will be included to store key system data such as fault history, built-in test (BIT) results, etc. Input and output devices will be provided to interface the machine to its peripherals. At a higher level the processor will be contained within an avionics line replaceable unit (LRU), which forms a

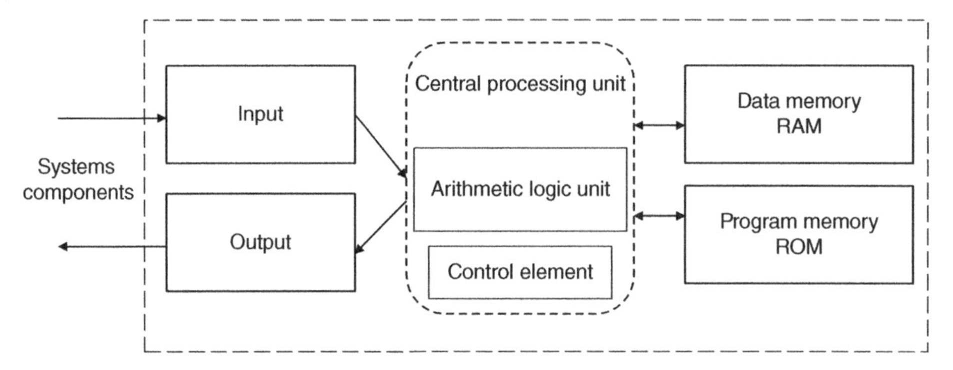

- Input/output: reads data into the machine and outputs results to system components.
- Control: sequences the series of instructions which constitute the computer program (application) held in the program memory.
- Arithmetic logic unit: executes arithmetic and logical operations
- Data memory: stores the intermediate and final results of the operations

Figure 9.10 General architecture of a microcontroller.

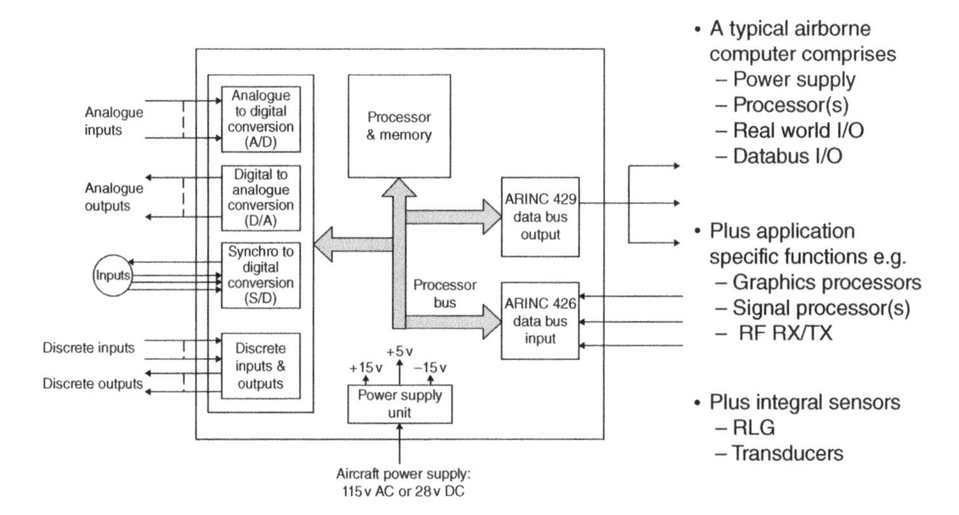

Figure 9.11 Typical LRU architecture.

convenient method to be able remove units from the aircraft when they have failed. These units may be rack mounted in the equipment compartment of the bay or they may sometimes be mounted directly to the aircraft structure as described in the electrical wiring section. Functionally, the unit contains the following elements, as shown in Figure 9.11:

- A power supply or power supplies to convert aircraft AC or DC power to provide stabilised power for the electronics: ±15 V, +5 V, +3.3 V.
- Input/output modules to interface with the aircraft sensor: analogue inputs and outputs, discrete inputs and outputs, and other specialised signals.

- The unit will also have data bus interfaces to enable the interchange of data with other aircraft systems. ARINC 429 digital data bus interfaces are shown in the figure but more recent aircraft are likely to have ARINC 664 buses at the aircraft level and CANbus within the system, as already described in the ICD section.

For a system to operate as a functional entity all the hardware and software elements need to be working in harmony and unison, transferring valid data sets around the system, such as the operational software programme.

The correct handling of data is performed in different ways and perhaps the easiest way to describe this is to give a few examples:

- For the ARINC 429 data bus interface shown, the data set is fixed by the appropriate aircraft equipment ARINC specification. All the expected data scope, scaling, accuracy, and refresh rates are specified such that all the equipment designed to meet that specification will be identical. This has the advantage that a specific item of aircraft equipment, for example a VHF transmitter receiver (as defined by ARINC 566), produced by different equipment suppliers will be interchangeable. The disadvantage is that it removes flexibility from the system designer in terms of how to configure the system data.
- In the example of the MIL-STD-1553B data buses that are extensively used on military aircraft, the system designer is given the freedom to specify all the data set himself, provided that the data bus operates correctly in accordance with the 1553B protocol. In this case the designer assumes the burden of specifying and maintaining a coherent data set. This would include the specification of which remote terminal (RT) addresses and sub-addresses are utilised, and defining the procedure that allows the processor and data bus elements within the LRU to communicate. Finally, the system designer has responsibility for designing, coding, and testing the bus controller software, which acts as the data scheduler for the entire system.
- In the ARINC 664 example, typical of the aircraft level data buses used on aircraft like the Airbus A380 and the B787, the integration task is more complex and is worthy of examination in more detail. In particular, the A380 introduced a modular avionic architecture using newly common core modules interfacing to the avionics full duplex switched (AFDX) internet (ARINC 664) whilst still using legacy equipment with somewhat dated, but nevertheless effective, ARINC 429 data buses.

9.8.1 Airbus A380 Example

The A380 was the first example of the introduction of an aircraft-wide set of common processor elements – common processor input/output modules (CPIOMs) – tied together using a version of ARINC 664. In Airbus terminology this is called AFDX internet and it is a twin copper wire bus transmission system using COTS 100BaseT technology (100 Mb/s data passed over twisted wire pairs), technology originally designed for switched packet digital telephone exchanges. The data buses are dual redundant in implementation and connect the core integrated module architecture (IMA) architecture together via a series of AFDX switches, as shown in Figure 9.12.

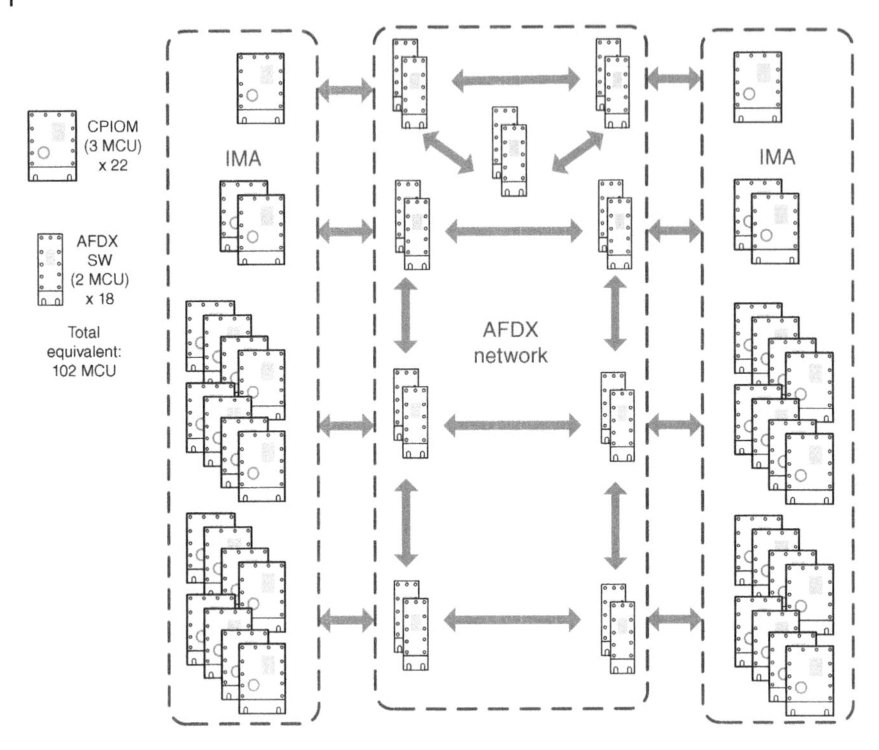

Figure 9.12 Airbus AFDX/IMA architecture.

- Within the central spine of the aircraft is the AFDX switching network – a total of 18 × 2 modular concept unit (MCU)-sized switches connecting key areas of the aircraft together in a dual redundant architecture.
- The IMA consists of a total of 22 × 3 MCU CPIOMs which provide the processing core of the aircraft system processing. In reality, due to the differing input/output requirements, there are a total of seven different CPIOM types termed CPIOM A through to CPIOM G. As an example, CPIOM G is specifically designed for the aircraft landing gear whereas CPIOM F is tailored to the fuel system requirements. Despite there being seven different types of CPIOM the processor within each is common and all types are supported by a common software development environment and tool set. An enormous degree of commonality is achieved compared to previous architectures.

This arrangement worked well for the new 'core' systems, but Airbus were still left with the problem of how to interface existing or legacy systems from previous Airbus family models, typically later A320 models and A330/A340. Much of this equipment was interfaced using ARINC 429 data buses and was perfectly functionally adequate for the A380, yet to embed these functions within the new IMA core would have been a risky and expensive undertaking. The solution that was adopted was to leave these legacy systems untouched, essentially peripheral to the central core and relying upon the existing ARINC 429 data buses to integrate the elements together. The key attributes of this solution are described in the following text.

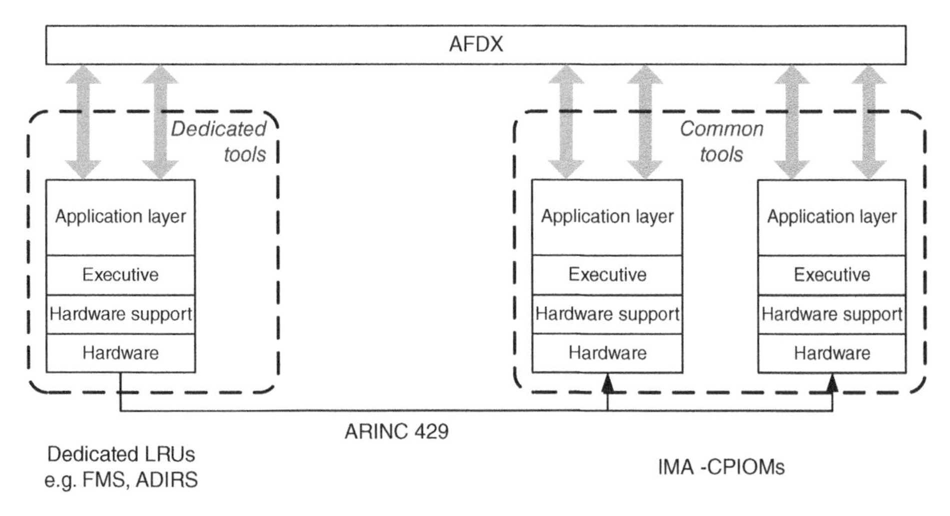

Figure 9.13 AFDX equipment example.

In Figure 9.13 typical legacy systems such as the flight management system, air data and inertial reference system, etc. are shown on the left. These systems do not have the capability to interface directly with the aircraft level AFDX buses. They do have existing ARINC 429 data buses inherited from former architectures that permit them to communicate with the CPIOMs. The development tools associated with the hardware/software configuration of these units would have been fixed by the development strictures at the time of the earlier Airbus models.

The IMA common core elements are shown on the right. These elements have the capability of interfacing directly with the AFDX switches and therefore the aircraft-level 100 Mbit/s data buses. The hardware/software combination was developed using common tools and state-of-the-art development methodologies that are outside the scope of this book. Suffice to say that it was developed using 'three-layer-stack' technologies which serve the dual purpose of isolating the hardware implementation from the effect of obsolescence whilst also providing partitioned software with a high degree of portability between different applications.

The back-door ARINC 429 data links enable the dedicated legacy LRUs to communicate with the IMA/CPIOM core.

The final issue to resolve is how the data communications are defined and controlled for the AFDX network. The answer is in the use of configuration and commutation tables, as illustrated in Figure 9.14:

- Configuration tables reside within the equipment and determine the data input/output and the format of that data.
- The commutation tables reside within the AFDX switches and control the movement of data around the network between the LRUs and the CPIOMs.

Only when a coherent set of configuration tables and commutation tables are embodied will the system data flows be correct. Correct configuration control is vital to ensure that this is achieved.

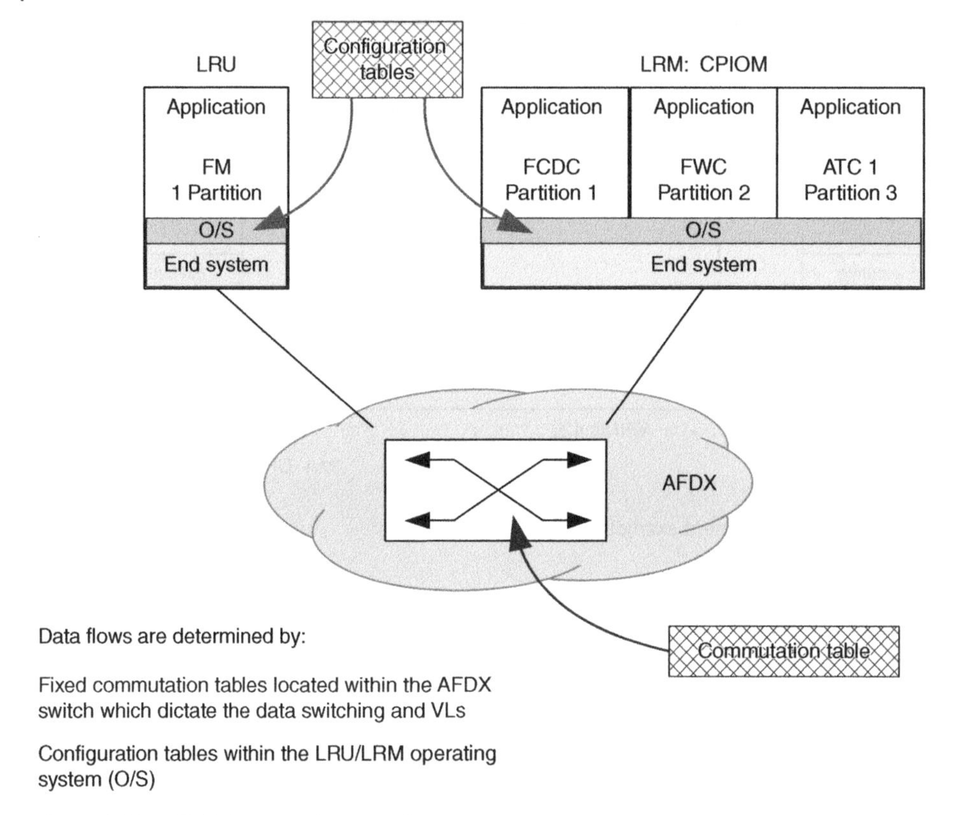

Figure 9.14 AFDX data transfer control.

9.9 Interface Control

The control of interfaces between system components owned by various stakeholders is essential. Rigorous mechanisms need to be in place to define the interfaces and to record ownership and agreement between owners.

9.9.1 Interface Control Document

The increasing use of modern COTS technology offers greater system functionality and performance but with an accompanying increase in complexity. All of the aircraft system interfaces have to be defined and bounded. Every aircraft system interacts with others so the aircraft is truly a system-of-systems. This is illustrated in Figure 9.15.

To define and control the system interfaces an ICD is used which defines all of the electrical interfaces. To illustrate this point a notional system is portrayed consisting of four units. The example shown in Figure 9.15 could be typical of a fuel gauging and management system on a large transport aircraft. Units A and B represent fuel gauging

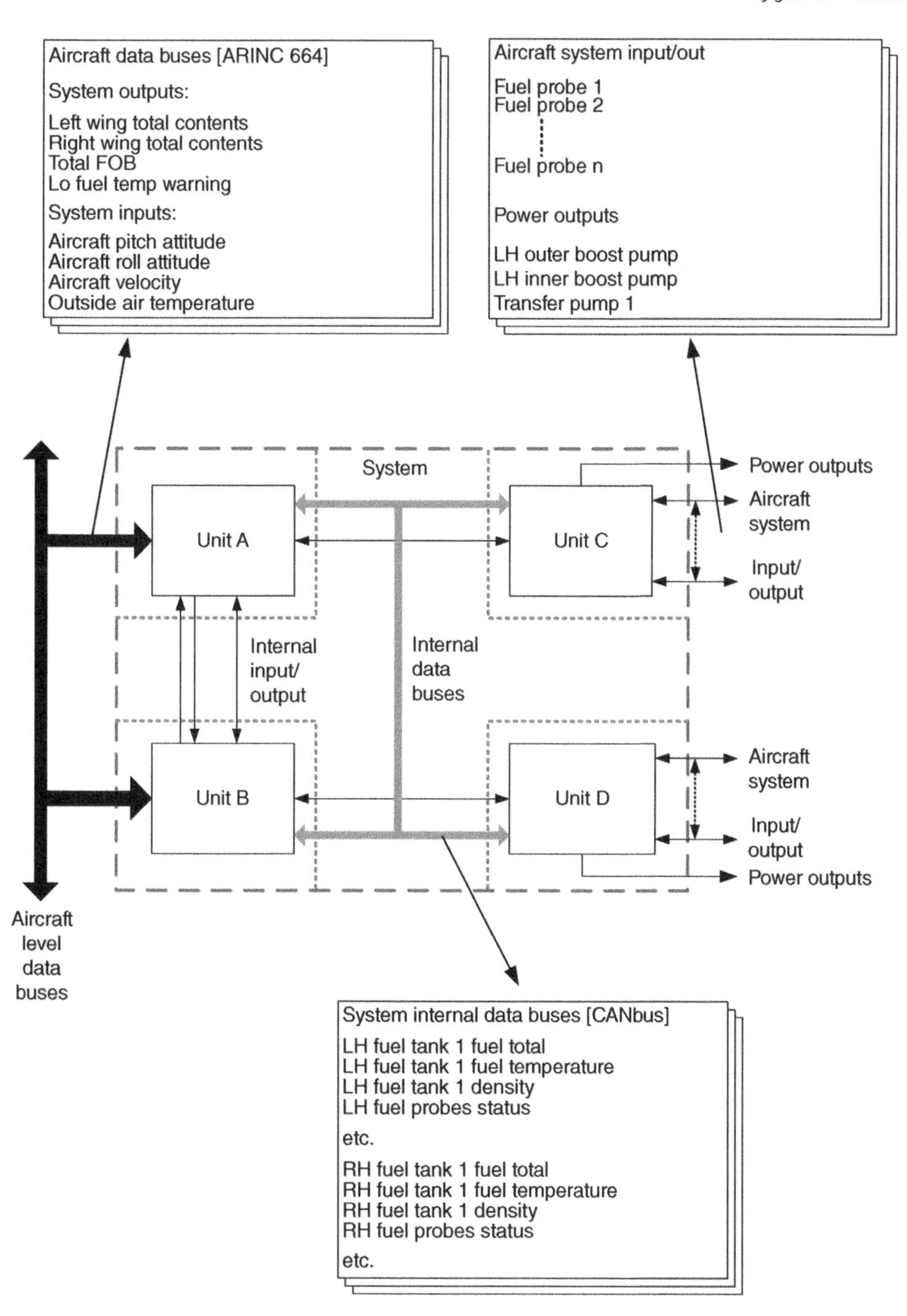

Figure 9.15 Aircraft ICD example.

and management computers whereas units C and D represent remote data concentrators interfacing directly with the components in the aircraft fuel tanks such as fuel measurement probes, temperature and densitometer sensors, and fuel pumps and valves. In the simplified system example chosen there are four major types of system interfaces:

- *Aircraft level data buses*: For an aircraft such as the Boeing 787 or the Airbus A380/A350 the aircraft-level data buses will be implemented in a form of ARINC 664 data buses. These aircraft-level data buses will be transmitting data typically at 100 Mbit/s using COTS technology which originated within the telecommunications industry using either conventional twisted wire pairs or fibre-optic technology.
- *System internal data buses*: Within the system digital data will need to be exchanged between units at a lower bandwidth. A COTS data bus called CANbus, developed by Bosch and originally intended for automobile automatic braking systems (ABS), is commonly used, albeit in a deterministic and ruggedised form. Typical data rates are of the order of 1 Mbit/s.
- *Internal system input/output signals between the system units*: Hardwired signals between units to compare data, synchronise the operation of the system computers, and establish which computer/channel is in control.
- *System internal interfaces in which the remote data concentrators interface with the components within the aircraft fuel tanks*: Key issues include the provision of electrically intrinsically safe interfaces where the power allowed into the tank to feed a fuel probe is constrained at miniscule energy levels to ensure that the system is inherently safe.

9.9.2 Aircraft-level Data Bus Data

Aircraft-level data will include top-level aircraft data which is useful to the flight crew in terms of operating the aircraft. In many cases this will be data needed for other aircraft systems or that needs to be displayed to the flight crew. In the example in Figure 9.15 typical data presented by the system could include the total fuel on-board (FOB) or the contents of individual fuel tanks. Warning and advisory data would also be provided.

Inputs to the fuel system would include aircraft attitude information in order that the fuel contents may be accurately calculated and corrected for aircraft velocity, and outside air temperature (OAT), which is of particular interest in understanding cold fuel issues during prolonged cold soak at altitude.

9.9.3 System Internal Data Bus Data

Many systems use an internal system data bus to exchange system-specific data. In the example shown in Figure 9.15 the fuel probe and other sensor data are exchanged. System BIT and other sensor health-related data will be included. The system will also have in-built monitors to ensure that hazardous events do not occur and ensure that the flight crew are kept fully informed of any failures and advised of what remedial action to take.

9.9.4 Internal System Input/Output Data

There will be a number of hardwired interfaces between the system units that are not appropriate to be passed over the internal data buses.

9.9.5 Fuel Component Interfaces

The ICD defines and controls all of the parameters defining:

- electrical characteristics
- wire sizes and types
- bonding and screening
- termination and matching
- data resolution and accuracy
- data rates and refresh rates
- power levels
- EMI categorisation.

9.10 Control of Day-to-Day Documents

A lot of day-to-day communication in the office takes place by electronic media, as discussed in Chapter 8. Discipline is essential to ensure that this communication is recorded and controlled in some way. It is all too easy to send a document out for review and incorporate the comments without recording that the document has changed. This is made easier by the use of multiple devices for document compilations, aided by easily available memory drives. For example, a report can be compiled using a company desktop machine, a portable laptop, a tablet or a smart phone. A minimum set of rules needs to be established and used. These rules should include the following:

1) A unique number shall be registered and clearly incorporated in the document title page and preferably the header.
2) A title shall be included on the title page.
3) The names of the author and preferably a check and authorisation name shall be included.
4) The date of first issue of the document shall be included.
5) Each change in status of the issue shall be recorded as, for example, Revision (Rev 1, Rev. 2, etc.), new issue (Iss1, Iss 2, etc.) and draft or full issue recorded.
6) This change of status shall be included in the document title page and header and in the file name and covering email.
7) Preferably a page should be included that records each stage of the configuration record.

In Figure 9.16 a report is written and first issued as Draft A under a Word file name **Title IssueA.doc**. It is issued to a number of people to obtain their comments. The comments are received and the report is updated and re-issued as **Draft A Rev 1**. The file name is changed to **Title Iss A Draft rev1.doc.** The report is issued again to allow people to check

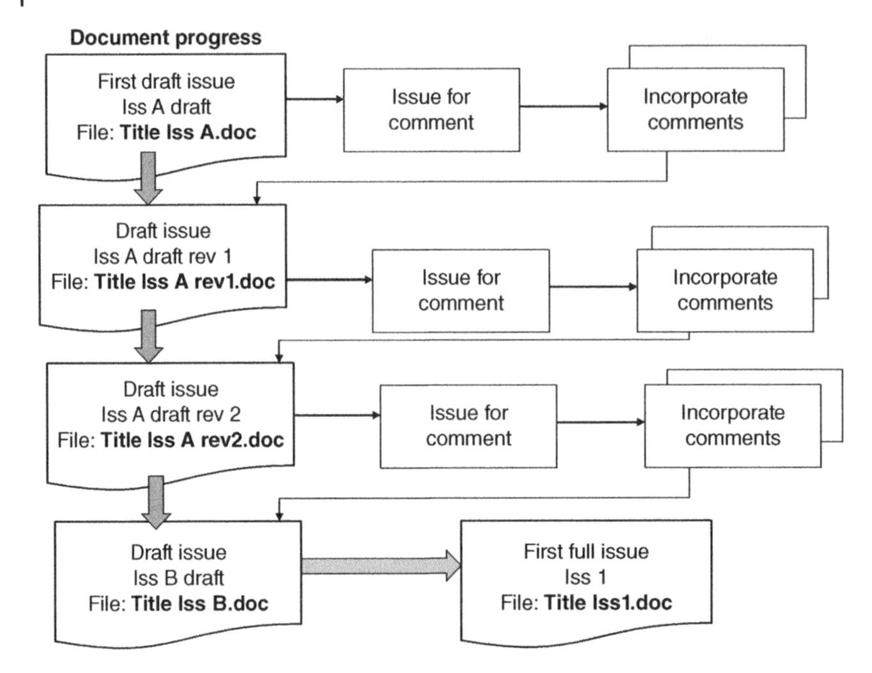

Figure 9.16 Progress of a document from first draft to first full issue.

that their comments have been interpreted correctly and to comment further if necessary. If more comments are received the report is updated again and issued as **Title Issue Draft Rev 2** and the file is renamed **Title Iss A rev2.doc.** After further comments or agreements the report is raised to **Iss B Draft** and filed is renamed **Title Iss B.doc.** After formal review or project approval the report is deemed suitable for use and is released formally as **First Full Issue Iss 1** under the file name of **Title Iss1.doc.**

These rules are not simply a matter of adherence to a process for the sake of it, but a measure to ensure that information used in the project design process has been carefully compiled, rigorously checked, stored in the correct place, and is easily identifiable as being trusted data.

Exercise

1 Consider the scenario where a drawing or document has been issued and circulated for comment. Comments have been received and incorporated, and the document has been released for circulation but has not had its issue updated. List some implications of this as work proceeds and the document becomes ready for updating again. How do think this situation has affected the recipients?

10

Aircraft System Examples

10.1 Introduction

The objective of this chapter is to provide an overview of a specific system to allow the reader to better understand and assimilate the content of the foregoing chapters by placing system issues within an overall context.

A useful example to illustrate the interaction of several systems is to examine the interrelationship of the systems on a modern civil aircraft. A civil aircraft is an interesting and perhaps almost unique subject as many of the systems are required to meet high levels of integrity in order that the aircraft can successfully complete a flight. At the same time these same systems have to perform safely and reliably in adverse environments of low or high temperature and often under conditions of high vibration. The need for the aircraft to meet performance goals places additional constraints on weight and volume, therefore these issues have to balanced and satisfied so that the aircraft can perform its mission both safely and economically.

Three major systems that contribute to aircraft operation are shown in Figure 10.1. The aircraft structure comprising the wings fuselage and empennage provides the lift and control surfaces, and the passenger cabin. The aircraft systems comprising propulsion system, flight control, fuel, hydraulic and environmental control systems provide the means of flying the aircraft. The avionics systems represent the 'brains' of the aircraft, providing navigation, communications, autopilot, and display functions. In this book the aircraft systems and avionics systems will be discussed. Companion volumes will enable the reader to research these topics in more detail to aid their understanding of the principles developed within this chapter (Moir and Seabridge 2008, 2013).

10.2 Design Considerations

The design of an aircraft follows the approach shown in Figure 10.2.

The first items to be defined are the mission requirements: what is the aircraft required to achieve in terms of payload, speed and range, and operating cost? This specifies the role of the aircraft. Clearly a long-range large passenger aircraft designed to fly from London to

Design and Development of Aircraft Systems, Third Edition. Allan Seabridge and Ian Moir.

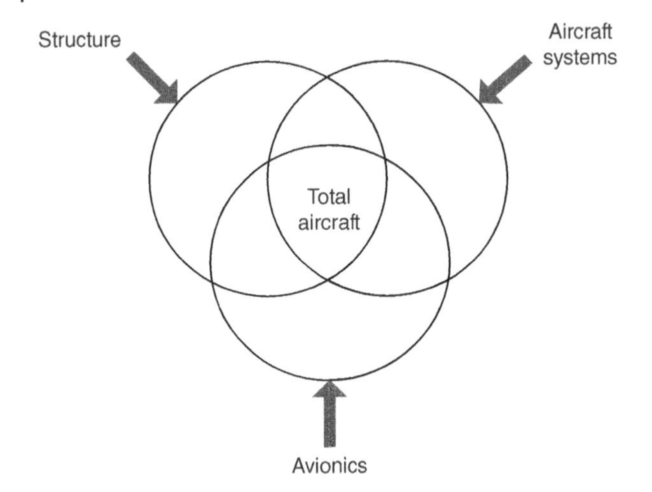

Figure 10.1 Major aircraft systems overview.

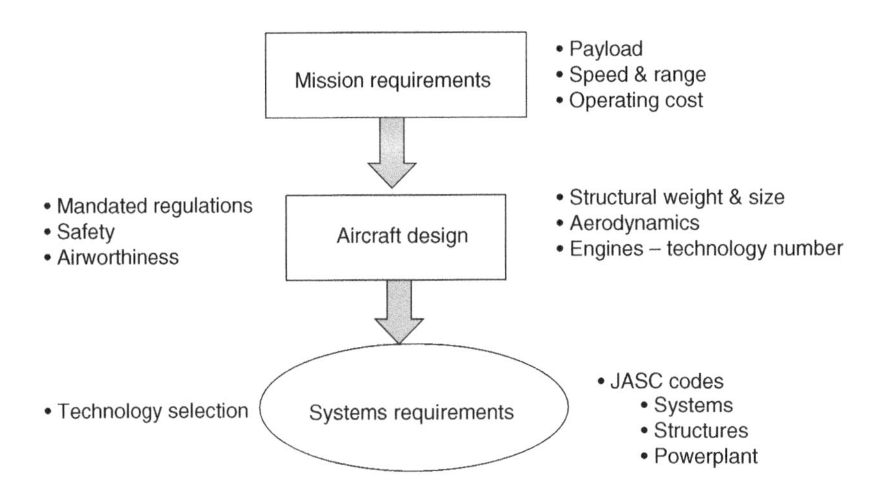

Figure 10.2 Top-level design process.

Chicago will have different characteristics from a short-range feeder liner that will be used to fly from Chicago to Grand Rapids.

The mission requirements will dictate the overall aircraft design, determining structural weight and size as well as the aerodynamic characteristics. The number and type of engines will also be defined. These physical parameters will be also dictated by the mandated regulations that apply to the aircraft design and by safety and airworthiness considerations.

Finally, the aircraft design will flow down into the detailed aircraft system requirements: the technology used and the type and capability of aircraft systems. In the civil field the functional aircraft systems are defined by FAA/JASC (2002) codes that categorise each system type. In this system categorisation 2400 always addresses electrical power, 2700 – flight controls, 2900 – hydraulic power and so on.

10.3 Safety and Economic Considerations

A key requisite is to balance the safety and economic requirements, as shown in Figure 10.3.

Naturally, safety has to be a paramount consideration in any form of transportation and nowhere is this more true than in air travel. Several factors influence safety:

- System function: the task the system has to perform. In a fuel system this may involve moving fuel around the aircraft to maintain the centre of gravity in the correct position to reduce trim drag and increase aircraft range.
- Performance: this relates to the true system performance; for certain types of manoeuvre or procedure the aircraft may need to be flown with a higher level of navigation accuracy than for others.
- Integrity: the inherent nature of the systems architecture that ensures the system is sufficiently robust to continue operating safely following one or more system failures; ensuring that the passengers are safely delivered to their destination.
- Reliability: the inherent ability of a system or component to continue to operate correctly, thereby ensuring that the system function and design level of integrity are maintained.
- Dispatch availability: this relates to the ability to dispatch the aircraft on a flight with known system defects. The aircraft still has to be able to meet the design levels of performance and integrity even given the fact that some defects may be present at the beginning and throughout the flight. The operation of aircraft by an airline or charter operator must be carried out economically or else the company will fail financially and be unable to continue to provide a service. The economics of operation are affected by the following factors:
  - Function, performance, reliability and dispatch availability can all affect the economy of operation as they can all affect system cost. Only integrity cannot be compromised by economy for obvious reasons.
  - Maintainability: this relates to the ease of maintaining the system in a fit state to be able to deliver the correct function. It can include the ease of component replacement, repair, and test.

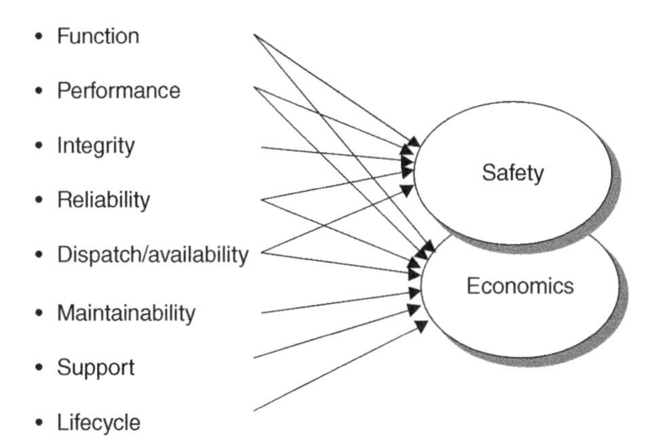

Figure 10.3 Safety and economic considerations.

- Support: this is the ability to provide an infrastructure of spare components, documentation, training, and expertise to support the system during the period of operation of the aircraft.
- Lifecycle: this relates to the cradle to grave concept of defining, designing, developing, manufacturing, and supporting the system at all stages until it reaches the end of its useful life.

10.4 Failure Severity Categorisation

In the aerospace industry levels of failure severity are universally specified in an unambiguous way, as defined in Table 10.1.

There are four main categories of failure severity. The most serious is a catastrophic failure which would result in the loss of the aircraft and passengers. The probability of such an event occurring is specified as extremely improbable and in analytical or qualitative terms it is directed that a catastrophic failure should occur less than 1×10^{-9} per flight hour. That is less than once per 1000 million flying hours. Other less significant failures are hazardous, major and minor; in each case the level of risk is reduced and the probability of the event occurring correspondingly increased, therefore a minor failure – perhaps the failure of a navigation light – can be expected to be reasonably probable, with such an event occurring less than 1×10^{-3} per flight hour or less than once every 1000 flying hours.

During the initial aircraft design all of those failures that can cause the various levels of failure severity are identified and used to modify the aircraft systems design accordingly, therefore long before an aircraft is built all of these conditions are identified, appropriate design steps taken, and the quality of design assured. This process helps to define the system architecture, the number of control and power channels, level of redundancy, etc. It also specifies a design assurance level according to what the effects of a failure might be.

10.5 Design Assurance Levels

Table 10.2 lists the design assurance levels for the US Radio Technical Committee Association (RTCA) documents DO-178B, which is used to specify software design procedures and DO-254, which specifies the design procedures for hardware. Other relevant

Table 10.1 Definition of failure severity.

Failure severity	Probability	Analytical
Catastrophic	Extremely improbable	Less than 1×10^{-9} per flight hour
Hazardous	Extremely remote	Less than 1×10^{-7} per flight hour
Major	Remote	Less than 1×10^{-5} per flight hour
Minor	Reasonably improbable	Less than 1×10^{-3} per flight hour

Table 10.2 Design assurance levels.

Design assurance level (DO-178B Software/ DO-254 Hardware)	Definition
A	Design whose anomalous behaviour as shown by the SSA would cause or contribute to a failure of a system function resulting in a *catastrophic* failure condition for the aircraft
B	Design whose anomalous behaviour as shown by the SSA would cause or contribute to a failure of a system function resulting in a *hazardous* failure condition for the aircraft
C	Design whose anomalous behaviour as shown by the SSA would cause or contribute to a failure of a system function resulting in a *major* failure condition for the aircraft
D	Design whose anomalous behaviour as shown by the SSA would cause or contribute to a failure of a system function resulting in a *minor* failure condition for the aircraft
E	Design whose anomalous behaviour as shown by the SSA would cause or contribute to a failure of a system function resulting in a *no-effect* failure condition for the aircraft

SSA, system safety analysis.

documents are ED 12 and ED 80, which are the European Organisation for Civil Aviation Equipment (EUROCAE) equivalents to DO-178B and DO-254, respectively.

These design assurance levels are categorised from A through to E according to the definition in the right-hand column. It can be seen that the failure conditions for catastrophic, hazardous, etc. failures mirror the failure severity conditions listed in Table 10.1. Therefore the highest design assurance level – level A – relates to the catastrophic failure severity that in turn should occur less than 1×10^{-9} per flight hour. This logic continues for design assurance levels B through to D and it can be seen than an additional level E applies when a failure has no effect.

In this way the system design is scrutinised to ensure that every system meets the necessary design goals in order to meet the necessary levels of integrity. Furthermore, the process specifies increasingly more stringent levels of design assurance as the effects and the impact of system failure becomes more severe for the aircraft and passengers.

The documents that support this design process have been evolved by the most experienced designers across the aerospace industry, initially as industry best practice, then adopted and mandated for all design processes. In this way the industry has set uniformly high standards that everyone in the global aerospace community has to invoke during systems design.

10.6 Redundancy

The complex nature of modern air transport aircraft systems means that special design rules need to be employed. These methodologies are described elsewhere in this volume and are a crucial part of the development process. Many of the systems that are vital to flying the aircraft are required to preserve the safety and well-being of the flight crew and passengers. In the parlance of the aerospace community these are flight critical systems.

During the engineering design phase the system architect devises system concepts employing various levels of redundancy that provide the necessary levels of system performance, availability, and safety. These architectures are carefully crafted and reviewed using an industry-wide series of methodologies, tools, and techniques that allow the provisional system design to be evaluated and to ensure that it meets the necessary requirements.

Within the aerospace community these tools provide a range of possible architectures that may be invoked to meet the system design requirements. The provision of redundant channels of control bears a different burden. Additional channels cost more due to the additional provision of hardware and are also less reliable as there are now more channels to fail. Modern technology – particularly in terms of electronics/avionics – is a help because as time goes on avionics technology becomes more reliable and more rugged in terms of use in an aerospace concept. Reduced cost and development risk, sadly, does not reduce commensurately.

In this community the diversity of redundancy varies between a single channel implementation (simplex or times 1) through to a fourfold channel implementation (quadruplex or times 4). There is an obvious practical limit as to what levels of redundancy may be sensibly employed. In practice, quadruplex implementations have only been used in specific military applications.

The following description broadly outlines the main candidate architectures and implementations, although in practice there may be considerable subtleties between specific implementations.

10.6.1 Architecture Options

The main architectures to be outlined include:

- simplex (illustrated in Figure 10.4)
- duplex (illustrated in Figure 10.4)
- dual/dual (illustrated in Figure 10.5)
- triplex (illustrated in Figure 10.6)
- quadruplex (illustrated in Figure 10.6).

Examples of each of these architectures will be provided and the implication of various failures examined and explored. The choice of which architecture to select is subject to a rigorous examination using the design tools already referred to. These techniques analyse risk per flight hour and the level of redundancy is chosen accordingly.

10.6.1.1 Simplex Architecture

Many control systems within an aircraft will be relatively simple and their loss will not be of great consequence. Such systems are likely to be implemented in a simplex form in terms of sensors and control, and if there is a failure then the control function will be lost. Failures will be detected by built-in test (BIT) functions that may be continuous or interruptive in nature. BIT is not perfect, however, and conventional wisdom has it that the effectiveness

Figure 10.4 Simplex and duplex architectures.

Simplex

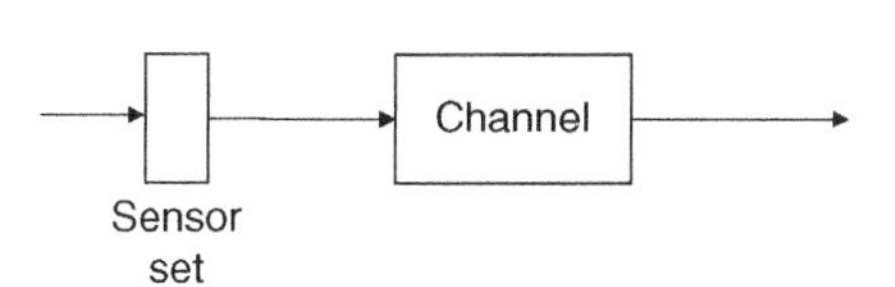

Dual

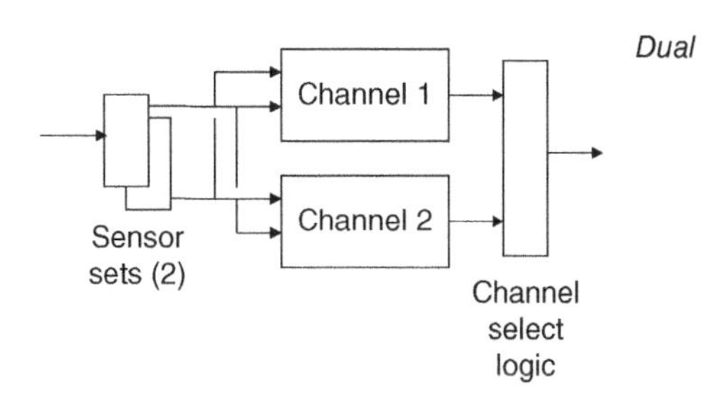

Figure 10.5 Dual duplex architecture.

Dual/dual

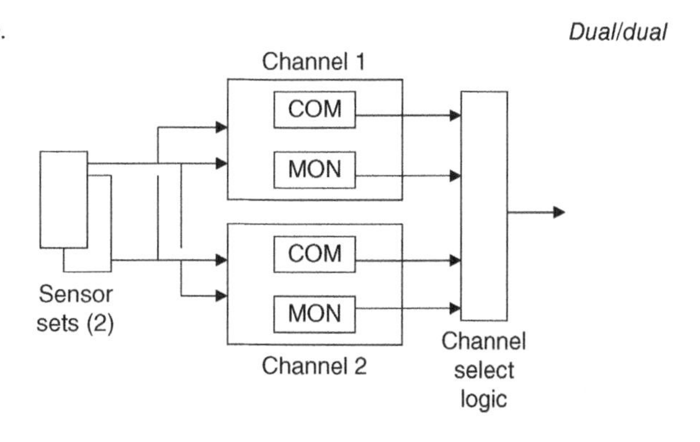

Triplex

Quadruplex

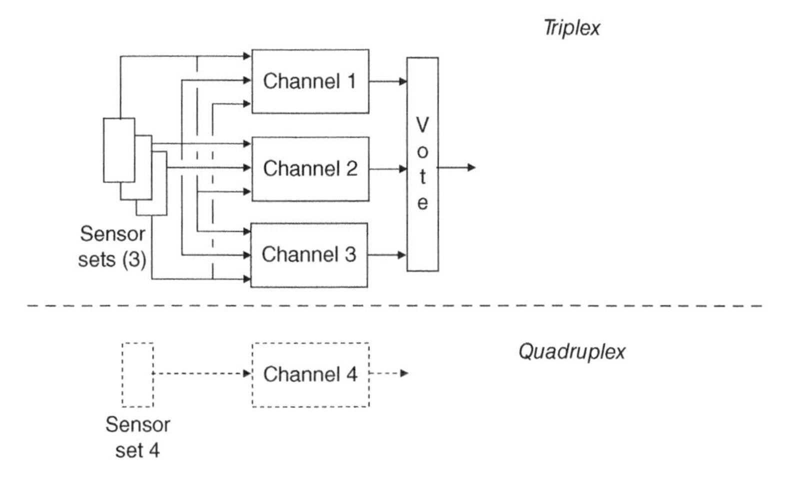

Figure 10.6 Triplex and quadruplex architectures.

of BIT is around 90–95%. There is a possibility that the control system may be configured to revert to a known, safe value or state and some limited control may still be possible.

Nature of control: **Fails to safe value**

10.6.1.2 Duplex Architecture

For more complex systems a dual channel implementation may be preferred. The sensor set and the control channel are replicated and if a sensor or control channel fails there is an alternative available. Such architectures will also use channel control logic to select which channel is to be in control. A cross-monitor compares the output of both channels after the processing has been carried out. This method offers close to 100% coverage and has the advantage that should one channel fail, the system may continue to operate in a simplex mode albeit with reduced safety margins.

A drawback of this system is that as the channels are identical, other means are necessary to determine which channel has failed, e.g. channel BIT or operator intervention.

Nature of control: **Fail safe**

10.6.1.3 Dual/Dual Architecture

A more sophisticated arrangement is dual/dual architecture, which is often implemented in a COM/MON fashion, that is, each channel has a command (COM) and a monitor (MON) lane within it, the command lane being in control and the monitor lane checking for correct functioning of the command element. The command and monitor lanes may be similar or dissimilar in terms of implementation. There will be a cross-monitor function associated with this arrangement. One weak point is that the cross-monitor function which allows the MON lane to arbitrate whether the COM lane has failed can itself be subject to a failure. In such a case the channel will fail even if the COM and MON lanes are themselves fully serviceable.

This is a very common architecture that is used widely within the civil community for the control of major utilities systems and for full authority digital engine control (FADEC) applications.

Nature of control: **Fail operational**
Fail safe

10.6.1.4 Triplex Architecture

Higher levels of integrity lead to higher levels of redundancy – triplex architectures where there are three independent sets of sensors and control. Arbitration in such an architecture is typified by a voter/monitor channel where the output of all three channels is compared and if one channel should deviate from the others then it is discounted and 'voted out'.

Nature of control: **Fail operational**
Fail safe

10.6.1.5 Quadruplex Architecture

In an extreme design situation the sensors and control channels may be replicated four times, leading to a quadruplex architecture as shown in Figure 10.6. Such architectures are usually utilised for flight control implementations in aircraft which are fundamentally unstable and where only the highly redundant flight control enables the aircraft to fly. Examples of this type of architecture are the Eurofighter Typhoon and the Northrop B-2 Spirit stealth bomber.

After a first fault the system degrades to triplex. After the second fault the system degrades to dual. In operational use a Typhoon pilot would be able to continue the mission after a first failure whereas after the second he would be well advised to curtail the mission and land the aircraft at the first convenient opportunity.

Nature of control: **Fail operational**
Fail operational
Fail safe

10.6.2 System Examples

As has already been described the architecture employed will be related to the levels of integrity (redundancy) that the system requirements dictate. Two examples are given below:

- a system example based on a major systems effect (shown in Figure 10.7)
- an example based on a flight critical event (shown in Figure 10.8).

10.6.2.1 Major Systems Event

A major systems event is one in which the occurrence is considered to be remote and which may cause a significant effect for the flight crew and passengers without necessarily

Requirement

Severity	Probability	Requirement (per flight hour)
Catastrophic	Extremely improbable	no more frequent than 1 in 10^{-9}
Hazardous	Extremely remote	no more frequent than 1 in 10^{-7}
Major	Remote	no more frequent than 1 in 10^{-5}
Minor	Reasonably probable	no more frequent than 1 in 10^{-3}

Architecture

No of lanes	Failure of a single lane	Failure of all lanes
1	$P = 1 \times 10^{-3}$	$P = 1 \times 10^{-3}$
2	$2P = 2 \times 10^{-3}$	$P^2 = 1 \times 10^{-6}$
3	$3P = 3 \times 10^{-3}$	$P^3 = 1 \times 10^{-9}$
4	$4P = 4 \times 10^{-3}$	$P^4 = 1 \times 10^{-12}$

Figure 10.7 Probability of a major system event.

Requirement

Severity	Probability	Requirement (per flight hour)
Catastrophic	Extremely improbable	no more frequent than 1 in 10^{-9}
Hazardous	Extremely remote	no more frequent than 1 in 10^{-7}
Major	Remote	no more frequent than 1 in 10^{-5}
Minor	Reasonably probable	no more frequent than 1 in 10^{-3}

Architecture

No of lanes	Failure of a single lane	Failure of all lanes
1	$P = 1 \times 10^{-3}$	$P = 1 \times 10^{-3}$
2	$2P = 2 \times 10^{-3}$	$P^2 = 1 \times 10^{-6}$
3	$3P = 3 \times 10^{-3}$	$P^3 = 1 \times 10^{-9}$
4	$4P = 4 \times 10^{-3}$	$P^4 = 1 \times 10^{-12}$

Figure 10.8 Probability of a flight critical event.

endangering them. A typical example might be an aircraft pressurisation failure, which may necessitate an emergency descent procedure. However, once this has been successfully executed the passengers and flight crew are no longer at risk.

Such a system is typically implemented as a dual/dual system in order to enable the 1×10^{-5} per flight hour requirement to be satisfied.

10.6.2.2 Flight Critical Event

A more flight critical event would be the total loss of aircraft flight control, termed a catastrophic event. In this situation the aircraft would be lost along with all the occupants.

As might be expected, the analysis for a flight critical outcome is more demanding than that for a major occurrence. In order that the 1×10^{-9} per flight requirement may be satisfied a triplex architecture is demanded.

Such replication is not without cost. The greater the number of active channels then the higher the equipment cost and the lower the reliability. In reality additional hardware may be provided to improve aircraft dispatch performance when the aircraft is allowed to safely depart carrying system faults.

In order to assess the levels of integrity that a system will be capable of it is usual to perform a probability analysis, as shown in Figure 10.9.

- Items that are linked together in series to form a channel or a lane are added together to give a total failure rate. In this case the failure of P_A or P_B or P_C.
- Elements that are organised in parallel are multiplied together, the loss of function is caused by the loss of P_1 or P_2 or P_3. In order to sustain total function failure all three elements must fail.

A sub-system might comprise a number of items linked together in series to form a channel or 'lane'

P_A — P_B — P_C

The probability of channel or lane failure is
$P = P_A + P_B + P_C$ ....

A complete system in likely to consist of a number of channels or lanes arranged in parallel

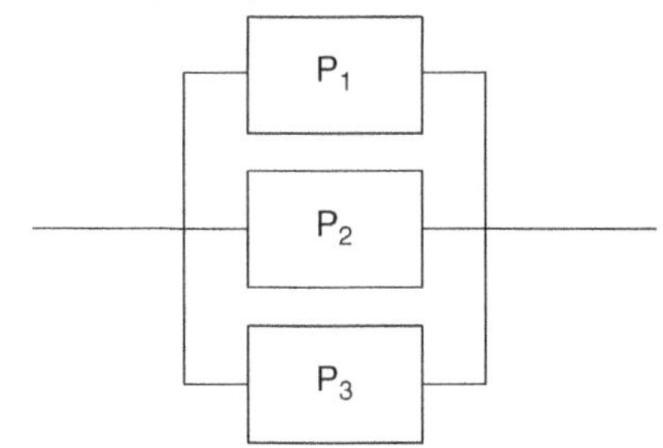

The probability of sub-system failure is
$P = P_1 \times P_2 \times P_3$ ....

Figure 10.9 Probability analysis.

Sources of hydraulic power – an example:

- Engine-driven pumps
- Air-driven pumps
- AC motor pumps
- DC motor pumps

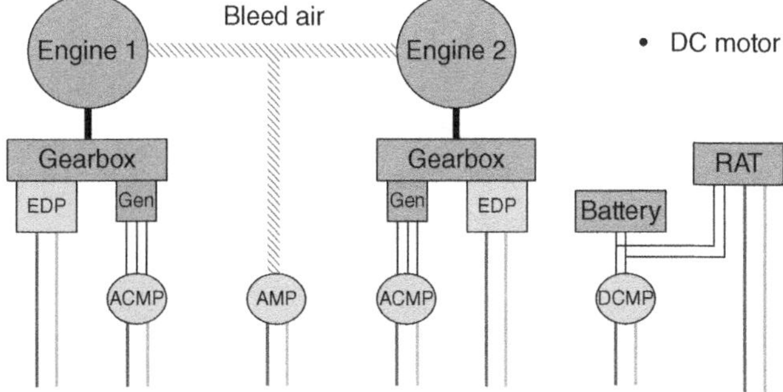

Figure 10.10 Alternative hydraulic power sources. EDP, engine drive pump; Gen, generator; AMP, air motor pump; ACMP, AC motor pump; DCMP, DC motor pump; RAT, ram air turbine.

The discussion so far has concentrated on the loss of the control function. However, systems such as flight control also need 'muscle' in the form of hydraulic and electrical systems to enable them to function. In the example shown in Figure 10.10 there are sources of hydraulic power derived from:

- engine-driven pumps (EDPs)
- air-driven pumps using bleed air (ADPs)
- AC-driven pumps (ACMPs)
- DC-motor pumps (DCMPs)
- ram air turbines (RATs) or even batteries.

10.7 Integration of Aircraft Systems

A number of aircraft systems contribute towards the correct functioning of a modern civil airliner. In the example chosen the following systems will be described:

- an engine control system: in the most recent aircraft this will use a FADEC system
- a flight control system: usually these days a fly-by-wire (FBW) system
- an attitude system to detect the aircraft attitude movement in pitch, roll, and yaw
- an air data system that provides the aircraft with information about its movement through the air: airspeed, altitude etc.
- an electrical system to provide electrical power for the systems computers
- a hydraulic system to provide the 'muscle' for the actuators to enable the pilot to fly the aircraft.

These systems are all required to fly the aircraft (see Figure 10.11).

All of these systems contribute to the aircraft function. The total loss of any one system will deny the correct operation of the others, therefore as well as contributing to the total aircraft function there is an inter-dependence between them, as shown in Figure 10.12.

The flight control system is irrelevant if the aircraft does not have the motive power or a propulsion system to allow the aircraft to reach flying speed. The flight control system cannot operate without electrical power to activate the flight control computers or hydraulic power to provide the actuator 'muscle'. Without aircraft attitude and air data information the flight crew will be unable to fly the aircraft safely within the flight envelope, and will be

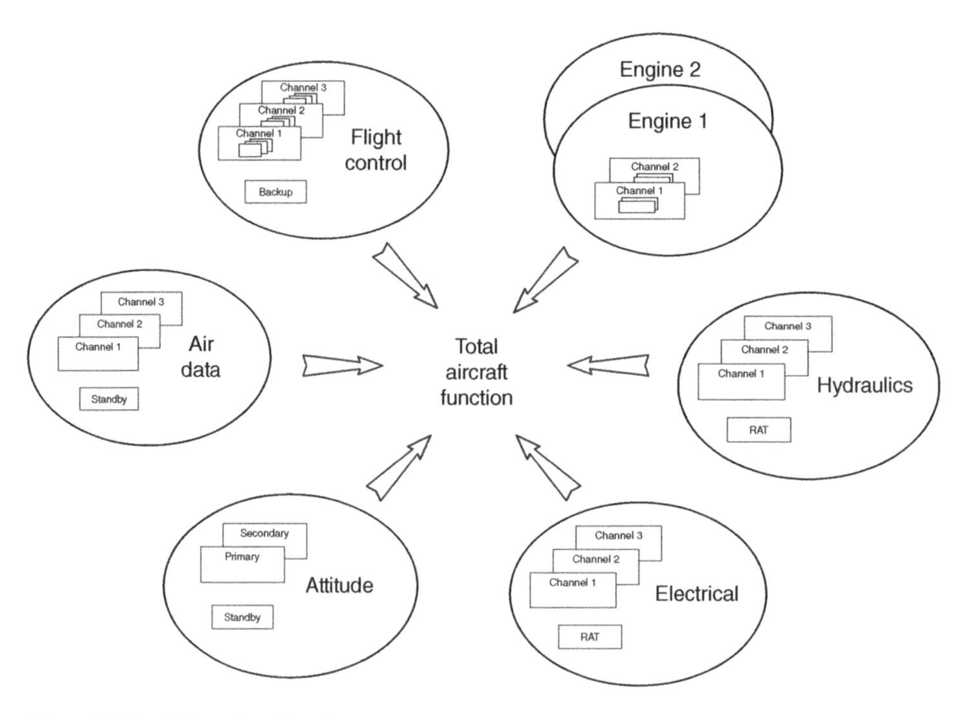

Figure 10.11 Major aircraft systems.

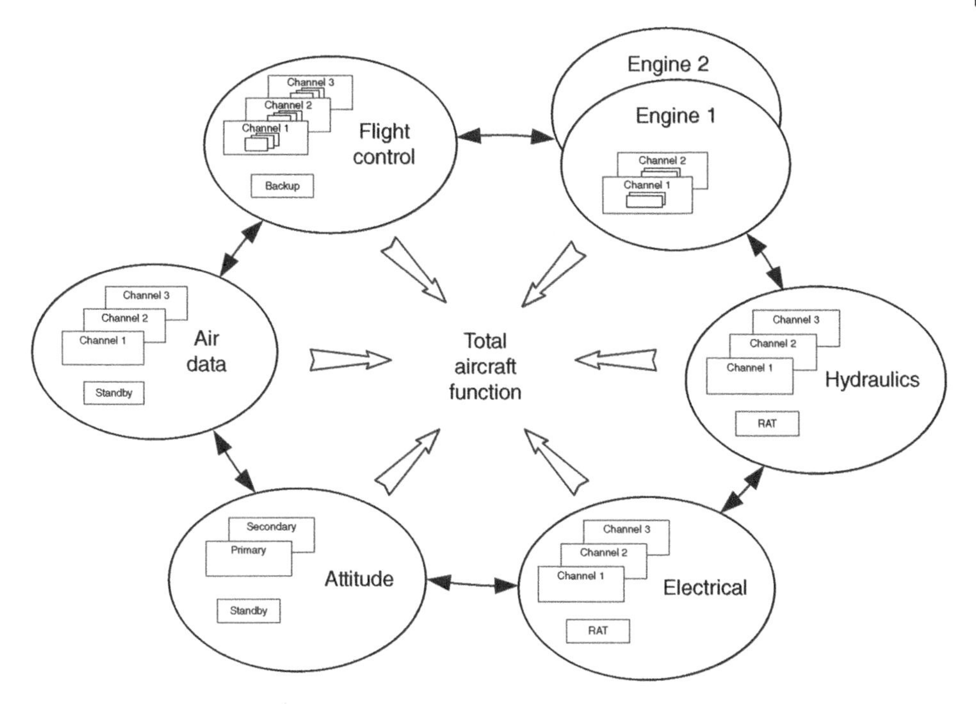

Figure 10.12 Contribution to total aircraft function and inter-dependence.

unable to determine in which direction the aircraft is travelling. Furthermore, air data relating to the aircraft's passage through the air is required to enable the flight control and engine system control laws to be correctly executed, therefore even this simplified portrayal indicates how important the interplay is between many of the aircraft systems. There are others such as the fuel system, flight deck displays, and avionics system which are equally important and some of these will be addressed later.

Each of these systems has been developed to meet high levels of integrity. For any of the systems described a total catastrophic failure could result in the loss of the aircraft and all will have been developed according to the highest levels of design assurance outlined above. As these systems have evolved each of them has developed its own architecture which the certification authorities, such as the Federal Aviation Authority (FAA) in the USS or the Joint Aviation Authority (JAA) in Europe have deemed suitable to meet the demanding system integrity requirements.

In the following sections each of these systems is examined in turn for a civil aircraft and the typical architectures that may be expected for each one are identified:

- engine control system
- flight control system
- attitude measurement system
- air data system
- electrical power system
- hydraulic power system.

Whilst the principles of operation and levels of redundancy for these systems are very similar for most aircraft – effectively representing an industry-wide response to a common design statement – specific implementations are not discussed. The cardinal issue within the context of this chapter is to view each of these systems at the top level. For more detailed system architectures relating to specific aircraft or technologies recourse will need to be made to Moir and Seabridge (2008, 2013).

Aerospace Recommended Practice (ARP) documents that may be applied during the system design process can be found in SAE ARP 4754 and 4761. They are not mandated, but they contain essential advice and systems designers who choose to ignore their contents may experience great difficulty in achieving certification for their system.

10.7.1 Engine Control System

Many modern aircraft have two engines and this is a common propulsion system configuration for short-, medium-, and some long-range aircraft. The example shown in Figure 10.13 assumes a two-engine turbofan propulsion system. Modern turbofan engines are usually controlled by a FADEC mounted on the engine it controls.

The FADEC architecture usually comprises two identical channels, channel A and channel B, each of which is capable of fully controlling the engine. Each channel consists of two elements: a control lane and a monitor lane. The control lane exercises control of the engine whilst the monitor lane carries out cheques on the control lane to ensure that it is performing correctly. In the event of a failure of the control lane control is passed to the other channel, which still has a fully functional control/monitor lane pair.

Each FADEC channel (control and monitor lanes) is independently supplied with electrical power provided by a small dedicated generator called a permanent magnet aternator (PMA) located on the engine and driven directly off the accessory gearbox.

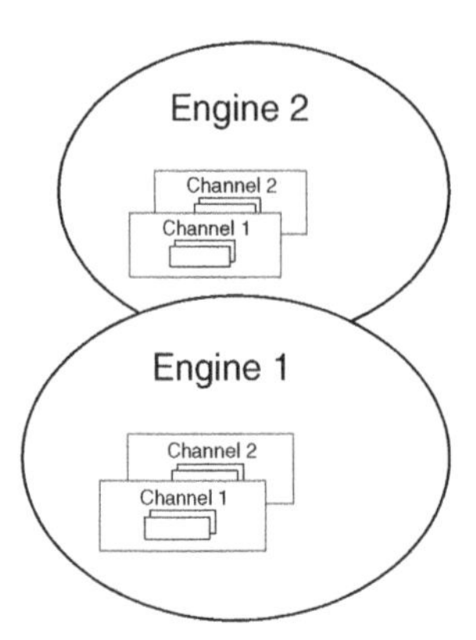

Figure 10.13 Each engine is controlled by an electronic engine controller (EEC). The EEC comprises a dual/dual architecture and is powered by dual sources of engine-derived electrical power. Each engine and EEC are independent of one another, although single point failures can occur, for example severe hail or volcanic ash ingestion can cause both/all engines to fail simultaneously.

Each FADEC control and monitor lane is therefore fed by a dedicated electrical supply derived on the engine, thus the FADEC is not dependent upon the aircraft electrical power system for uninterruptible power and correct operation. This feature gives the engine control system independence of operation should there be a total failure of the aircraft electrical power. In order to fully implement the engine control laws the FADEC needs to be supplied with air data from the aircraft air data system.

In some cases it may be possible to dispatch the aircraft for a limited period with a single monitor lane – but not a control lane – inoperative. This is possible because a risk assessment has demonstrated that for a limited period of operation the loss of this monitor lane, in effect the loss of one engine control lane amongst the set of four lanes in the aircraft, represents an acceptable risk for a short period, perhaps to enable the aircraft to be ferried to a maintenance base so that a repair can be carried out.

10.7.2 Flight Control System

Whilst the philosophies adopted by individual aircraft manufacturers may differ, most FBW systems use multiple lanes of redundancy in the various computing channels. The example shown in Figure 10.14 illustrates the trio-triplex configuration employed on the Boeing 777 flight control system. Airbus have adopted a multi-channel philosophy employing five independent command/monitor channels across the pitch, roll, and yaw axes. The in-built redundancy is such that it is usually possible to dispatch the aircraft safely with a number of lanes inoperative; the precise details depend upon the aircraft type and system architecture, and the duration of the intended flight.

In the event that all the FBW computing fails it is usually possible to operate the aircraft in a direct electrical link mode as a get-you-home function. Even if this direct link were inoperative all FBW systems operating today also have a direct mechanical link in the pitch and yaw channels though this potentially offers reduced control authority.

In order to execute the control laws the flight control computers need information from the aircraft attitude measurement system and the air data system. Electrical power is required to power the system computers and hydraulic power is required for most of the flight control system actuators.

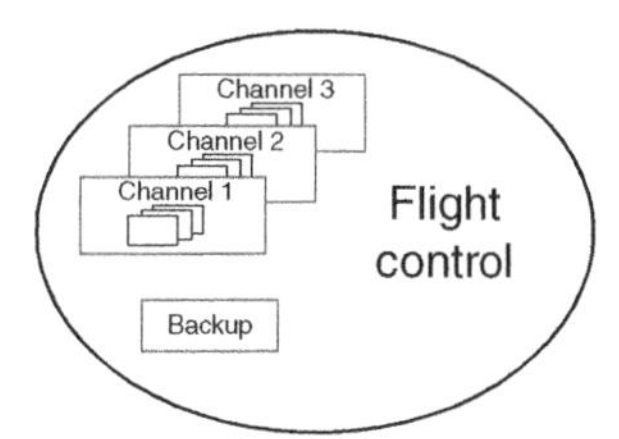

Figure 10.14 Flight control architecture will have multiple lanes of control depending upon the degree of authority. In the example shown the B777 flight control system has three channels of triplex computing in pitch roll, and yaw and a mechanical backup in pitch and roll. FCS also relies heavily upon the other systems for data: air data, aircraft attitude and body rates and power for control and actuation, electrical and hydraulic power.

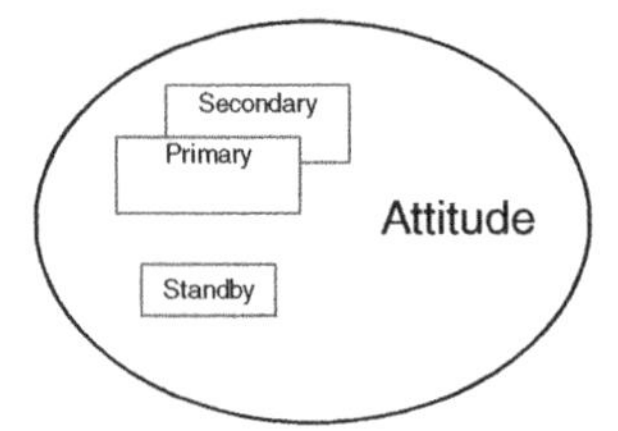

Figure 10.15 The aircraft requires multiple sources of aircraft attitude information to provide the pilot with attitude guidance and to support the needs of many subsystems. Normally primary, secondary, and standby sources of attitude are provided: primary attitude source (inertial navigation system or air data and inertial reference system), secondary attitude source (attitude and heading reference system), standby source of attitude to stand-alone backup instruments.

10.7.3 Attitude Measurement System

The aircraft requires pitch, roll, and yaw attitude information in order to be flown safely. Attitude systems are required to display pitch and roll information on the primary flight display (PFD) and yaw/heading information on the navigation display (ND), which both pilots have as part of their electronic flight instrument system.

In addition, three-axis velocity and acceleration data relating to aircraft motion are required for the FBW and autopilot systems. Usually aircraft have primary and secondary attitude systems with some form of standby system, as illustrated in Figure 10.15; the exact implementation depends on the system architecture and the technology employed. On most aircraft the inertial reference system (IRS) provides the primary attitude source whilst a secondary attitude and heading reference system (AHARS) yields an alternative source of attitude information.

In recent years improved technology using micro-inertial sensing devices and packaging techniques has resulted in small 2ATI instruments providing an integrated standby instrument system (ISIS) that provides an independent source of attitude information to the pilot in the event that the primary and secondary sources are not available.

10.7.4 Air Data System

The air data relating to the aircraft speed and height whilst passing through the air is of critical importance. A combination of pitot and static probes senses the total and static pressure as the aircraft moves through the air as a fluid medium. In the simplest form this information can be combined to derive airspeed, altitude, and rate of climb and descent. The addition of other sensors, such as total air temperature probes and air-stream direction detectors, allows even more useful derived data to be calculated. By using the digital computing capability, air data computers (ADCs) or air data modules (ADMs) are able to calculate other more useful parameters, such as indicated airspeed (IAS), true airspeed (TAS), Mach etc.

Due to the level of criticality of air data it is usual to have three independent channels with a standby channel, as illustrated in Figure 10.16. In the past, the standby channel has been composed of dedicated small standby instruments fed by their own dedicated pitot-static system. The availability of the ISIS technology described above now means that two

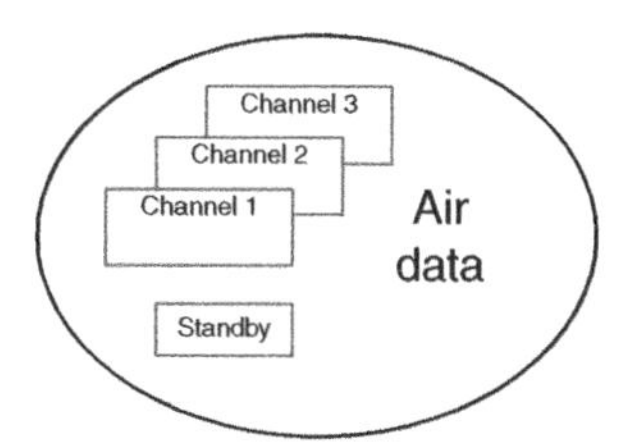

Figure 10.16 The availability of air data is critical for safe flight. Air data computers or air data modules provide triple redundant sources of pitot and static air data information. Calculations convert basic pitot and static pressure into more meaningful parameters such as indicated air speed, true air speed, Mach number, barometric altitude etc. Air data are extensively used by many aircraft systems. Backup data are provided to standby systems.

multi-function, solid-state backup instruments can be used in the place of three dedicated mechanical instruments. The reduction in the number of standby instruments is enabled in part due to the increase in reliability by changing to solid state instruments rather than conventional instruments as well as the multi-function display capability of the newer instruments.

10.7.5 Electrical Power System

As has been mentioned earlier, the electrical power system is a key system. Most, if not all, aircraft systems depend upon the supply of electrical power either as 115VAC 3-phase or 28VDC supplies derived from transformer rectifier units (TRUs).

A common electrical power generation configuration employed is to have three generators, one driven by each of the aircraft engines and a third driven by the aircraft auxiliary power unit (APU). The original purpose of installing the APU was to provide a source of electrical power and medium pressure air whilst the aircraft was on the ground prior to departure. The APU could also be used as a source of pneumatic power to start the aircraft engines. In many configurations in service today, the APU can be used to provide a third electrical power generation source by starting the APU in flight. In some systems this can be accomplished at cruising altitude in excess of 30000ft; in others the aircraft has to descend to medium altitude – say 20000ft – in order for the APU to be in the start envelope. Still other systems run the APU continuously during the cruise when operating in an extended twin operations (ETOPS) configuration, that is operating a twin-engine aircraft such as a Boeing 737 over 60 minutes away from a diversion airfield. Special considerations apply to ETOPS operation (see Airworthiness Circular AC 120-42A 2008).

Due to the importance of electrical power a further backup power system is provided. In smaller aircraft this may be provided by sizing the aircraft batteries such that they can provide adequate power for short-term use (up to 30minutes) in the event that primary and secondary sources are lost. Other aircraft rely on the use of a RAT to provide the aircraft with an emergency power source. The RAT is an air-driven turbine normally stowed within the aircraft fuselage. When required it is released into the airstream it is rotated by the airflow and drives a small embedded electrical generator. Once deployed the RAT cannot be re-stowed in flight and must be reset by a maintenance action once the aircraft has landed. As will be seen, the RAT can also be used as a source of emergency hydraulic power. This is illustrated in Figure 10.17.

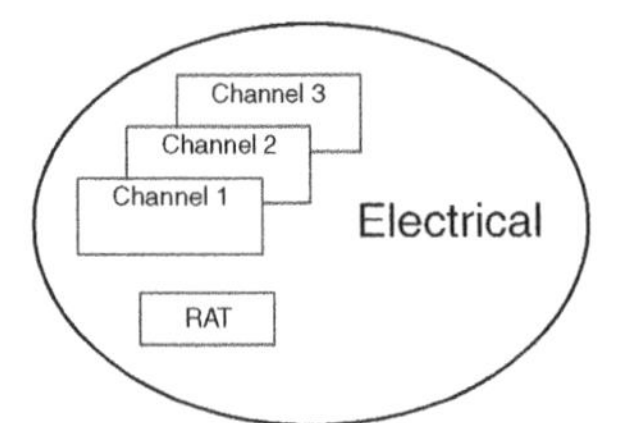

Figure 10.17 Aircraft usually possess three independent channels of electrical power: left main generator, right main generator, and auxiliary power unit generator. A battery is also provided which can provide a short-to medium-term alternative energy source. An emergency source such as a ram air turbine supplies power when all other power is lost.

10.7.6 Hydraulic Power System

The need for hydraulic power to provide the muscle to drive the flight control and other hydraulic systems has already been described. Normally hydraulic power is derived from EDPs mounted on the engine accessory gearbox. However, hydraulic pumps may also be electrically driven by either AC or DC electrical motors. On Boeing's wide-body aircraft – the Boeing 747/767/777 – some hydraulic pumps are air driven. The reason for using these dissimilar means of generating hydraulic power is to achieve additional levels of segregation and redundancy as well as meeting diverse system demand (flow rate) conditions.

A typical aircraft will have three independent hydraulic systems driven by a combination of the aircraft EDPs and electrical and/or ADP. These systems will be isolated such that a major system failure, such as total loss of hydraulic fluid within one system, does not cascade from one hydraulic system into another. Some systems use units called power transfer units (PTUs) which can transfer energy from one system to another whilst at the same time preserving segregation between them.

As has been mentioned, another source of hydraulic power can be provided by a RAT driving a dedicated hydraulic pump to provide a short-term source of power in an emergency situation. This is illustrated in Figure 10.18.

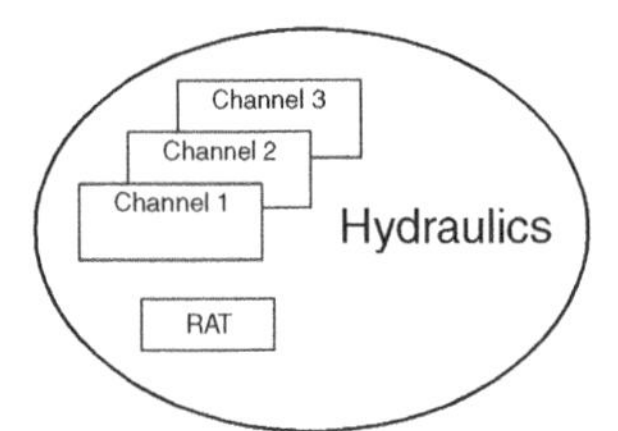

Figure 10.18 Usually three independent channels of hydraulic power are provided. Hydraulic power is derived from engine-driven pumps, AC motor pumps, and in some cases air-driven pumps. Hydraulic accumulators provide a short-term backup. Emergency power is supplied from a ram air turbine.

10.8 Integration of Avionics Systems

The previous discussion has related to some of the aircraft systems, but on a modern aircraft the avionics system is equally important. The avionics system is – as has previously been described – essentially the brains of the aircraft, helping the flight crew negotiate the busy airspace of today with precision and in safety. The key elements of the avionics system are shown in Figure 10.19. There are four main functional components:

- navigation: flight management, inertial navigation, satellite navigation, navigation aids, and terrain avoidance warning system (TAWS)
- flight control: air data, attitude systems, FBW, autopilot, and auto-throttle
- communications: high frequency (HF), approach aids, very high frequency (VHF), mode S and traffic collision avoidance system (TCAS), and satellite communications (SATCOM)
- displays: primary displays including an electronic flight instrument system (EFIS) providing flight information and an engine indicating and crew alerting system/electronic centralised engine monitor (EICAS/ECAM) providing system synoptic and status displays, aircraft system overhead panels, multi-purpose control and display units (MPDCUs), equipment control panels, and standby instruments.

These functions have significant interaction between them and with the aircraft systems already described.

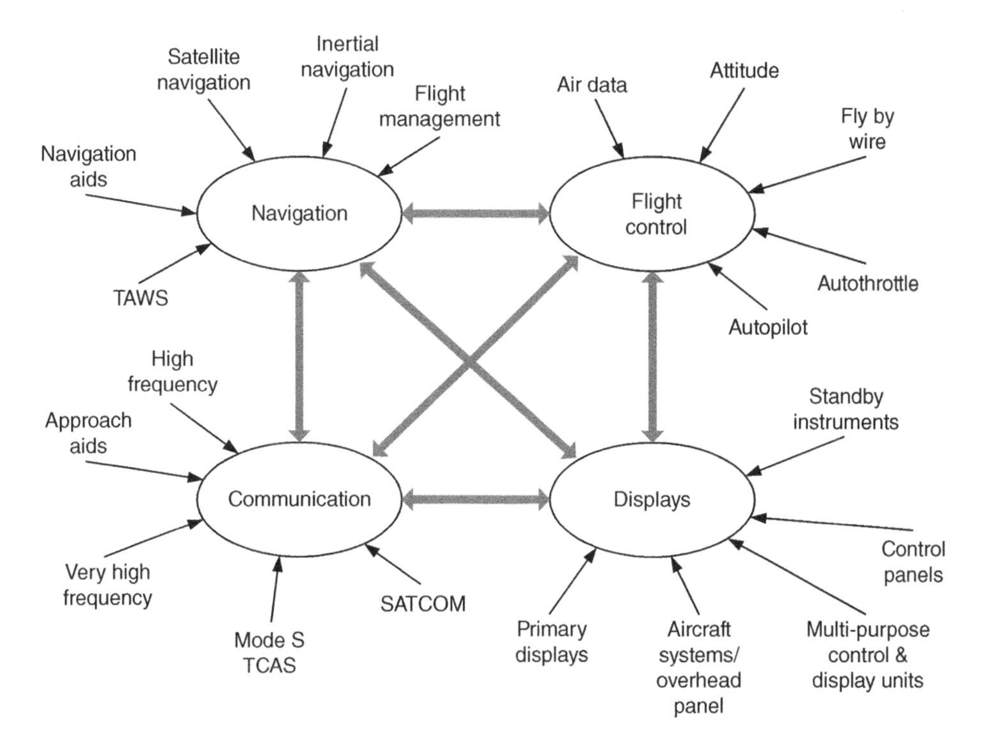

Figure 10.19 Major avionics system functions.

Some of these systems and their inter-reactions can be neatly described as a series of nested control loops. Other inter-reactions are more subtle. Figure 10.20 shows the interaction between FBW, the autopilot and the flight management system (FMS) as a simplified series of nested control loops.

These control loops, starting from the simplest on the inside and working outwards towards the most complex, are as follows:

- *Primary flight control – ATA Chapter 27*: This is used to control the *aircraft attitude*. Inputs from the pilot's controls are fed through the FBW system control computers to the flight control actuators that modify the aircraft attitude in response to the pilot's commands. The aircraft dynamics and attitude sensors result in feedback to the pilot either visually or by means of the EFIS displays.
- *Autopilot – ATA Chapter 22*: Once the autopilot is engaged the aircraft is controlled by the next control loop, which controls the *aircraft trajectory*. By means of selecting speeds, height, heading and speed datums, and navigation and approach aids the pilot is able to accurately manage the aircraft trajectory during flight with much reduced work load. Control of the autopilot is achieved by the use of a flight control unit (FCU) or mode control panel situated on the flight deck glare-shield.
- *FMS – ATA Chapter 34*: The FMS effectively assists the flight crew in achieving the *aircraft mission*. For a civil airliner this involves executing departure and arrival procedures, and negotiating the navigation of the aircraft on a series of way points as the aircraft flies from the departure airport to the destination airport. The primary flight crew interface with the FMS is the multi-purpose control and display unit (MCDU) of which there are three units on the flight deck.

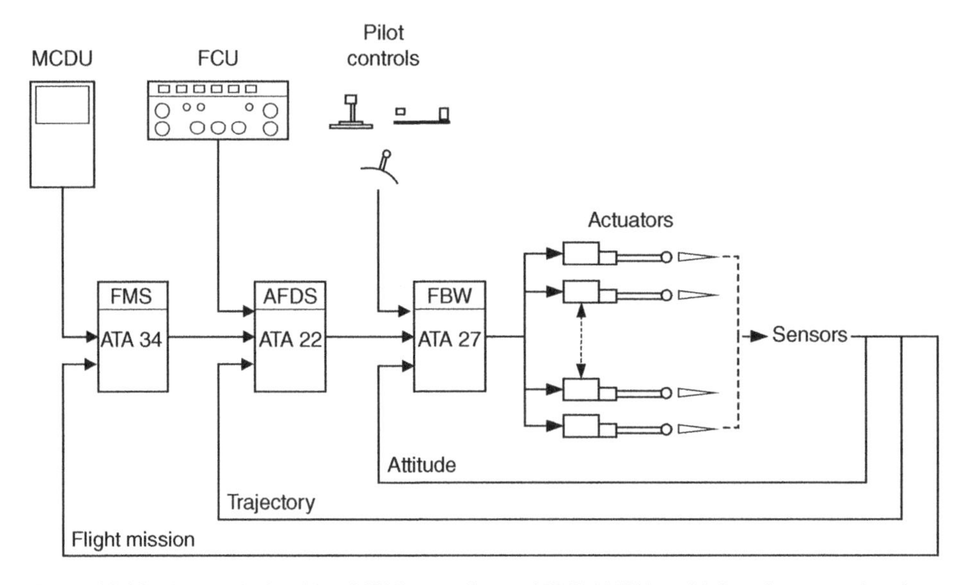

Figure 10.20 Inter-relationship of FBW, autopilot and FMS. MCDU, multi-function control and display unit; FCU, flight control unit; FMS, flight management system; AFDS, auto-pilot and flight director system; FBW, fly by wire.

It will be recognised that the co-ordination of the overall specification of all of these systems, including a range of equipment produced in different timescales with different technology baselines, is a mammoth task. The industry approaches this problem by using a series of equipment and technology specifications derived for and by the industry to be used worldwide. These specifications are controlled by the Air Radio Inc. (ARINC) organisation in the USA and the hierarchy of specifications as they presently exist at the time of writing is shown in Figure 10.21.

There are four series of ARINC specifications:

- *ARINC 400 Series*: These are termed the design foundation and relate to the earliest avionics specifications. Typical amongst this series are ARINC 429, which was used to specify the first civil serial avionics data bus, and ARINC 404A, which was an early packaging standard. Presently 21 specifications in the ARINC 400 series are still in use.
- *ARINC 500 Series*: As electronics became increasingly used in aircraft equipment a series of specifications was derived to deal with analogue equipment used in the older analogue – now called 'classic' – aircraft such as the DC-9, MD-10, A300, and early Boeing 737 and Boeing 747 aircraft. Typical examples include ARINC 578 and 579, which were used to define the characteristics of instrument landing system (ILS) and VHF omni-range (VOR) equipment, respectively Some 21 specifications in this series still exist.
- *ARINC 600 Series*: When the scale of the application of digital electronics to civil aircraft became clear this series became the vehicle for specifying enabling digital technologies. Examples are ARINC 629, which defines a 2 Mb/s digital data bus used on the Boeing

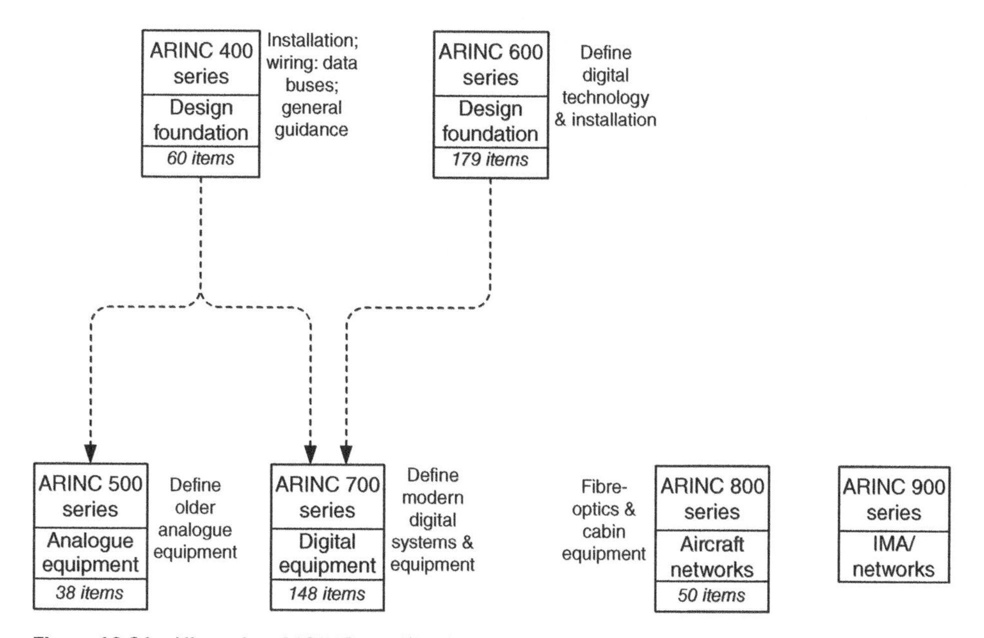

Figure 10.21 Hierarchy of ARINC specifications.

777, and ARINC 600, which specifies advanced packaging techniques beyond the earlier ARINC 404A standard. Around 70 ARINC 600 series are presently in use.

- *ARINC 700 Series*: To define the new digital equipment used in the present modern generation of digital aircraft the ARINC 700 series evolved. Typical examples include ARINC 708 for a digital weather radar and ARINC 755 for a multi-mode receiver (MMR) incorporating GPS, ILS, and other RF receivers in a single unit. Some 66 ARINC 700 series specifications are presently in use and doubtless others are in evolving in draft form for the specification of new and emerging equipment. Details on these specifications made be found on the ARINC website.
- *ARINC 800 Series*: Specifications and reports define enabling technologies supporting the networked aircraft environment. Amongst the topics covered in this series is fibre optics used in high-speed data buses, A802–1 for fibre optic cables, and A808–1 for cabin distribution systems.
- *ARINC 900 Series*: ARINC characteristics define avionic systems in an integrated modular and/or networked structure. They include detailed functional and interface definitions.

References

Airworthiness Circular AC 120-42A (2008) *Extended Range Operation with Two-Engine Airplanes*, 30 December 1988.

FAA/JASC(2002). *FAA Joint Aircraft System/Component code tables and definitions.* Issued February 2002. Google JASC Codes PDF. Accessed March 2019.

Moir, I. and Seabridge, A. (2008). *Aircraft Systems*, 3e. Wiley.

Moir, I. and Seabridge, A. (2013). *Civil Avionics Systems*, 2e. Wiley.

RTCA DO-178B, (1992) *Software Considerations in Airborne Systems and Equipment Certification.*

RTCA DO-254, (2000) *Design Assurance Guidelines for Airborne Electronic Hardware.*

SAE ARP 4754, *Certification Considerations for Highly-Integrated or Complex Aircraft Systems*, Society of Automobile Engineers Inc.

SAE ARP 4761, *Guidelines and Methods for Conducting the Safety Assessment Process on Civil Airborne Systems*, Society of Automobile Engineers Inc.

11

Integration and Complexity

The Potential Impact on Flight Safety

11.1 Introduction

So far in this book, as in other aerospace series books by Moir and Seabridge, the broad assumption has been made that aircraft systems will be designed as individual entities. This is very often what happens in large organisations, generally as a result of their departmental or organisational structures, which are often formed around individual system responsibilities. The subject of 'interconnectedness' or interactions between systems has been introduced as a warning to designers to be aware of unintended consequences arising from the performance of their systems and the interactions that occur, directly or indirectly, with other systems in the total aircraft design.

Chapter 6 introduced the topic of systems integration as a normal part of the continuing development of the systems process; Chapter 5 introduced the topic of integrated systems architectures and a number of different systems examples were given. The content of those chapters is still relevant. In this chapter the results of the integration of physical systems, functional aspects and functions associated with automation, and potentially autonomous unmanned operations is examined. The purpose of the chapter is to discuss whether or not such continued integration of functions is likely to pose a threat to the continuing safe operation of the aircraft. The reason for examining this is not to impose a limitation on such innovations but to cause the managers of system integration – chief engineers, system integrators, systems managers, and systems engineers – to reflect on the process and to act to ensure that no such threat is posed.

What follows is a number of observations and examples, some based on experience, some based on published material, together with some warnings intended as guidance for systems engineers and integrators.

11.2 Integration

Integration can be characterised by the performance of many functions by a small number of units. It is more commonly applied to avionics-type systems, but is also found in major structural and mechanical systems such as aircraft powerplant and propulsion systems and even the airframe itself. In an integrated system the individual systems interact with each other and display forms of 'interconnectedness' that make the larger system more than the sum of its constituent sub-systems. Interconnectedness of systems is achieved by:

- data communication networks, conductive or optical
- direct physical interconnections by design, e.g. welded, fastened or glued connections
- direct mechanical interconnections by drive shafts, air motors or actuators
- indirect physical interconnections, e.g. thermal, vibration, electrical conduction
- physical amalgamation at component level
- functional interconnection in processors by shared computer architectures, memory locations or data use
- serendipity – emergent properties or unexpected behaviours in one system that affect one or more others.

There is little doubt that integration has brought advantages to the design of aircraft systems. There is better inter-system communication and greatly improved functional performance, and innovative technologies have been introduced. Data is available for on-board use by the crew in real time, whilst the accident data recorder, maintenance management system, and flight test instrumentation provide data for real-time or post-flight analysis. There has been a great improvement in the design of cockpits and flight decks with a consequently reduced crew workload.

System integration has been achieved by design and has been implemented by the application of readily available technologies by means of which functionality has been incorporated using a combination of:

- computing systems based on distributed computers
- high-order languages
- high-speed data buses
- multi-function displays and controls
- touch, voice, and switch controls.

All this has been enhanced over the years by the use of colour screens, high-resolution images, easy access to on-board databases for intelligence and maintenance data, access to ground-based systems by data link, and incorporation into the air transport management system. The cockpit or flight deck is now roomy, uncluttered, comfortable, and a safe and clean place in which to work. The pilot is now integrated with the whole aircraft and its systems. This looks likely to continue with predicted technologies such as gesture control, synthetic vision, and neural networks.

Some mechanical transfer functions such as those previously provided by mechanical devices such as cams and springs or hydro-mechanical devices have been incorporated in software and electrical effectors, thereby blurring the distinction between mechanical and avionics functions.

There are issues associated with each of these characteristics that need to be understood:

- *Computing systems based on distributed computers*: the functions in individual systems may be conducted in real time in a cluster of computers, and are part of a complex network of more clustered computers. All of these will generally operate asynchronously.

- *High order languages*: the early choice of languages such as Pascal and Ada was based on their inherent rigid structure and the use of limited instruction sets to form a deterministic structure essential for the acceptance of safety critical software. As time progressed this structure was seen as limiting in terms of speed, even cumbersome in its inherent inelegance of software design, and the computer games industry began to spawn languages more suited to graphics, visual effects, and high-speed operations. The favoured language became C++, which is still prevalent today and has been adopted for aerospace applications, after early resistance. There are potential issues with its use (unless a limited instruction set is mandated) which can lead to non-deterministic structures.
- *High-speed data buses*: a similar sequence of events led from the initial choice of MIL-STD-1553B because of its simple and deterministic nature, towards the widespread use of high-speed data bus structures which can also exhibit non-determinism if not correctly designed.
- *Multi-function displays and controls*: these have released panel space in cockpits and improved pilot awareness by the use of colour and easily understood formats for data and information. Many of these advances have come from the world of desktop computing and games. The perceived need to give air crew information only when it is needed for a particular phase of flight means that data may only be accessed after several pages of information (screens) have been viewed. Cummings and Zacharias describe this as *glass displays are not always guaranteed to promote optimal human performance ... because there are many more sources of information, often hidden in many layers of hierarchical software menus. It can be difficult to incorporate all the needed information in a dynamic fashion that follows the changing needs of the operator while avoiding the clutter of unneeded information.* This can be an issue in emergencies, where clarity of information in decision making is essential. The same authors also cite an accident in which the crew mistook one alarm for another whilst busy trying to sort out a problem, noting that humans find it difficult to either remember or to clearly distinguish alarms when multiple cues are present.
- *Touch, voice, and switch controls*: these have increased the opportunity for overlaying visual screen information with other cues such as voice and aural tones to signify events. They have also increased the opportunity for instinctive reactions by the pilot in response to these cues. In a controlled environment this is seen as an advantage, but in emergencies it may result in an overload of information which is not permanently displayed – it needs to be remembered. Instinctive reaction may also lead to a response being less considered than a switch press action where the switch is on a panel and can be checked by a second pilot, rather than being a 'twitch of the finger'.

The result of this is a perceived obscuration of the end-to-end behaviour of the system. This may be seen as an inability to be able to trace the route of an input signal to an output effector in a simple manner – once the way in which system behaviour was checked. There are severe implications of this obscuration:

- It is difficult to envisage the correct behaviour of any one individual system and the whole system for an observer and, more seriously, for a reviewer checking and signing-off for correctness of operation.

- It is difficult to design a series of tests that go deeper than merely checking that an appropriate combination of inputs gives rise to an expected output and does not give rise to an unexpected output.
- It is almost impossible to verify the soundness and segregation of systems and to detect any unexpected interactions.
- It is difficult to determine the correct operation of the whole system, not simply the fact that functions have been performed correctly, but in understanding the mechanics of this.
- It is very possible that serious issues will remain dormant until a particular combination of events, demands, data structure, and software instructions leads to their initiation.

This all leads to uncertainty in the complete design unless something is done to improve the visualisation and understanding of the design, its implementation, and the results of testing.

11.3 Complexity

It certainly appears that modern aircraft systems have become more complex, both as individual systems and as total systems. Some of the reasons for this include the following:

- Advances in technology have made it easier to implement more functions in software in a high-speed processing architecture.
- The desire for engineering advancement: the simple need for engineers and designers to incorporate new functions and to apply new technology.
- The need for more automation in a bid to reduce crew workload.
- A desire for autonomous operations for unmanned vehicles and in the event of further reductions in flight deck crew complement.
- The desire to get more functional performance from fewer items of equipment.

As computing technology makes it easier to include new functions, so customer requirements are enhanced by the results of this and their needs become more complex. This may be spurred on by pilots flying new types and learning new techniques, thus making more demands based on their experience. Engineers are usually more than happy to entertain this scenario as their natural inclination is to apply new technologies.

This desire is not solely restricted to the flying side of operations. Airline and military fleet maintenance and logistics support are also under pressure to improve turnaround times. One method of doing this is to improve the accurate location of failures and to provide advance notification to the ideal repair and replacement bases. This has led to the introduction of more sophisticated algorithms to enable monitoring and on-board analysis together with data link download of information. It has also led, incidentally, to scenarios in which faults can be 'carried' in redundant systems until the most cost-effective repair base is reached.

This has the result of making the systems associated with control of the aircraft and mission become closely connected to the maintenance function, further increasing complexity.

Complexity is almost never a good thing in engineering (hence the adage 'keep it simple') and it has also been singled out for comment (Maier and Rechtin 2002) with a warning of the impact of complexity on inter-system interconnectedness:

- *When architects and builders are asked to explain cost over-runs and schedule delays, by far the most common ... explanation is that the system is much more complex than originally thought.*
- *... As systems become more complex, the inter-relationships among the elements increase far faster than the elements themselves.*
- *It is generally accepted that increasing complexity is at the heart of the most difficult problems facing today's architects and engineers.*

There are two important messages (at least) here. The first is that despite the process of requirements capture, analysis, careful design, and robust reviews, many projects fail because of cost and time over-runs. Much of this has to do with the impact of complexity. Complexity adds time to the design and test process as re-work is required and because the task has been under-estimated. It is important to try to assess or measure any trends towards increasing complexity, and to focus attention on their solutions or to amend the estimate.

The second message is to do with the increase in relationships between elements of the system. This is bad enough if the complexity is confined to a bounded system, but when the impact of integration blurs the boundaries of individual systems then the situation becomes difficult to control. Hence the widening scope of integration to encompass external systems such as maintenance and air traffic management which requires more diligence in design.

It is good then to make an assessment of a new system design to see if it has become more complex than its predecessors. This assessment can be used for a number of purposes:

- to determine what additional design and test work will be required
- to determine what further investment will be needed in new design tools
- to determine what new or additional skills will be required
- to determine what training will be needed
- to ensure that any work or cost estimates carried forward from a previous project are appropriately factored
- as an aid to determining the feasibility of the project.

Indications of the growth in complexity can be estimated by examining the current project design and comparing it with predecessor projects. Figure 11.1 is an extract of work done in 1986 by the author in trying to estimate what was ahead for a project that was to become the experimental fighter aircraft EAP. The work was carried out to examine the feasibility of introducing an integrated computing system for the control of the aircraft systems.

It was essential to determine how complex the systems were to make sure that the proposed integrated solution was indeed feasible. The task started with the human machine interface as a suitable indicator of new functions in the aircraft. Figure 11.1a shows that the number of display parameters was increasing, Figure 11.1b shows that the same was true for number of instinctive switch controls required in the throttle handle, and Figure 11.1c shows the number of switches and controls anticipated for the cockpit. It was clear that this

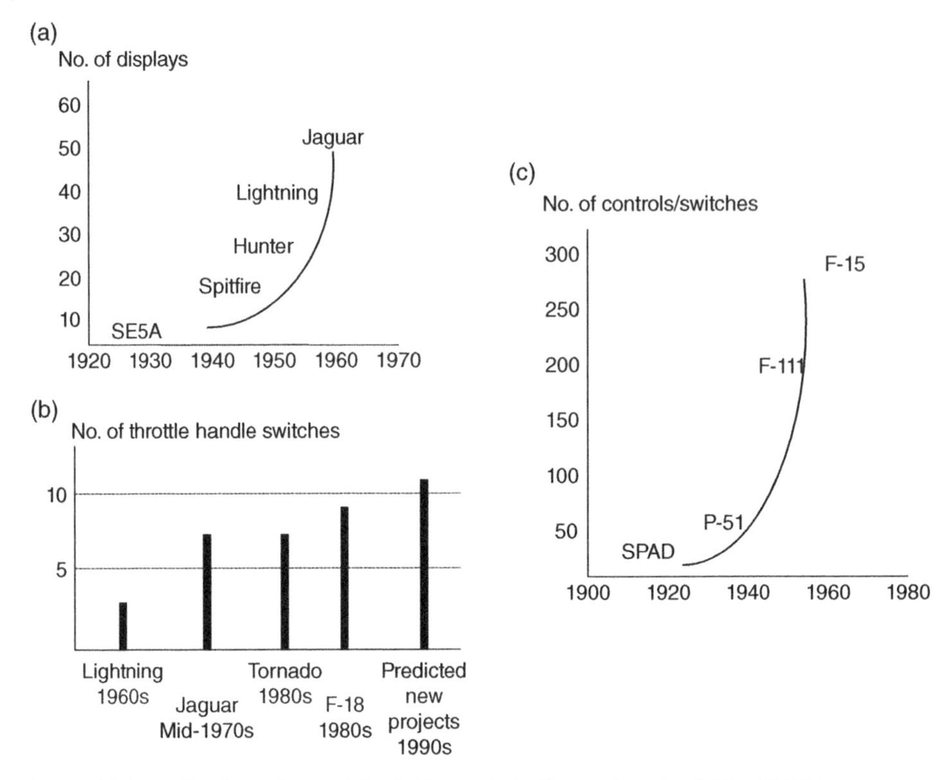

Figure 11.1 Indications of growth in similar projects. *Source:* from Seabridge (1984).

growth could not be accommodated in the future cockpit real estate unless a significant new technology for displays, controls, and computing power emerged – there would not be sufficient panel area available. Although these were simple enough measurements the questions for future projects included:

- How many discrete switches are going to be needed and what impact will this have on the control panel area?
- How many new items for display are needed and what impact will this have on the panel area?
- What alternative technologies are emerging that might lead to even more displays and switches?

These were important questions that eventually led to the decision to make use of colour display screens and 'soft' or multi-function switches, as technology made them available of course. Over time this led to the introduction of flat-screen technology and touch screens.

Since that time there have been significant changes in the technology employed which have resulted in reductions in air-crew complement. This has resulted in a reliance on the aircraft systems to perform some routine tasks of automatic flight, systems performance, and failure monitoring.

Another example is to look at the growth in power demand. Figure 11.2 was introduced in the second edition of *Civil Avionics Systems* (Moir et al. 2013). It shows that power

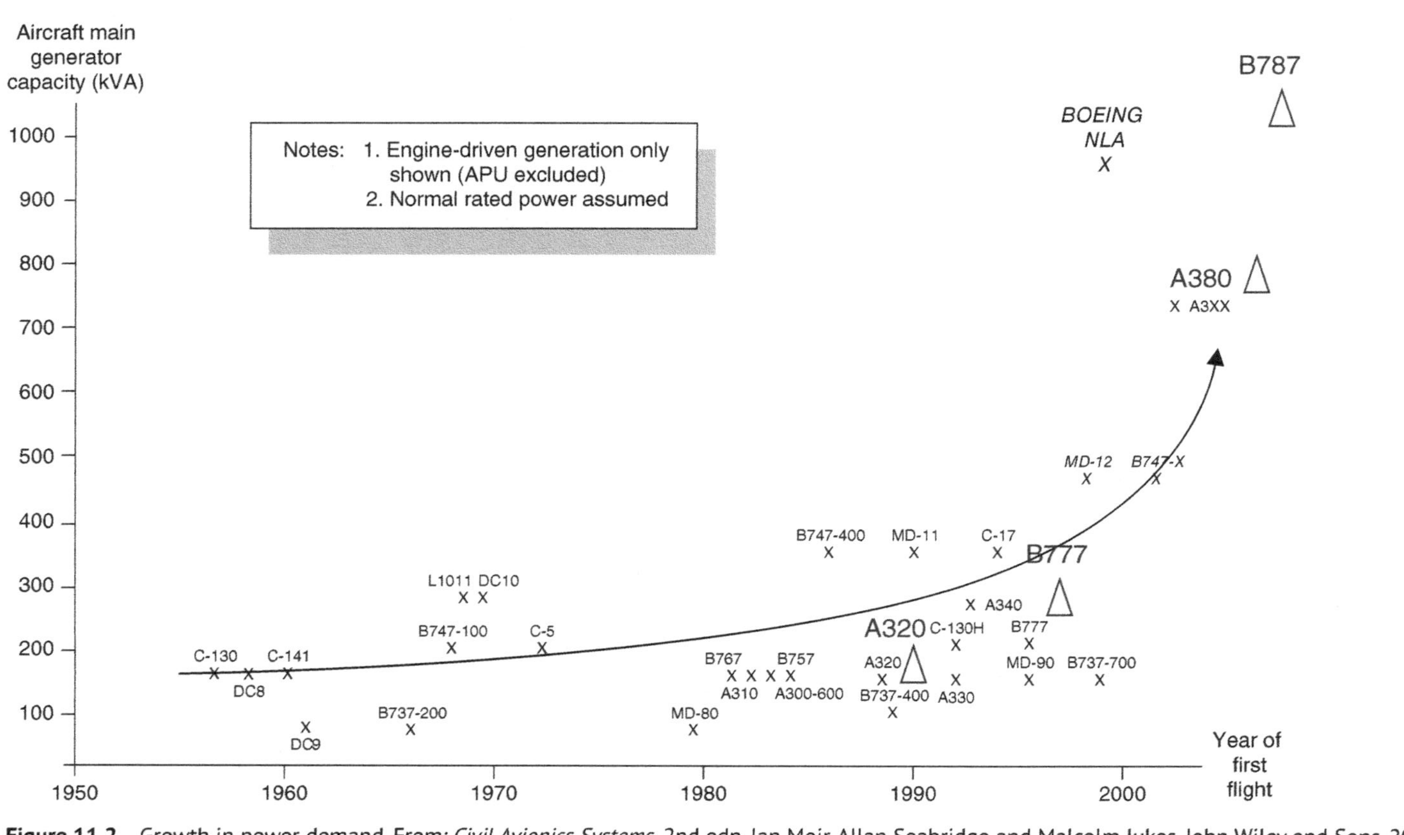

Figure 11.2 Growth in power demand. From: *Civil Avionics Systems*, 2nd edn, Ian Moir, Allan Seabridge and Malcolm Jukes, John Wiley and Sons, 2013.

demands started to increase as integrated avionics systems were introduced and increased further as the demand for more passenger entertainment systems grew. It is also true that larger aircraft and more passengers stimulated a demand for more power. An interesting phenomenon is the large increase for the B787. This came about from that aircraft's strategy for improving engine performance by removing the variable bleed air demand and replacing it with external air pressurised by electrically driven compressors and also with electrical anti-icing.

For similar projects today there are other measures that can be used, including:

- the size of software loads in thousands of lines of code (kLoC)
- the number functions implemented in software
- the number of pages of displayed information
- how many 'layers' of screen are needed to reach a particular display
- the number of routine warnings, alerts, and displays
- the number of safety critical warnings, alerts, and displays
- what combinations of graphic display, text display, voice, and aural tone are needed
- crew workload analysis

There is work in progress today to pursue this issue further. The Department of Trade/Federal Aviation Administration (DOT/FAA 2017) have produced a final report that was commissioned with the goal of defining an appropriate definition of complexity and coming up with a measure to estimate the complexity of avionic systems.

11.4 Automation

The integrated system has enabled functions to be added to the basic design that can be seen to offer an improvement. The basic flying task has been developed over many years from completely hands-on through the introduction of autopilot and flight director to today's fully automatic operation of flight control, flight management, and landing. This has resulted in a situation where pilots are able to do a complete mission 'hands off' from take-off to landing. This has led to serious questions being asked about unmanned operations or, if this is unacceptable, single-pilot operation. To do this successfully will probably require an artificial intelligence (AI) input to aid the single pilot and provide an independent check function.

One reason for introducing automated functions is that it appears to reduce the crew workload. It is partly the introduction of increasing automation that has led to the removal of the flight engineer from the flight deck and to the reduction in the number of pilots to a commonly found complement of two only. However, it has been noted that workload reduction is not always the right result (Billings 1991): *While automation brings with it many advances in terms of pilot workload, it does not always reduce workload and in most cases merely changes the nature of the task.*

Figure 11.3 shows how automation is a likely solution to the conflict of reducing crew complement in the face of a persistent increase in the complexity of the flying task. This situation was predicted in the early concept phase of a UK fighter programme (Seabridge and Skorczewski 2016) by looking at trends in past projects and looking forward.

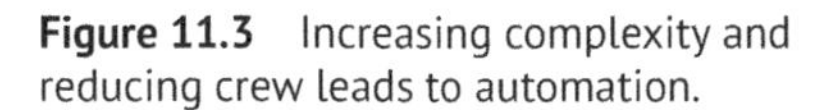

Figure 11.3 Increasing complexity and reducing crew leads to automation.

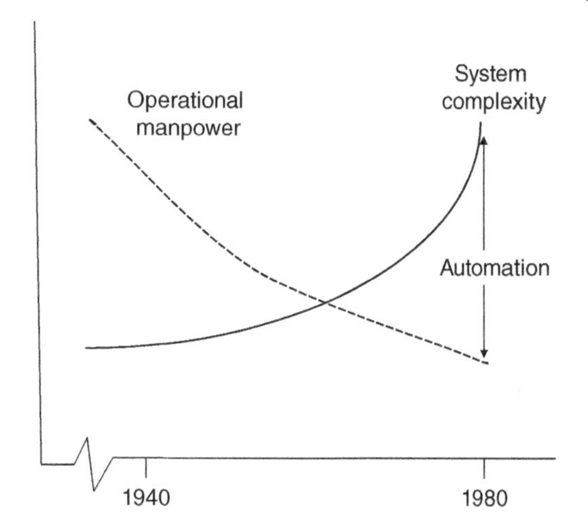

This project was to be designed to meet a particularly stringent set of operational requirements with a single crew member. This would lead to an increasingly complex system and sensor architecture in a multi-role platform where the pilot's task would be to concentrate on prosecuting the mission. The traditional tasks of monitoring and managing every system on the aircraft, whilst flying safely and within limits, were predicted to lead to unprecedented levels of automation in order to alleviate pilot workload. This was likely to lead to deeper integration of systems and more complex software. This risk was accordingly built into the design task.

An important point to note in all discussion of automation is that there will be occasions when the automated system will fail and the pilot must take over. Notable modern examples of this include the descent into the Hudson (Langewiesche 2009) and the 2010 Qantas A380 incident (ATSB 2010). Designers must note that the automated system must be configurable to enable this to happen. It must be clearly comprehensible and must not conflict with the actions of the pilots.

It is important to avoid the situation in which air crew place too much trust in automation or become dependent on it. Is it possible, or even wise to design an automated system that asks for human assistance, or does this add to complexity and cast doubt upon the system?

11.5 Impact on Flight Safety Discussion

There is the potential for system behaviours to arise that may affect flight safety. Although no such incidents have been positively identified to date there is a growing awareness of unusual behaviours and reports are beginning to emerge. These describe incidents and accidents in which there has been access to aircrew and accident data recorder records to explain the incident. Unfortunately, there have been some incidents where there has been no explanation as the aircraft has never been recovered. Some published reports are described in the following notes.

Note 1 The National Transportation Safety Board (NTSB) has reported on the nature of its own task to keep pace with future developments.

The nature of NTSB investigations and the agency's future workload will be shaped by changes in the aviation environment, in particular by increasing technological complexity and growth in general and commercial aviation air traffic, and by important changes in the composition of the air transport fleet. The growth in aircraft system complexity is exponential in many areas, with the most significant trend being the interconnectedness of systems. Current-generation aircraft operate as highly integrated systems with extensive cross-linking. As system complexity grows, so does the concern about hidden design flaws or possible equipment defects. Accidents involving complex systems and events present investigators with new and different failure modes that multiply the number of potential scenarios they must consider. The historically common causes of accidents are occurring less frequently, leaving more challenging accidents to diagnose. In response, the NTSB must develop new investigative processes and training procedures to meet the challenges that the rapid growth in systems complexity presents'.

Note 2 Cummings and Zacharias have described human supervisory control as *the process by which a human operator intermittently interacts with on-board computers receiving feedback and providing commands to a controlling process or task environment which is connected to the computer through actuators and sensors. This means that controls, displays and computers separate the crew from the aircraft and sub-systems they are meant to manage.* (Cummings and Zacharias 2010)

Note 3 Complexity in the human–machine interface has been discussed by Lintern et al. (1999), linking some of the issues with designing for new technology rather than the technology itself.

There is an emerging concern that modern glass cockpits induce information overload. This is sometimes thought to be an inevitable result of the increased complexity and the need for automation that accompanies the transition to high technology. We argue here that the human performance problems created by glass cockpits are not an inevitable consequence of increased hardware complexity or of automation but, instead, are a result of non-functional design that increases complexity at the cockpit interface. The essential danger with computerized interfaces is that many physical design constraints are removed and designers are permitted unheralded opportunities for new information and control formats. Low technology forces the use of functional properties at the interface, but computer technology does not. On the other hand, computer technology does not preclude functional design. Computer technology may offer far broader opportunities for functional design by releasing designers from many physical constraints.

Note 4 An example of complexity and crew interaction with an automated system. In this example it is clear that the aircraft system had a large number of modes of operation.

The ASIANA B-777 incident in 2013 prompted a comment in *Aviation Week & Space Technology* in July 2015: *Complications and distractions aside, over-reliance on automation systems appears to have trumped basic flying skills and crew resource management.*

- *Avionics manufacturers are working to simplify the complex and often confusing human-machine interfaces that hinder rather than help pilots.*

- *There were also 'mode surprises', certain conditions in which modes will transition without the pilot's knowledge.*
- *Rockwell Collins is working on a project to reduce the number of federated automatic flight control modes added to the flight deck. By aligning auto-flight modes with pilot 'goals' – arriving at a certain point at a certain time with a given amount of fuel – researchers were able to design a prototype model that reduces mode choices from 38 to 7.*

Note 5 One point that is worth noting is that modern integrated systems and a high degree of automation have made automated flight path adherence more precise. It should be noted that aircraft flying routes repetitively and automatically will tend towards flying the same route precisely. The accuracy of modern navigation systems is such that a heading and a height will be precisely flown without the variability of the human pilot. There is an increased likelihood of collisions because of this in aircraft flying in free skies mode in close proximity.

Note 6 An example of crew interaction with an automated system in which confusion arises and there is a risk of an accident.

In any system that is both highly automated but also allows operator intervention during various phases of operation, there is a significant risk of operator 'mode confusion' (Billings 1997; Sarter and Woods 1994). As evidenced by numerous airline crashes in the 1990s, during the transition to highly automated flight control and navigation systems mode confusion occurs when an operator attempts to take control of a highly automated system, but does not understand the current mode of automation, i.e. the goals or objectives the automation is trying to achieve. This lack of understanding can cause catastrophic human-system failure due to confusion over who is in control – the human or the system, especially when the desired goal of the operator differs from that of the automation.

An example is given:

...the fatal A320 crash near Strasbourg Entzheim airport in 1992 in which the crew entered a glideslope angle command into the computer which was in vertical descent mode, causing the aircraft to descend into the ground (Hughes 2005). (Cummings and Zacharias 2010).

Note 7 Another example in which integration of systems plays a part was discussed in Chapter 6. This is the issue associated with contaminated cabin air, a complex and as yet unconfirmed issue, said to be caused by an integration of engine, lubricants, and the air-conditioning system. It is alleged that pilots have suffered from neurological disorders leading to enforced retirement and reported incidents of deaths. The impact on flight safety arises because some pilots reported conditions experienced during flight such as fatigue, loss of concentration, bad temper, and dizziness.

Note 8 A number of incidents have been described by Masako Miyagi (2005) in a series of accidents, some fatal, to do with pilots switching modes and in conflicts between an automatic system and reflexive operation by the pilots. These incidents concerned the MD-11 and A300 between 1991 and 1994, thus there were very early indications of serious issues. An interesting observation is that this book contains section titles such as 'Advances in science and technology and new kinds of danger' and 'The pitfalls of computer control'.

The examples given above are by no means comprehensive; more will found by research, still others may have not been recognised and thus have not been reported. There are sufficient warnings to cause designers to be cautious.

As aircraft systems move towards closer integration with external systems and consideration is given to autonomous operations, there is a need for increasing rigour in design and system certification.

11.6 Single-pilot Operations

Since the 1950s, the commercial aviation flight crew has reduced in number from five to the current pilot flying (PF) and pilot non-flying (PNF) personnel. This process, called 'de-crewing' by Harris (2007), has been enabled through advances in technology through the years and it is now deemed acceptable for two pilots to perform what five people used to do in the past. The tasks of the flight engineer, navigator, and radio operator are now mostly ensured by the high level of automation. Harris reported that the historic crew reduction events in commercial aviation have not posed threats to flight safety, and now some in the aviation community believe that the concept of single-pilot operations is possible.

Currently, there is a mandatory requirement, underwritten by the Code of Federal Regulations, which states that all large commercial aircraft must be flown by a flight deck crew of not less than two pilots. Nevertheless, these regulations also specify that all aircraft must be capable of operation by a single pilot from either seat. This has been taken by some to mean that the current flight deck design is already implicitly ready for single-pilot operation. Another constraint for the current regional aircraft is that they must be able to operate not only into and out of big airports but also smaller airports, which might have fewer navigation and landing aids to support automated landings. Moreover, an important operational policy for two pilots to be able to fly a commercial aircraft is crew resource management (CRM). Indeed, it is required that the two pilots coordinate, cooperate, and communicate with respect to CRM policies in order to ensure safe flight procedures.

The concept of single-pilot operation of passenger transport aircraft by scheduled airlines has therefore become a serious topic of debate. It features in the Cranfield University master's degree curriculum as part of the course project for the Air Vehicle Design MSc course. Students attending the course have included it in their group design project. In response to comments made at the project preliminary design review a concept of operations was produced to explain the rationale for a proposed solution. This document and the course material described a potential solution for detecting pilot incapacitation and for safely recovering the aircraft. A potential process for this, from which a set of requirements could be derived, is as follows:

1) Equip the aircraft with a system for monitoring the pilot's vital signs using non-intrusive sensors to monitor, for example, heart rate, breathing, pupil movements or to provide a regular stimulus which demands a pilot response.
2) Measure these inputs and provide a warning should any deterioration be detected using an existing on-board system such as on-board maintenance system.
3) The warning alerts the cabin crew and a senior member is authorised to over-ride the flight deck door security and enter the flight deck to confirm incapacitation and alert air traffic control (ATC) using an agreed code.
4) The warning instructs the flight management system (FMS) to hold the current flying conditions.

5) ATC acknowledges receipt of the warning, identifies the aircraft and, using appropriate security, takes control via the FMS in order to direct the aircraft to a suitable airport where it is able to land.

This process pre-supposes a number of additions to the current method of operation:

1) The set of non-intrusive sensors and an on-board system can be designed to declare the pilot's incapacitation reliably with a very low false alarm rate. Modern leisure wear designed to measure the health of athletes and health enthusiasts may provide a suitable solution.
2) This system must be declared safety critical and suitable architecture needs to be declared.
3) The aircraft must be equipped with a system to allow ATC intervention for remote operation.
4) ATC will need access to a workstation and an uplink to allow remote operation as well as trained (and possible licenced) staff.
5) In the period between the detection of pilot incapacitation and ATC take-over the aircraft is effectively an unmanned air vehicle, and therefore will need to be certificated as such.
6) The acceptance of the regulatory authorities, the airlines, passengers, and insurance underwriters will be needed.
7) The system will need to be secure and terrorist proof.
8) Training will be required for pilots, cabin crew, ATC, and airports likely to receive such aircraft.

11.7 Postscript: Chaos Discussion

The current situation of aircraft systems is that they have reached a state of integration that is illustrated in Figure 11.4. This illustrates how the 'core' or essential systems of the aircraft are themselves tightly integrated and are becoming an integral part of advances in

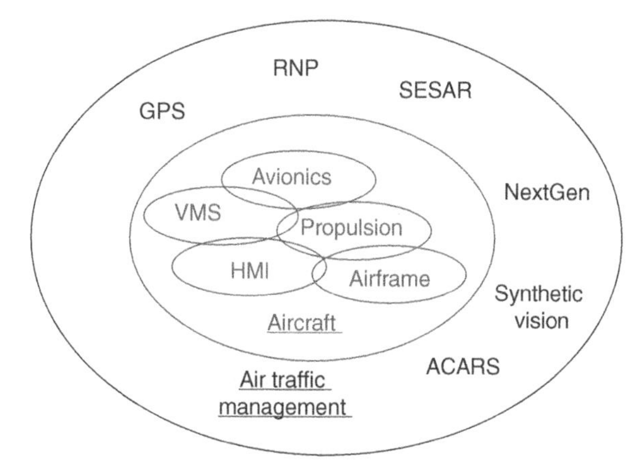

Figure 11.4 Current state of integration of the aircraft 'core' systems. GPS, global positioning system; RNP, required navigation performance; SESAR, Single European Sky air transport management; NextGen, next generation; ACARS, aircraft communications addressing and reporting system.

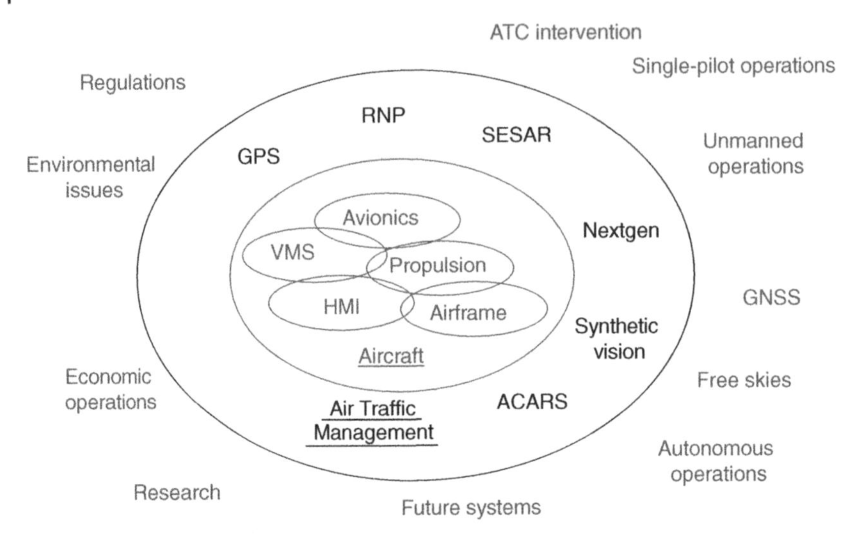

Figure 11.5 Increasing integration. GPS, global positioning system; RNP, required navigation performance; SESAR, Single European Sky air transport management; NextGen, next generation; ACARS, aircraft communications addressing and reporting system; GNSS, global navigation satellite system.

navigation and air traffic management systems aimed at providing efficient navigation. In part this is being driven by environmental issues aimed at reducing fuel burn and potential pollution of the atmosphere.

Figure 11.5 takes this further to illustrate how advances in external systems, many of them ground-based and in continuous operation, are also embracing the aircraft. In this way the controlled operation of systems, with regular and relatively frequent power-up and power-down phases, are now a part of a continuously operating aviation system. Thus there are fewer opportunities for system clocks to be reset and for memories to be refreshed.

The 'core' systems can be represented by their own individual sub-system architectures and can be visualised as a combined architecture. This illustration can be extended to encompass a complete aircraft architecture to give an example of the complexity of modern systems, as shown in Figure 11.6.

This extension can be illustrated as a federated architecture or an integrated modular architecture. Each functional block in the extension should be considered as performing one or more functions which can be implemented in individual avionics boxes or integrated into some form of computing architecture with interconnecting data bus links. The functions are implemented in software and interconnected with some form of data bus. The message is the same no matter how it is implemented – it is complex. The system should be perceived as many functions performed in many computers, with instructions and data in software, and inter-functional data embedded in data bus messages. These systems will probably run asynchronously, there will many data items and many non-linearities. It is likely that some processing structures and some data bus mechanisms will be non-deterministic, which may lead to variability in system timing.

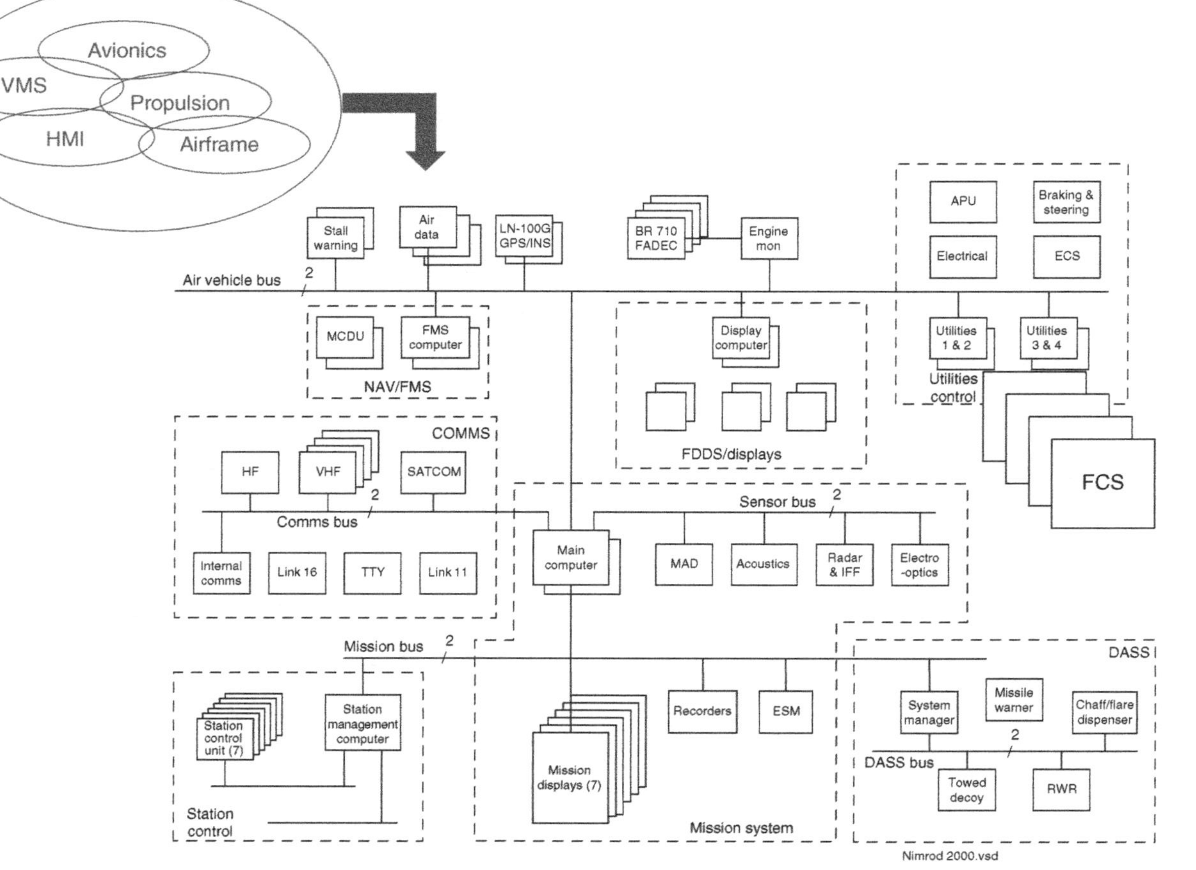

Figure 11.6 Extension to a complete aircraft system. APU, auxiliary power unit; DASS, defensive aids sub-system; ECS, environmental control system; ESM, electronic support measures; FADEC, full authority digital engine control; FCS, flight control system; FDS, flight deck display system; FMS, flight management system; HF, high frequency; IFF, identification friend or foe; MAD, magnetic anomaly detector; MCDU, multi-function display unit; NAV/FMS, navigation/flight management system; RWR, radar warning receiver; SatCom, satellite communications; TTY, teletype; VHF, very high frequency; ATC, air traffic control; GNSS, global navigation satellite system; NextGen, next generation.

The system is also the recipient of random inputs from the human operators and maybe from external sources.

This situation leads one to pose a question about complex real-time systems: Are there any conditions in which the system can become chaotic? Is there a risk that the whole system can enter a state of chaos situation?

At one time the aircraft systems operated for a short time only in any one flight and were then powered down, and thus all conditions were re-set. For fast military jets the power-up time might only be an hour or so, for long-haul commercial airliners it might be 8 hours. Now with air-to-air re-fuelling even fast jets are powered up for many hours, and airliners routinely remain powered up for days at a time. This situation arises on long-haul flight of up to 20 hours where the systems are not shut down during the aircraft turnaround interval. They are operating within an air transport management system that itself operates continuously.

As a result of some unexplained behaviours it has been necessary to impose a mandatory re-boot for the B787 and A350. This was reported in http://MRO-Network.com under the header 'EASA orders periodic reset of A350 internal clock'. The report describes the issue of an EASA airworthiness directive AD 2017-0129 mandating that operators power-down the aircraft systems after a continuous period of operation of 149 hours. This was issued as a response to reports of loss of communications between avionic systems and networks. According to EASA *different consequences have been observed by operators from redundancy loss to a specific function hosted on the remote data concentrators and core processing I/O modules*. Shutting down and powering-up (power re-boot) has had some success in previous incidents. In 2015 an FAA directive directed a re-boot of all B787s that had been powered-up continuously for more than 248 days to prevent a computer internal counter overflow. A year later, software issues were reported to have re-surfaced when flight control modules were found to reset automatically after 22 days of continuous power-up.

There are a number of reasons why the question about chaos arises: aviation systems have moved from relatively short time-scale durations to a situation in which some parts of the systems are in use continuously, and now airborne systems have moved to long duration power-up. In this way they have closely approached the operating scenarios for ground-based commercial systems, which have been known to produce unusual results.

The complexity of systems means that it is difficult to guarantee that some or all of the precursor conditions for chaos will not arise. In its 'normal' state the total system operates with many transactions, iterations, rates, and asynchronous conditions. Inputs to the system will be provided by the operators on board such as first pilot, second pilot, and cabin crew. There are situations in which some use of non-deterministic software and data bus applications may arise. When installed in the aircraft there are many interactions between software and data buses, and many systems contain non-linearities. These are all conditions in which chaos can arise.

- Is it possible for a transition to occur that will disturb the 'normal' state?
- Could this lead to unexpected system behaviour?
- Consider the MH370 incident. Could this be an example of an unexpected event?

Exercises

1 This chapter provides some examples of how the complexity of systems has increased over successive generations. Look at the current literature and see if you can add to these. Do you think that this trend is likely to diminish in the near future? Give your reasons.

2 This chapter has postulated that complex avionic systems may be verging on chaotic behaviour. What is your view? How would you challenge or defend this view?

3 Take a systems architecture that you are familiar with. Can you envisage any circumstances in which conditions for chaos will emerge in the system?

4 Examine the literature surrounding the issue of cabin air quality. What is your own view? How do you think this controversy can be resolved?

References

ATSB. (2010). *Investigation report AO-2010-089. In-flight uncontained engine failure, Airbus A380–842, VH-OQA, overhead Batam Island, Indonesia, 4 Nov 2010*. Available: atsb.gov.aus/publications/investigation reports/2010.

Aviation Week & Space Technology, July 15 2013.

Billings, C.E. (1991). *Human-Centred Aircraft Automation Philosophy: A Concept and Guidelines*. Moffet field, CA: NASA.

Cummings, Mary L & Zacharias, Greg L. (2010). Aircraft Pilot and Operator Interfaces. In: *Wiley Encyclopaedia of Aerospace Engineering, Vol 8, part 38*, Chapter 398. Wiley.

DOT/FAA. (2017). *Definition and measurement of complexity in the context of safety assurance, Final Report DOT/FAA/TC-17/26*.

Harris, D. (2007). A human-centred design agenda for the development of single crew operated commercial aircraft. *Aircraft Engineering and Aerospace Technology* 79 (5): 518–526.

Langewiesche, W. (2009). *Fly by Wire: The Geese, the Glide, the Miracle on the Hudson*. Penguin.

Lintern, G., Waite, T., and Talleur, D.A. (1999). Functional interface design for the modern aircraft cockpit. *International Journal of Psychology* 9 (3): 225–240.

Maier, M.W. and Rechtin, E. (2002). *The Art of Systems Architecting*. ORC Press.

Miyagi, M. (2005). *Serious Accidents and Human Factors*. AIAA.

Moir, I., Seabridge, A., and Jukes, M. (2013). *Civil Avionic Systems*, 2e. Wiley.

Sarter, N.B. and Woods, D.D. (1994). Decomposing automation: autonomy, authority, observability and decision fidelity in the supervisory control of multiple UAVs. *Presence* II (4): 335–351.

Seabridge, AG. (1984). *Study of an Integrated Computing System for the Control and Management of Aircraft Utility Systems*, M. Phil. thesis Lancaster University.

Seabridge, A. and Skorczewski, L. (2016). *EAP: The Experimental Aircraft Programme*. BAE Systems Heritage Department.

Further Reading

The following publications have been consulted to assist the writing of this chapter. There are many more sources and any reader with an interest in the topics discussed here will do well to read further.

Ballin, MG, Sharma, V, Vivona, RA, Johnson, EJ and Ramiscal, E. (2002). A flight deck decision support tool for autonomous airborne operations. In *Proceedings of the AIAA Guidance, Navigation and Control Conference, Monterey, CA, USA.*

Billings, C.E. (1996). *Aviation Automation: The Search for a Human-Centred Approach.* Lawrence Erlbaum Associates Publishers, November 1996.

CAA (2012). *CAP 722: Unmanned Aircraft System Operations in UK Airspace – Guidance.* Civil Aviation Authority.

Comerford, D, Lachter, J, Feary, M, Mogford, R, Battiste, RV and Johnson, W. (2012) *Task allocation for single pilot operations: A role for the ground.*Technical Report, NASA.

Croft, J. (2013). Asiana crash puts focus on training, automation. *Aviation Week and Space Technology* http://www.aviationweek.com/Article.aspx?id=/article-xml/AW07152013p22-596107.xml.

EASA. (2013, May) Air ops and amc/g annex iv part-cat. EASA. Available: http://easa.europa.eu/agency-measures/docs/agency-decisions/2012/2012-018-RAnnex%20to%20ED%20Decision%202012-018-R.pdf.

EASA. (2017) Airworthiness directive CE 2017–0129.

Gleick, J. (1998). *Chaos: Making a New Science.* Vintage Books.

Hansman, J. (2012). *Single pilot operation: Motivation, issues architectures and con-ops.* NASA. Available: http://humansystems.arc.nasa.gov/groups/FDDRL/SPOdownload/presentations/JohnHansman-Single%20Pilot%20Operations.pdf.

Hersman, DA. (2012) *Safety recommendation.* NTSB. Available: http://www.ntsb.gov/doclib/recletters/2012/A-12-050-051.pdf.

Hughes, D. (2005). Incidents reveal mode confusion. *Aviation Week & Space Technology,* January 30, 2005.

ICAO (2005). *ICAO Annex 2 Rules of the Air.* International Civil Aviation Organization.

Longworth, JH. (2014). *Test Flying in Lancashire, Vol 3.* BAe Systems Heritage Department.

Petroski, H. (1992). *To Engineer Is Human.* NY: Vintage Books.

MRO-Network. https://www.mro-network.com/safety-regulatory/easa-orders-periodic-reset-a350-internal-clock. Accessed March 2019.

NASA. (2012). *Single pilot operations technical interchange meeting.* Available: http://human-factors.arc.nasa.gov/groups/FDDRL/SPO/agenda.php.

NTSB: *Emerging aviation trends: potential impact on aircraft accident investigations, 38 – improving the safety and efficiency of aviation.* Available: www.rand.org\pubs\monograph_reports MR1122.1.

Schutte, PC, Goodrich, KH, Cox, DE et al., (2007). *The naturalistic flight deck system: an integrated system concept for improved single-pilot operations,* Technical Report. NASA, Hampton, VA.

Seabridge, A. and Morgan, S. (2010). *Air Travel and Health: A Systems Perspective.* Wiley.

12

Key Characteristics of Aircraft Systems

12.1 Introduction

This chapter provides a simple description of typical aircraft and vehicle systems and will emphasise for each the key factors in the design that affect interfaces, integration, design drivers, and opportunities for modelling. This information will be provided in a table for each system and is intended as a guide for systems engineers. The reference row in the table refers to sources of further information at the end of the chapter. It should be noted that the field of publications is constantly being refreshed by new material or new editions. Students are encouraged to seek out further information as required. Table 12.1 provides a model for the way in which system characteristics will be presented.

Also provided is a description of a process that will enable students to 'size' a system approximately. This will be of use to students involved in project work that requires them to model aircraft projects in terms of mass, power requirement, and dissipation in order to trade-off different designs. This will provide approximate, but sufficient, quantitative data to obtain a first-order approximation of the impact of a system on the whole aircraft. More detailed information must be sought from suppliers of equipment. Since the aerospace industry supplier base is constantly changing through mergers and acquisitions, an internet or library search is recommended.

To help the reader understand the inter-relationship between the flight deck and the major aircraft systems described in this chapter, Figure 12.1 provides an overview, albeit at a very top level.

The aircraft systems are in general controlled by a series of switches and push-buttons grouped on a system-by-system basis on the overhead. Basic system configuration and status information is displayed on the overhead panel but more information can be displayed on request (with the exception of engine displays) as system synoptic displays on the two centre multi-function displays. The synoptic displays may be used by the flight crew for operational purposes; they also permit access to more detailed system information for maintenance activities.

The input to the avionics and mission systems are provided by a series of control panels and display units located on the centre pedestal between captain and first officer. These include the flight management system (FMS) control and display unit (CDU), the

Design and Development of Aircraft Systems, Third Edition. Allan Seabridge and Ian Moir.

Table 12.1 System characteristics.

System title	The name by which the system is usually known
Purpose of system	A brief description of the purpose of the system.
Description	Brief description of the system's physical and functional characteristics.
Safety/integrity aspects	Impact on flight safety or mission availability and redundancy considerations. See notes on dispatch criteria (1, 2). For this reason some of the definitions offered are over-simplified.
Key integration aspects	Opportunities and reasons for integration with other systems.
Key interfaces	Physical, functional or human–machine interfaces.
Key design drivers	Those design drivers having a major impact on systems engineering decisions.
Modelling	Tools available to model the system. Typical characteristics and limitations of the application.
References	See tables in this chapter.
Future considerations	Notes indicating future trends or new requirements arising from integration or environmental considerations.
Notes	1) For civil aircraft some of these criteria vary greatly by vehicle type and systems, route to be flown, and other operational issues and appropriate limitations which may apply. These are defined by the aircraft MMEL as defined by the FAA (2018). 2) Military aircraft will have a similar MMEL equivalent for airworthiness considerations but in addition the availability of mission sensors will dictate whether the allocated mission may be prosecuted or not.

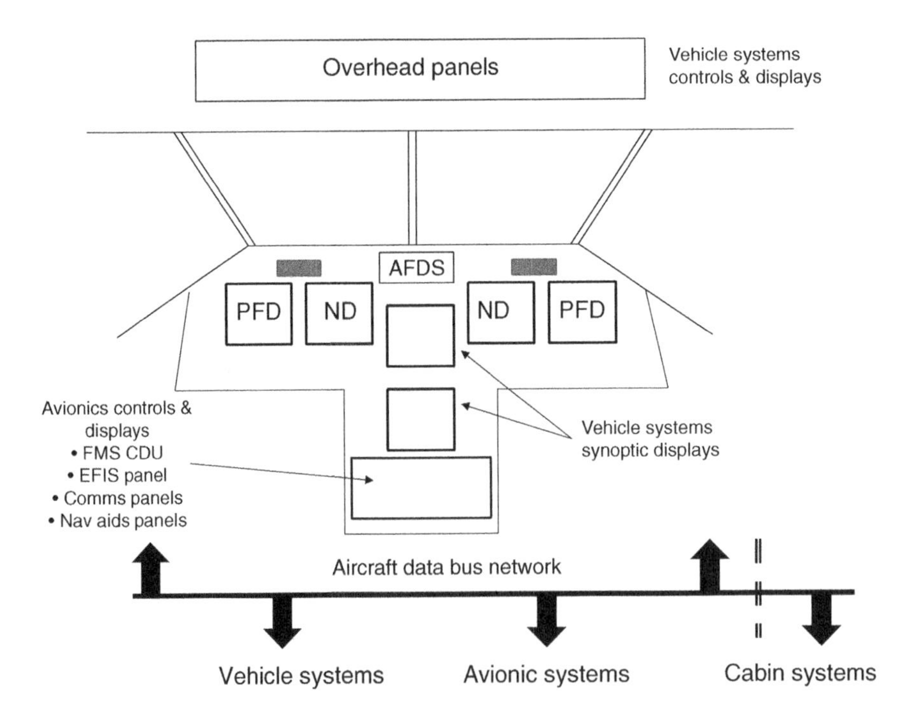

Figure 12.1 Interaction of flight deck and major aircraft systems. AFDS, autopilot and flight director system; Comms, communications; EFIS, electronic flight instrument system; FMS CDU, flight management system control and display unit; Nav Aids, navigations aids; ND, navigation display; PFD, primary flight display.

electronic flight instrument system (EFIS) control panel, and other controls for the communications systems and navigation aids.

Flight crew inputs to the autopilot and flight director system (AFDS) are managed by a dedicated control panel located centrally just below the glare-shield.

Emergency warnings and annunciators are usually situated in a prominent position above the primary flight display (PFD) and navigation display (ND).

All the aircraft and avionics systems are linked by a series of data bus networks to facilitate data exchange.

Some avionics data are provided to the passenger services via a firewall to segregate flight and mission critical data from the less important task of providing data to the passengers.

12.2 Aircraft Systems

Tables 12.2–12.31 show the characteristics of various aircraft systems.

Table 12.2 Propulsion system characteristics.

System title	Propulsion
Purpose of system	To provide thrust for the vehicle and to provide a source of power off-take for electrical power generation, hydraulic power generation, and bleed air for pneumatic systems and environmental cooling system.
Description	Main propulsion units, propulsion control system, interfaces with intake and airframe, air and mechanical power off-takes.
Safety/integrity aspects	Safety critical.
Key integration aspects	Total integration of propulsion unit with intake, nacelle, and jet pipe/nozzle. Integration of off-take drives to prevent vehicle loads having an impact on the engine. May be integration with the flight control system in a highly agile aircraft.
Key interfaces	Airframe installation, thrust bearings, synoptic displays, and throttle and reverse thrust controls.
Key design drivers	Aircraft performance: military – thrust, handling, range/endurance civil – thrust, economy, reliability and availability cost, operating costs.
Modelling	Propulsion test rigs, altitude test facility.
References	Moir and Seabridge (2002, 2008), MacIsaac and Langton (2011), Langton (2006, 2010), (Jackson (2010), Schutte Jeff et al. (2016), Bomani and Hendricks (2016), Pornet (2016), Agarwal (2016).
Sizing considerations	Throttle levers as part of flight deck, treat engine control unit as part of engine. Cooling required for engine oil and impact on fuel system.
Future considerations	Environmental (green) issues are driving consideration of alternative means of propulsion such as electric, open rotor, and geared turbofan. Alternative fuels are also being considered such as hydrogen, liquid/natural gas, and bio-fuels. These will influence installation interfaces and cockpit indications. Noise reduction measures may lead to design to reduce noise or operational restrictions to limit noise in the vicinity of airports.

Table 12.3 Fuel system characteristics.

System title	**Fuel system**
Purpose of system	To store fuel in tanks and to transfer fuel from tank to tank whilst measuring the quantity of fuel on board, and to provide a continuous flow of fuel to the engines. Fuel is often used as a thermal sink for aircraft heat loads, both on- and off-engine, e.g. fuel-cooled oil cooler.
Description	A collection of fuel tanks, fuel gauge probes, interconnecting pipes, and couplings, together with pumps, valves, fuel gauge probes, and level sensors.
Safety/integrity aspects	Safety critical system. Some architectures may dictate multiple transfer paths and multiple lane control electronics. Intrinsic safety to be considered because of fire or fuel vapour explosion risk leading to the need for nitrogen inerting systems, particularly on composite airframes. Safety considerations apply to ground refuelling and maintenance to reduce the risk of fuel/air vapour explosion.
Key integration aspects	Control can be integrated in the utility/vehicle management system or the IMA. Integration with FCS for management of aircraft centre of gravity. Heat exchangers to make use of fuel as a heat sink for engine oil, hydraulic loads, and avionics cooling.
Key interfaces	Propulsion system, ground refuelling, air-to-air refuelling, pilot's displays, and warning system.
Key design drivers	Range/endurance, gauging accuracy, safety.
Modelling	3D modelling (e.g. Catia) to model tanks shapes. Computational fluid dynamics to model fuel flow and slosh characteristics. FloMaster, Bond Graph models available, Matlab.
References	Moir and Seabridge, (2008), Langton et al. (2009, 2010), Langton (2010), Schutte Jeff et al. (2016), Bomani and Hendricks (2016), Freeh (2016).
Sizing considerations	Main components boost pumps, transfer pumps, gauge probes, transfer and shut off valves, fuel pipes, tanks, fuel mass. Fuel as a source of cooling for engine oil, hydraulic fluid, and avionics cooling.
Future considerations	Environmental (green) issues are driving consideration of alternative fuels such as hydrogen, liquid/natural gas, and bio-fuels. These will influence installation interfaces and cockpit indications as well as storage at airfields. Impact on intrinsic safety. In-flight refuelling has been discussed to reduce take-off mass and fuel burn.

Table 12.4 Electrical power generation system characteristics.

System title	**Electrical power generation and distribution**
Purpose of system	To provide a source of regulated AC and DC power to the aircraft systems via bus-bars and circuit protection devices.
Description	AC generators powered by engine off-take, generator control units, batteries, bus bars and feeders, and load protection devices (fuses, circuit breakers, electrical power controllers).
Safety/integrity aspects	Safety critical, multiple redundant system, failure propagation protection.

Table 12.4 (Continued)

System title	Electrical power generation and distribution
Key integration aspects	Pilot overhead panel and synoptic displays. Integration with engine power off-take loads.
Key interfaces	Electrical ground power supply.
Key design drivers	Total electrical load, electrical power quality, safety, reliability.
Modelling	Electrical load analysis by phase of flight – spreadsheet. SABER to model system. Power generation test rig.
References	Moir and Seabridge (2008), Pallett (1987), Moir (2010), Pornet (2016), Xue et al. (2016), Agarwal (2016).
Sizing considerations	Generators and control units, batteries, TRUs, bus bars, distribution panels, contactors.
Future considerations	Future generators integrated into engine shaft. Thermal scavenging devices to produce power and hydrogen fuel cells. Higher demands as a result of the use of non-engine bleed environmental control systems and electro-hydrostatic actuation. Higher in-flight entertainment loads. Advances in battery technology.

Table 12.5 Hydraulic system characteristics.

System title	Hydraulic system
Purpose of system	To provide a source of high-pressure motive energy for actuation mechanisms.
Description	A collection of hydraulic pumps, reservoirs, accumulators, pipes, and couplings.
Safety/integrity aspects	Safety critical system. Redundancy will match that of the highest integrity system – usually the FCS. Hydraulic system redundancy is generally triplex. Diversity of pump power sources is important – engine gearbox, electrical, pneumatic, and emergency sources.
Key integration aspects	Control and monitoring can be integrated in the utility/vehicle management system. Can also be integrated into vehicle domain of IMA.
Key interfaces	Propulsion system power off-take, pilot's overhead panel and synoptic displays, and warning system.
Key design drivers	Actuator power and rates, safety.
Modelling	Matlab/Simulink, hydraulic test rig. Iron Bird rig.
References	Moir and Seabridge (2008), Hunt and Vaughan (1996) Seabridge (2010).
Sizing considerations	Hydraulic pumps, reservoirs, valves, power transfer units, piping, accumulators, heat exchangers.
Future considerations	Reduced need for hydraulic power generation, more use of electro-hydrostatic and electric actuators.

Table 12.6 Secondary power system characteristics.

System title	Secondary power
Purpose of system	Starting of main propulsion system, provision of air and electrical power during ground operations with no engines operating to provide autonomous operation – rapid turnaround.
Description	APU, starter, and connections to airframe systems.
Safety/integrity aspects	Mission critical.
Key integration aspects	Integration with ground facilities.
Key interfaces	Pilot's overhead panel and synoptic displays, secondary sources of electrical and hydraulic power, and cooling air circuits.
Key design drivers	Mass, cost, efficiency, noise.
Modelling	Test rig.
References	Moir and Seabridge (2008), Freeh (2016).
Sizing considerations	APU, fire protection, intake/exhaust hatches and actuation mechanism.
Future considerations	More emphasis on in-flight operable APU. Ground APU contribution to airport noise and pollution.

Table 12.7 Emergency power system characteristics.

System title	Emergency power
Purpose of system	Provision of electrical and/or hydraulic power during period of failure of main propulsion system.
Description	Emergency power unit, for example monofuel (hydrazine) or air operable APU, RAT, electro-hydraulic pumps, hydraulic accumulators, one-shot battery.
Safety/integrity aspects	Part of safety critical analysis – must operate when required.
Key integration aspects	Integration with airframe for optimum intake performance, deployment of RAT for optimum energy extraction from air flow.
Key interfaces	Interfaces with secondary sources of electrical and hydraulic power.
Key design drivers	Availability, effective operation.
Modelling	3D modelling (Catia).
References	Moir and Seabridge (2008), Giguere (2010).
Sizing considerations	Power unit and source of energy.
Future considerations	Fuel cells.

Table 12.8 Flight control system characteristics.

System title	Flight controls
Purpose of system	To translate the pilot's commands into a demand for power to drive primary and secondary control surfaces, to respond to auto-pilot demands for automatic control and stability. For unstable military aircraft to ensure that demands are acted upon rapidly, to limit demands to a safe operating envelope, and constantly react to external aerodynamic conditions.
Description	Demand input sensors, computing system, actuators, position and rate feedback sensors.
Safety/integrity aspects	Safety critical.
Key integration aspects	Integration with air data system, auto-pilot, flight management, propulsion, and landing aids to complete guidance and control integration. Integration with fuel system for cg control in unstable aircraft.
Key interfaces	Electrical system, hydraulic system; air data, and inertial sensors; pilot's effectors, autopilot, FMS and pilot's displays: PFD, ND, synoptic displays, and overhead panel.
Key design drivers	Safety, structural limitations, flight envelope, and performance.
Modelling	Control loop modelling, Iron Bird.
References	Moir and Seabridge (2002, 2008), Lloyd and Tye (1982), Bryson (1994), Raymond and Chenoweth (1993), Pratt (2000), Langton (2006), Weller (2018).
Sizing considerations	Flight control computers, actuators, control column on flight deck, redundancy aspects.
Future considerations	Full electric actuation, integration of controls into IMA.

Table 12.9 Landing gear system characteristics.

System title	Landing gear
Purpose of system	To enable the aircraft to be mobile on the ground, includes nose wheel steering.
Description	Nose gear, main gear, oleos, retracting mechanism, doors, locks, and position monitoring devices.
Safety/integrity aspects	Safety critical – usually provided with a mechanism for manual lowering of the gear if the normal means fails.
Key integration aspects	Integration with airframe to provide for efficient stowage of gear. Weight on wheels signals for other systems, cockpit warning system for indication of safe gear positions.
Key interfaces	Landing gear installation with airframe. Hydraulic and electrical systems, pilot controls and synoptic display.
Key design drivers	Mass, aircraft all-up weight, aborted take-off mass, airfield condition (runway LCN and braking conditions).
Modelling	3D (Catia) modelling of extension and retraction of gear. Iron Bird test rig.
References	Moir and Seabridge (2008), Conway (1957), Currey (1984).
Sizing considerations	Gear, attachments, wheels, brakes and tyres, braking and rejected take-off loads.
Future considerations	All electric actuation.

Table 12.10 Brakes/anti-skid system characteristics.

System title	Brakes/anti-skid
Purpose of system	To allow the aircraft to be decelerated on the ground, to absorb braking energy, and to prevent loss of wheel traction during braking.
Description	Brake discs and pads, braking control system, anti-skid control system, sensors.
Safety/integrity aspects	Safety critical.
Key integration aspects	Highly dynamic integration within the high bandwidth/brake control.
Key interfaces	Interface to brake pedals, weight on wheels sensors, hydraulic and electrical systems.
Key design drivers	Aircraft all-up weight, maximum rejected take-off clearances, landing characteristics, dissipation of brake energy for ramp departure – cooling fans.
Modelling	Dynamic landing test rig.
References	Moir and Seabridge (2008).
Sizing considerations	Braking system and energy source, energy during braking, energy dissipation, and cooling mechanism.
Future considerations	All electric actuations. Use of brakes heat for scavenging devices.

Table 12.11 Steering system characteristics.

System title	Steering
Purpose of system	To provide a means of steering the aircraft under its own power or whilst being towed.
Description	Steering tiller or pedals, actuator acting on nose wheel.
Safety/integrity aspects	Safety affected – failure to steer correctly at high speeds can lead to departure from runway or taxiway.
Key integration aspects	Human factors. Hydraulic system, pilot's displays, including video and wheel-monitoring cameras (some models).
Key interfaces	Integration with flight control to ensure correct hand-over from rudder steering during landing run.
Key design drivers	Taxy way curve radius, landing speeds.
Modelling	CAD.
References	Moir and Seabridge (2008).
Sizing considerations	Steering mechanism and source of energy.
Future considerations	

Table 12.12 Environmental control system characteristics.

System title	Environmental control system
Purpose of system	To provide heating and/or cooling air for passengers, crew, and avionics equipment.
Description	Heat exchangers, cooling systems, air distribution.
Safety/integrity aspects	Safety affected – loss of all cooling can lead to equipment malfunction.
Key integration aspects	Ability to extract air without affecting engine performance.
Key interfaces	Interfaces with engine air off-take. Controlled by environmental control system. Pilot's overhead panel and synoptic displays.
Key design drivers	Crew and passenger comfort, ambient operating conditions – regional or world-wide. Good filtration to reduce biological contamination risk and to produce clean cabin air.
Modelling	Modelling of air flow in ducting using CFD.
References	Moir and Seabridge (2008), Lawson (2010).
Sizing considerations	Volume of cabin, number of occupants, pressurisation, air intakes (drag), air distribution system, cold air units, filters, redundancy, emergency air supply.
Future considerations	Reduced need for engine bleed, trend towards electrical compressors, increased AC power demand.

Table 12.13 Fire protection system characteristics.

System title	Fire protection
Purpose of system	To detect fire or overheat in engine or secondary power bays, and to provide a source of extinguishant.
Description	Overheat or UV detectors installed in a bay to provide wide area coverage, fire extinguisher fluid, and spray nozzles.
Safety/integrity aspects	Major – dormant system with limited test coverage, must operate when required.
Key integration aspects	Local system integration.
Key interfaces	Cockpit warning system.
Key design drivers	Rapid and unambiguous detection mechanism.
Modelling	Simple simulation.
References	Moir and Seabridge (2008), Giguere (2010).
Sizing considerations	Detection loop and control unit, extinguishers.
Future considerations	

Table 12.14 Ice detection system characteristics.

System title	Ice detection
Purpose of system	To detect entry into icing conditions that may lead to the accretion of ice on leading edges of wing, empennage or intake lips.
Description	Ice detector probe.
Safety/integrity aspects	Major.
Key integration aspects	Integration with ice protection system.
Key interfaces	Cockpit warnings.
Key design drivers	Aircraft operating envelope and operating conditions.
Modelling	Simple simulation.
References	Gent (2010).
Sizing considerations	Detector, control unit.
Future considerations	

Table 12.15 Ice protection system characteristics.

System title	Ice protection
Purpose of system	To prevent the build-up of ice and/or to remove ice already formed.
Description	Electrically or hot air heated surfaces, usually air-heated leading edge inner surface, inflatable rubber boots.
Safety/integrity aspects	Safety involved – must work when required or aircraft must rapidly leave icing conditions.
Key integration aspects	Integration with ice detection system.
Key interfaces	Avionics for static air temperature calculations.
Key design drivers	Mass, electrical load, drag.
Modelling	Simple simulation.
References	Moir and Seabridge (2008), Gent (2010).
Sizing considerations	Type of ice protection mechanism and potential electrical load.
Future considerations	Reduced dependence on engine bleed air, may be trend towards electric anti-icing (re B-787).

Table 12.16 External lighting system characteristics.

System title	External lighting
Purpose of system	To ensure that the aircraft is visible to other airspace users and to provide lighting for landing and taxying. Also to provide lighting of company logos.
Description	Wing tip high intensity strobe lights, fuselage strobe lights or anti-collision beacons, logo lights. Military users will include formation lights and air-to-air refuelling probe light.
Safety/integrity aspects	Safety involved.
Key integration aspects	Structure.
Key interfaces	Pilot's overhead panel.
Key design drivers	Regulations, visibility to other aircraft.
Modelling	Simple simulation.
References	Moir and Seabridge (2008).
Sizing considerations	Lamp types, installation.
Future considerations	

Table 12.17 Probe heating system characteristics.

System title	Probe heating
Purpose of system	To provide a means of heating the pitot-static and temperature probes on the external skin of the aircraft to ensure that they are kept free of ice.
Description	Electrical heater built into the probes.
Safety/integrity aspects	Safety critical. Failure of heaters will affect accuracy of air data sensing and will affect cockpit indications and flight and propulsion control system input data.
Key integration aspects	Flight control system, cockpit displays and controls.
Key interfaces	Air – ground/weight-on-wheels signals to other systems.
Key design drivers	Accuracy of air data for flight control and navigation – may be driven by minimum height separation requirements on airways.
Modelling	Simple simulation.
References	Moir and Seabridge (2002).
Sizing considerations	Electrical load.
Future considerations	

Table 12.18 Vehicle management system characteristics.

System title	Vehicle management systems
Purpose of system	To provide an integrated processing and communication system for interfacing with vehicle system components, performing built-in test, performing control functions, providing power demands to actuators and effectors, and communicating with cockpit display.
Description	A number of interfacing and processing units geographically dispersed in the airframe to reduce wiring lengths and a data bus to interconnect the units.
Safety/integrity aspects	Integrity depends on control functions – generally safety involved or safety critical.
Key integration aspects	Integration with avionics systems, displays, and controls.
Key interfaces	Vehicle systems components.
Key design drivers	Safety, availability.
Modelling	Integrated modelling across the systems.
References	Moir and Seabridge (2008), Lloyd and Tye (1982), Spitzer (1993), Principles of Avionics Data Buses (1995), Moir and Seabridge (2010).
Sizing considerations	Number of control and interface units, redundancy.
Future considerations	This function may be integrated into the aircraft integrated modular architecture as part of the utilities domain (see Table 12.50). Integration of FCS.

Table 12.19 Crew escape system characteristics.

System title	Crew escape
Purpose of system	Military – to enable crew to escape from the aircraft under a wide variety of conditions with minimum risk of injury or death, range from high altitude to zero speed, zero altitude.
Description	Rocket-assisted seat equipped with parachute and emergency oxygen.
Safety/integrity aspects	Safety critical – dormant system with limited test coverage, must operate when required.
Key integration aspects	Integration with canopy jettison or shattering mechanism.
Key interfaces	Pilot and personal equipment.
Key design drivers	Clear ejection lines, crew physiology, safety.
Modelling	3D modelling ejection clearance lines, rig test.
References	Moir and Seabridge (2008), Giguere (2010).
Sizing considerations	Seat or escape module.
Future considerations	

Table 12.20 Canopy jettison system characteristics.

System title	Canopy jettison
Purpose of system	To provide a means of removing or fragmenting the canopy material to provide a means of exit for escaping crew.
Description	Rocket-assisted jettison mechanism or miniature detonating cord embedded in canopy material.
Safety/integrity aspects	Safety critical. Danger to ground crew if not isolated on the ground.
Key integration aspects	Integrated with crew escape initiation.
Key interfaces	
Key design drivers	Must allow the crew to exit the aircraft without injury.
Modelling	Physical models or prototypes.
References	Moir and Seabridge (2008).
Sizing considerations	Canopy, jettison type, and mechanism.
Future considerations	

Table 12.21 Biological and chemical protection system characteristics.

System title	Biological and chemical protection
Purpose of system	To protect the crew from the toxic effects of chemical or biological contamination.
Description	Filtered air and oxygen supply, protective clothing and respirators, wash-down facility.
Safety/integrity aspects	Mission critical.
Key integration aspects	Human factors, operability of controls with gloves.
Key interfaces	Not available.
Key design drivers	Operator safety.
Modelling	Complex simulation.
References	
Sizing considerations	Threat substances, filters, air crew clothing, respirator mechanism.
Future considerations	New threat materials.

Table 12.22 Arrestor hook system characteristics.

System title	Arrestor hook
Purpose of system	To stop the aircraft by engaging a runway arrestor gear wire if the brakes should fail, normal method of stopping naval carrier borne aircraft.
Description	Arrestor hook stowed at rear of aircraft and deployed in emergency.
Safety/integrity aspects	Safety critical – must operate when required.
Key integration aspects	Airframe mass and speed (energy) requirements.
Key interfaces	Interface with in-service arrestor gear at military airfields and carriers.
Key design drivers	Safety, emergency operations.
Modelling	Stress calculation, 3D (e.g. Catia) modelling.
References	Moir and Seabridge (2008).
Sizing considerations	Energy requirement, hook, lock/release mechanism, attachment.
Future considerations	

Table 12.23 Brake parachute system characteristics.

System title	Brake parachute
PURPOSE OF SYSTEM	Used on military types and some commercial prototypes to decelerate the aircraft for ultra-short stopping distances or on short runways.
Description	Parachute normally stowed in a canister in the aircraft rear fuselage so that the parachute or canister can be jettisoned if required.
Safety/integrity aspects	Minor – dormant system with limited test coverage, must operate when required.
Key integration aspects	Single system – no opportunity for redundancy.
Key interfaces	Simple manual operation by pilot.
Key design drivers	Aircraft landing speed, stopping distance. Aircraft support – parachute repackaging.
Modelling	Simple simulation.
References	Moir and Seabridge (2008).
Sizing considerations	Energy requirement, hook, lock/release mechanism, attachment.
Future considerations	

Table 12.24 In-flight refuelling system characteristics.

System title	In-flight refuelling
Purpose of system	To enable military aircraft to obtain fuel from a tanker in flight and extend range/airborne capability.
Description	Receptacle for fuel hose from tanker – generally a retractable probe fitted to UK/European aircraft, and a receptacle mating with a tanker probe on US aircraft.
Safety/integrity aspects	Mission critical. Some safety aspects due to aircraft maintaining close formation.
Key integration aspects	Connection to fuel system to allow control of refuel to a recipient.
Key interfaces	Interface with tanker refuelling device – drogue/probe.
Key design drivers	Fuel quantity on offer, numbers of recipients on refuelling station(s), required transfer rates.
Modelling	Flight test.
References	Moir and Seabridge (2008), Purdy (2010).
Sizing considerations	Probe type, actuating mechanism, source of energy.
Future considerations	May be considered for civil aircraft applications for environmental reasons.

Table 12.25 Galley system characteristics.

System title	Galley
Purpose of system	To provide a safe and hygienic method of food preparation and cooking for passengers and crew. For refrigerated products very precise health and safety requirements must be applied.
Description	Storage, refrigeration, and cooking (heating and microwave) appliances.
Safety/integrity aspects	May be mission critical for long-range flights. Health and safety regulations, crew electrical shock, and fire risks to be minimised.
Key integration aspects	Interface with primary electrical system which includes precise fault protection schemes (the galley is an airline furnished item). The galley/passenger provision power requirements for a long-range passenger aircraft may equate to ~40–50% of overall connected load.
Key interfaces	Interfaces with standard airline provisions supplier for roll-on, roll-off modules and food packaging.
Key design drivers	Health and safety and customer comfort/preference.
Modelling	Load analyses performed by galley supplier.
References	Moir and Seabridge (2002).
Sizing considerations	Number of passengers and cabin areas, number of galleys, galley equipment and trolleys, electrical loads.
Future considerations	Reduced need for galley in short-haul budget flights.

Table 12.26 Passenger evacuation system characteristics.

System title	Passenger evacuation
Purpose of system	To allow safe evacuation of passengers from the cabin when the aircraft is on the ground or has ditched in water.
Description	Emergency exit doors, evacuation chutes, life vests, and fully equipped rafts.
Safety/integrity aspects	Must be available when required.
Key integration aspects	Door and slide operation. Flight deck awareness.
Key interfaces	Passenger and evacuation requirements and demonstration. Interface with airports for very large aircraft and high passenger numbers.
Key design drivers	Availability, passenger safety.
Modelling	Mock-ups and evacuation test rigs.
References	Giguere (2010).
Sizing considerations	Number of passengers, number of exits and escape equipment.
Future considerations	

Table 12.27 In-flight entertainment system characteristics.

System title	IFE systems
Purpose of system	To provide audio and video entertainment for passengers at their seats.
Description	Networked audio and video signals to cabin screens or seat-located devices.
Safety/integrity aspects	Dispatch critical for passenger preference reasons.
Key integration aspects	Large-scale integration of COTS system which needs a firewall between it and the avionics needed to fly the aircraft.
Key interfaces	Passengers, flight crew, and content providers.
Key design drivers	Passenger satisfaction, marketing appeal.
Modelling	Simulation and integration off-aircraft.
References	Moir and Seabridge (2002).
Sizing considerations	Number of seats, cabin class variations, electrical loads, impact on cabin heat load.
Future considerations	Provision of email and text.

Table 12.28 Telecommunications system characteristics.

System title	Telecommunications
Purpose of system	To allow passengers to make telephone calls and access the Internet in flight. Possible streaming video/TV.
Description	In-seat telephone handsets and personal computer/portable electronic device charging capability.
Safety/integrity aspects	None.
Key integration aspects	Aircraft communications antennas.
Key interfaces	Passenger seating, communications, cabin crew.
Key design drivers	Passenger satisfaction, marketing.
Modelling	Integrated with IFE.
References	Moir and Seabridge (2002).
Sizing considerations	Number of seats, cabin class variations, electrical loads, impact on cabin heat load.
Future considerations	Provision of email and text.

Table 12.29 Toilet and waterwaste system characteristics.

System title	Toilet and wastewater
Purpose of system	To provide hygienic management of toilets and water waste.
Description	Provision of flushing toilets, hot and cold water, and disposal.
Safety/integrity aspects	Dispatch critical due to the implications of the inability of passengers to use toilet facilities.
Key integration aspects	Human factors, cabin furnishings, safety.
Key interfaces	Ground waste disposal and water replenishment systems.
Key design drivers	Passenger satisfaction, hygiene, health, safety and environmental regulations.
Modelling	Simple simulation.
References	
Sizing considerations	Number of passengers, cabin class variations, health and safety.
Future considerations	

Table 12.30 Oxygen system characteristics.

System title	Oxygen
Purpose of system	To provide a source of breathable oxygen for crew members and passengers.
Description	Commercial – to cover descent to safe altitude in the event of pressurisation loss: bottled gaseous oxygen for pilots with quick-don masks. Oxygen masks for passengers and bottled oxygen or candles. Military – continuous pressure breathing from liquid oxygen or on board oxygen generation system.
Safety/integrity aspects	Commercial – must be available on demand to enable pilots to fly the aircraft to a safe altitude, must be available for passenger safety and comfort. Military – pressure oxygen must be available at all times in combat aircraft. Supply also available on ejection seat.
Key integration aspects	Commercial – integration with emergency system. Military – integration with ECS, human factors and crew escape systems.
Key interfaces	Human factors.
Key design drivers	Autonomous operation or availability of LOX or gaseous oxygen at remote sites.
Modelling	Simple simulation.
References	Moir and Seabridge (2008), Giguere (2010).
Sizing considerations	Number of passengers, type of breathing air supply, emergency sources.
Future considerations	

Table 12.31 Cabin and emergency lighting system characteristics.

System title	Cabin and emergency lighting
Purpose of system	To provide general lighting for the cabin and galley, reading lights, exit lighting and emergency lights to provide a visual path to the exits.
Description	General light in the cabin ceiling, reading lights with personal controls above each seat, emergency lighting.
Safety/integrity aspects	Must be available for emergency evacuation – dispatch critical.
Key integration aspects	Integration with other emergency systems.
Key interfaces	Normal and emergency power generation system and batteries.
Key design drivers	Human factors for lighting, safety, passenger satisfaction, health and safety regulations.
Modelling	Evacuation mock-up.
References	Moir and Seabridge (2002), Giguere (2010).
Sizing considerations	Size of cabin, number of exits.
Future considerations	

12.3 Avionics Systems

Tables 12.32–12.50 show the characteristics of various avionics systems.

Table 12.32 Cockpit displays and controls system characteristics.

System title	Cockpit displays and controls
Purpose of system	To provide the crew with information and warnings with which to operate the aircraft.
Description	The cockpit will be equipped with normal and emergency displays, control inceptors, and control switches to enable the crew to access and control all aircraft functions.
Safety/integrity aspects	Variable from safety critical to safety involved depending upon display/unit concerned and level of display redundancy.
Key integration aspects	Human factors. For military aircraft may need to be compatible with night vision goggles.
Key interfaces	Cockpit design and structure. Refer to Figure 12.1.
Key design drivers	Human factors, safety, pilot workload.
Modelling	Rapid prototyping, VAPS, altitude lighting test facility. Avionics integration rig.
References	Moir and Seabridge (2002), Jukes (2003), Pallett (1992), Rankin and Matolak (2010), Atkin (2010).
Sizing considerations	Number of display units, display computers, interfaces, redundancy, emergency displays, head-up displays.
Future considerations	Synthetic vision, more automation and integration. Gesture control.

Table 12.33 Communications system characteristics.

System title	Communications
Purpose of system	To allow two-way communication between the aircraft and air traffic control, other aircraft, and co-operating forces.
Description	Transmitting and receiving systems, antennas, personal equipment – headsets, mikes, speakers. For data link applications – terminals, crypto devices.
Safety/integrity aspects	Mission critical.
Key integration aspects	Antenna operability, drag, integration with FMS for auto-tuning.
Key interfaces	Structure – pressurisation sealing of antennas.
Key design drivers	All-weather communications, interface with emergency channels.
Modelling	Integrated with FMS.
References	Burberry (1992), Hall and Barclay (1980).
Sizing considerations	Types of radios, control panels, antennas, dissipation, electrical loads.
Future considerations	

Table 12.34 Navigation system characteristics.

System title	Navigation
Purpose of system	To provide world-wide, high accuracy navigation capability.
Description	Inertial, laser or global positioning system based.
Safety/integrity aspects	Mission critical with safety implications.
Key integration aspects	Integration with avionics and mission system.
Key interfaces	Structure, avionics.
Key design drivers	Accurate world-wide navigation – ATM (civil) or GATM (military).
Modelling	Avionics integration rig, mission system integration rig.
References	Moir and Seabridge (2002), Kayton and Fried (1997), Galotti (1998).
Sizing considerations	Navigation sensors, navigational aids, redundancy, antennas.
Future considerations	

Table 12.35 Flight management system characteristics.

System title	Flight management system
Purpose of system	To provide a means of entering and executing flight plans and allowing automatic operation of the aircraft in accordance with those plans.
Description	Flight management computers and control and display unit to enter and modify flight plans and tune navigation aids.
Safety/integrity aspects	Mission critical.
Key integration aspects	Navigation system and navigation aids, cockpit lighting, human factors.
Key interfaces	Cockpit location.
Key design drivers	Ease of use, accessibility, pilot workload, efficient route management.
Modelling	Integration rig.
References	Moir and Seabridge (2002), Cramer et al. (2010), Gradwell (2010).
Sizing considerations	Flight management control and display unit(s) on flight deck, redundancy.
Future considerations	

Table 12.36 Automated landing aids system characteristics.

System title	Automated landing aids
Purpose of system	To provide a means of automatic/assisted landing at airports world-wide.
Description	Ground-based antennas providing standard radio frequency beam at an angle and direction that facilitates a safe approach and landing pattern, associated beacons and markers. Airborne system to detect the beam and warn of deviations from the beam. Ground-based systems include ILS or MLS. Space-based systems using GPS are also used.
Safety/integrity aspects	Not safety critical.
Key integration aspects	Integration with flight management system, auto-pilot or flight director, ground-based landing system.
Key interfaces	Flight management system, flight control system.
Key design drivers	Safety and category of approach involving decision height and visibility.
Modelling	Avionics integration rig.
References	Moir and Seabridge (2002).
Sizing considerations	Type of landing aid, antennas.
Future considerations	

Table 12.37 Weather radar system characteristics.

System title	Weather radar
Purpose of system	Commercial – weather. Military – airborne or ground based targets, air, ground or sea surveillance, weather.
Description	Suitable antenna and radome, transmitter/receiver, radar processing, cooling system.
Safety/integrity aspects	Mission/dispatch critical.
Key integration aspects	Commercial – cockpit displays. Military – cockpit displays, mission system consoles, weapon system, mission computer.
Key interfaces	Radome with required transmission characteristics.
Key design drivers	Operational requirement, modes of search required.
Modelling	Avionics integration rig.
References	Moir and Seabridge (2002).
Sizing considerations	Antenna, transmitter/receiver, display.
Future considerations	

Table 12.38 Transponder system characteristics.

System title	Transponder, IFF/SSR
Purpose of system	To provide a response to ground interrogations which identify the aircraft and provide information relating to position and height. To provide a response to aircraft equipped with TCAS mode S transponders.
Description	Receiver, transponder, antennas. Known as IFF in military and ADS-B in civil applications.
Safety/integrity aspects	Mission critical – loss of operation will lead to air traffic violations. Military aircraft will be asked to leave the airways.
Key integration aspects	Integration with TCAS. Antenna may be shared with other RF devices using the same frequencies.
Key interfaces	Communication system, air traffic control.
Key design drivers	Identification of aircraft identification and height for air traffic control, for military aircraft – co-operative operations in combat zones.
Modelling	Avionics integration rig.
References	Moir and Seabridge (2002, 2006).
Sizing considerations	Antennas.
Future considerations	

Table 12.39 Traffic collision and avoidance system characteristics.

System title	Traffic collision avoidance system
Purpose of system	To reduce the risk of collision with other aircraft.
Description	Transponder-based control unit to interrogate aircraft within a certain spherical volume of the carrier aircraft and an indication and warning system.
Safety/integrity aspects	Dispatch critical for certain routes.
Key integration aspects	Cockpit displays, mission computing, navigation system, navigation aids, human factors.
Key interfaces	IFF/SSR mode S, cockpit displays.
Key design drivers	Safe operation in airport terminal areas and designated air lanes.
Modelling	Avionics integration rig.
References	Moir and Seabridge (2002).
Sizing considerations	Display type, control unit.
Future considerations	Automatic response to collision warnings.

Table 12.40 Ground proximity warning system/terrain avoidance warning system system characteristics.

System title	GPWS/TAWS
Purpose of system	To reduce the risk of aircraft flying into the ground or into high ground.
Description	Provides a series of advisory warnings for the flight crew when the aircraft is approaching a hazardous situation.
Safety/integrity aspects	Safety implications.
Key integration aspects	Cockpit displays, mission computing, navigation system, navigation aids, human factors.
Key interfaces	Radar altimeter, GPS and pilot's displays and warning systems.
Key design drivers	Reduce risk of accidents due to flight crew loss of situational awareness and subsequent controlled flight into terrain.
Modelling	Avionics integration rig.
References	Moir and Seabridge (2002).
Sizing considerations	Display type, control unit.
Future considerations	

Table 12.41 Distance measuring equipment system characteristics.

System title	DME
Purpose of system	To provide a measure of distance from a known beacon.
Description	Receiver tuned by flight management system to appropriate beacons along routes.
Safety/integrity aspects	May be mission critical.
Key integration aspects	Cockpit displays, mission computing, navigation system, navigation aids, human factors.
Key interfaces	Tuning by FMS where an integrated system is fitted.
Key design drivers	Navigational accuracy and location/availability of DME beacons.
Modelling	Avionics integration rig.
References	Moir and Seabridge (2002).
Sizing considerations	Control unit, antenna.
Future considerations	

Table 12.42 Automatic direction finding system characteristics.

System title	ADF
Purpose of system	To provide bearing from a known beacon.
Description	Antennas and control unit.
Safety/integrity aspects	Not safety critical.
Key integration aspects	Cockpit displays, mission computing, navigation system, navigation aids, human factors, communications.
Key interfaces	Tuning by FMS where an integrated FMS is fitted.
Key design drivers	Regulations, ease of navigation.
Modelling	Avionics integration rig.
References	Moir and Seabridge (2002).
Sizing considerations	Control unit, antenna.
Future considerations	

Table 12.43 Radar altimeter system characteristics.

System title	Radar altimeter
Purpose of system	To provide an absolute reading of height above the ground or sea.
Description	One or more antennas sends a signal to the surface and reads the return signal to calculate height above the surface. This is used for display or by other systems.
Safety/integrity aspects	Safety involved. Low flying, manoeuvrable aircraft will need antennas to be sited so that one antenna is always operable in high-g turns.
Key integration aspects	Cockpit displays, mission computing, navigation system, navigation aids, human factors.
Key interfaces	Structure – antenna.
Key design drivers	Accuracy of height measurement, independence from barometric conditions.
Modelling	Avionics integration rig.
References	Moir and Seabridge (2002).
Sizing considerations	Antennas, display.
Future considerations	

Table 12.44 Automatic flight control system characteristics.

System title	Automatic FCS
Purpose of system	To provide an automatic means of flying the aircraft during routine routes, automatic landing and to perform standard mission profiles and search patterns.
Description	Control unit and actuators connected to FCS and engine control. May be direct demands in FCS and engine control system.
Safety/integrity aspects	Primary flight control is safety critical. AFDS is mission critical.
Key integration aspects	FCS, engine control system, flight management system, human factors.
Key interfaces	Human factors.
Key design drivers	Pilot workload reduction, aircraft economy.
Modelling	Avionics integration rig, iron bird.
References	Pratt (2000).
Sizing considerations	Control panel, actuators, redundancy.
Future considerations	

Table 12.45 Air data system characteristics.

System title	Air data system
Purpose of system	To provide information to aircraft system on air pressures (total pressure and static pressure) and to convert these pressures into signals representing airspeed, altitude, and Mach number.
Description	Pitot probes and static vents (maybe combined) located in the airstream.
Safety/integrity aspects	Safety critical – used by flight control system, propulsion system, navigation and cockpit displays.
Key integration aspects	Integrated with navigation system, guidance and control, sole source of critical air data.
Key interfaces	Airframe, drag, probe heating.
Key design drivers	Air data accuracy.
Modelling	
References	Moir and Seabridge (2002, 2008).
Sizing considerations	Probes, electrical loads.
Future considerations	

Table 12.46 Accident data recording system characteristics.

System title	ADR
Purpose of system	To continuously record specified aircraft parameters for use in analysis of serious incidents.
Description	Data acquisition interfaces to relevant systems and continuous recording or solid-state bulk memory store. Locator beacon to aid recovery.
Safety/integrity aspects	Dispatch critical.
Key integration aspects	Data bus types.
Key interfaces	Relevant systems sensors.
Key design drivers	Regulations, crash survivable memory – impact, immersion and fire.
Modelling	Avionics integration rig.
References	Moir and Seabridge (2002).
Sizing considerations	Recording unit, special sensors.
Future considerations	

Table 12.47 Cockpit voice recording system characteristics.

System title	CVR
Purpose of system	To provide a continuous record of specified aircrew speech for use in analysis of serious incidents.
Description	Cockpit microphones and recording system.
Safety/integrity aspects	Dispatch critical.
Key integration aspects	Cockpit environment, communications, human factors.
Key interfaces	
Key design drivers	Regulations, crash survivable – impact, immersion and fire.
Modelling	Avionics integration rig.
References	Moir and Seabridge (2002).
Sizing considerations	Recording unit, microphones.
Future considerations	

Table 12.48 Prognostics and health management system characteristics.

System title	PHM
Purpose of system	To provide a continuous record of systems performance and failures. To use this information to determine trends and declining system health.
Description	Function connected to data buses and system LRIs to extract information and perform appropriate algorithms and output results to data storage or for transmission to the ground.
Safety/integrity aspects	Non safety critical.
Key integration aspects	All systems and ground aspect of maintenance management. Integration with data link for transmission of data to ground.
Key interfaces	All data bus and systems, ground aspect of maintenance, data link for download.
Key design drivers	
Modelling	Avionics integration rig.
References	Moir and Seabridge (2002), Srivastava et al. (2010).
Sizing considerations	Recording unit.
Future considerations	

Table 12.49 Internal lighting system characteristics.

System title	Internal lighting
Purpose of system	To provide a balanced illumination of cockpit panels to aid flight in poor or bright ambient lighting conditions and at night.
Description	Integral panel lighting, flood lighting, wander lights.
Safety/integrity aspects	Emergency lighting is required.
Key integration aspects	Integrated into cockpit design and lighting control system. May need to be compatible with night vision goggles.
Key interfaces	
Key design drivers	Human factors.
Modelling	By simulation or mock-up in lighting test facility. Altitude lighting test facility.
References	
Sizing considerations	Electrical loads.
Future considerations	

Table 12.50 Integrated modular architecture.

System title	IMA
Purpose of system	To provide a computing framework for the avionic systems, often divided into domains for cabin systems, energy management, and utility systems.
Description	Computing system with remote data concentrators to provide interfacing and control of systems based on a data bus, commonly ARINC 664.
Safety/integrity aspects	Segregation and redundancy as required to maintain integrity of redundant systems.
Key integration aspects	Performs as integrating medium and control for all avionics and utility systems.
Key interfaces	All systems analogue, discrete, digital for all avionics and utility systems.
Key design drivers	Safety, control.
Modelling	Modelling of individual systems and integrated model on test bench.
References	
Sizing considerations	Number of interfaces, processing requirement, throughput.
Future considerations	

12.4 Mission Systems

Tables 12.51–12.64 show the characteristics of various mission systems.

Table 12.51 Attack or surveillance radar system characteristics.

System title	Attack or surveillance radar
Purpose of system	To provide information on hostile and friendly targets for attack, airborne early warning or surface surveillance.
Description	A radar antenna and transmitter/receiver with appropriate displays. Attack aircraft house the antenna in the nose, whilst surveillance aircraft may have the antenna mounted in the nose, nose and tail, or in radomes mounted on the upper surface of the aircraft. Active sensor.
Safety/integrity aspects	Mission critical.
Key integration aspects	Integration with mission computing, display systems, weapon aiming systems. Power source for scanner – hydraulic or electric motor.
Key interfaces	Radome.
Key design drivers	Mission success, cost, performance.
Modelling	
References	Skolnik (1980), Schleher (1978), Walton (1970), Oxlee (1997), Airey and Berlin (1985), Stimson (1998), Rigby (2010), Moir and Seabridge (2006).
Sizing considerations	Antenna, antenna drive mechanism, radome, transmitter/receiver, cooling system, display.
Future considerations	Electronic scan (E-scan) will require significant power and may need cooling.

Table 12.52 Electro-optical system characteristics.

System title	EOS
Purpose of system	To provide passive surveillance of targets.
Description	Electro-optical sensors installed in a fuselage-mounted, steerable turret or in an underwing pod. Infra-red, ultraviolet, and TV sensors are able to provide images in poor visibility. Passive sensor.
Safety/integrity aspects	Mission critical.
Key integration aspects	Integration with mission computing and displays.
Key interfaces	Turret to fuselage or pod to pylon station.
Key design drivers	Mission success, cost, performance.
Modelling	Mission system test rig.
References	Moir and Seabridge (2006).
Sizing considerations	Sensor turret (drag), cooling system, deployment and steering mechanism.
Future considerations	

Table 12.53 Electronic support measures system characteristics.

System title	**ESM**
Purpose of system	To provide emitter information, range and bearing of hostile transmitters.
Description	A set of antennas to detect radar and RF transmissions, equipment to analyse the detected signals to determine their most likely source, and the ability to detect the direction of arrival of the signals. An on-board data base allows the signals to be analysed to determine the type of transmitter, and the most likely platform carrying the transmitters. Passive sensor.
Safety/integrity aspects	Mission critical.
Key integration aspects	Integration with mission computing and data link for access to remote intelligence data bases.
Key interfaces	Weapons systems operator; blanking of host RF equipment to avoid interference.
Key design drivers	Intelligence, self-protection.
Modelling	Mission system test rig.
References	Schleher (1999), Bamford (2001), Van Brunt (1995), Poisel (2003), Adamy (2003).
Sizing considerations	Antennas, workstation/displays.
Future considerations	

Table 12.54 Magnetic anomaly detector system characteristics.

System title	**MAD**
Purpose of system	To confirm the presence of large metallic objects under the sea (submarines) prior to attack.
Description	A sensitive magnetic sensor mounted clear of any items of fuselage likely to cause interference. Used to confirm the presence of a submarine by maritime patrol aircraft.
Safety/integrity aspects	Mission critical.
Key integration aspects	Mission computing and displays.
Key interfaces	Location so that there is no interference with the sensitive sensor.
Key design drivers	Mission success, cost performance.
Modelling	Mission system test rig.
References	
Sizing consideration	MAD sensor head, boom, display/chart recorder.
Future considerations	

Table 12.55 Acoustic system characteristics.

System title	Acoustic sensors
Purpose of system	To provide a means of detecting and tracking the passage of underwater objects.
Description	Passive and active sonobuoys are dispensed from the maritime patrol aircraft and provide a means of acoustic detection of submarines. Signals are transmitted back to the aircraft for analysis.
Safety/integrity aspects	Mission critical.
Key integration aspects	Integration with mission computing and displays.
Key interfaces	Sonobuoy dispensers in fuselage and potential depressurisation risk.
Key design drivers	Mission success, performance.
Modelling	Mission system test rig, acoustic test ranges.
References	Urick (1982, 1983), Gardner (1996).
Sizing considerations	Sonobuoy storage, sonobuoys (role fit), dispensers, workstation, antennas.
Future considerations	

Table 12.56 Mission computing system characteristics.

System title	Mission computing
Purpose of system	To collate the sensor information and to provide a fused data picture to the cockpit or mission crew stations.
Description	Suitable architecture computing and interfacing system, appropriate data transmission systems, recording, data loading.
Safety/integrity aspects	Mission critical.
Key integration aspects	Integration with avionic systems, cockpit, sensors. Human factors.
Key interfaces	Avionic and mission system data buses.
Key design drivers	Mission success, performance.
Modelling	Operational analysis modelling, mission system test rig.
References	Moir and Seabridge (2002), Jukes (2003).
Sizing considerations	Mission computer and recorders.
Future considerations	

Table 12.57 Defensive aids system characteristics.

System title	Defensive aids
Purpose of system	To provide a means of detecting missile attack and deploying countermeasures.
Description	A suite of sensors to detect missile approach, missile plume or missile homing radar, warning system and countermeasures such as chaff and flare, towed radar decoy, and active jamming.
Safety/integrity aspects	Mission critical.
Key integration aspects	Mission computing, cockpit, countermeasures.
Key interfaces	Structure.
Key design drivers	Mission success, self-protection.
Modelling	Mission system test rig.
References	Moir and Seabridge (2006).
Sizing considerations	Antennas, antenna pods, workstation/display, countermeasures dispensers.
Future considerations	

Table 12.58 Weapon system characteristics.

System title	Weapon system
Purpose of system	To arm, direct, and release weapons from the aircraft weapon stations.
Description	System for management of external or internal stores, fuselage, wing or bomb bay carriers or pylons for weapons carriage, and safe methods of emergency release.
Safety/integrity aspects	Mission critical. Weapon safety to prevent inadvertent release. Must meet ordnance safety standards.
Key integration aspects	Navigation, mission computing, aerodynamics, separation of wiring from all other wiring or sources of energy to prevent inadvertent release.
Key interfaces	Strong points on wing, fuselage and bomb bay, weapons loading and arming.
Key design drivers	Mission success, ordnance safety, probability of kill.
Modelling	
References	Rigby (2010).
Sizing considerations	Pylons (wing, fuselage or bomb bay), weapons (role fit), cockpit controls.
Future considerations	

Table 12.59 Station keeping system characteristics.

System title	Station keeping
Purpose of system	To provide a means of safely maintaining formation in conditions of poor visibility, especially for large transport aircraft.
Description	Detection system and separation warning.
Safety/integrity aspects	Safety involved.
Key integration aspects	Communications.
Key interfaces	
Key design drivers	Safety, safe operation of crew and aircraft, mission success.
Modelling	
References	
Sizing considerations	Display.
Future considerations	

Table 12.60 Electronic warfare system characteristics.

System title	Electronic warfare
Purpose of system	To detect and identify enemy transmitters, to collect and record traffic, and if necessary to provide a means of jamming transmissions.
Description	Antennas to detect a wide spectrum of signals for COMINT and identification of radars for SIGINT.
Safety/integrity aspects	Mission critical.
Key integration aspects	Antenna integration, mission computing, on-board intelligence database.
Key interfaces	
Key design drivers	Accuracy of detection and location, need to obtain intelligence on new emitters and current asset deployment.
Modelling	Mission system test rig.
References	Schleher (1999), Bamford (2001), Van Brunt (1995), Poisel (2003), Adamy (2003).
Sizing considerations	Antennas, antenna pods, receivers.
Future considerations	

Table 12.61 Camera system characteristics.

System title	Cameras
Purpose of system	To record weapon effects or to provide high resolution images of the ground for intelligence purposes.
Description	Cameras installed in the fuselage or in fuselage-/wing-mounted pods. Surveillance cameras will be high resolution with mapping ability for high-quality images for intelligence purposes (IMINT).
Safety/integrity aspects	Mission critical.
Key integration aspects	Alignment with aircraft axis, structure, mission system.
Key interfaces	Location of lenses, under fuselage/underwing pylons.
Key design drivers	Mission success, resolution of images.
Modelling	
References	Oxlee (1997), Airey and Berlin (1985).
Sizing considerations	Cameras, mountings, plane glass window.
Future considerations	

Table 12.62 Head-up display system characteristics.

System title	HUD
Purpose of system	To provide the crew with primary information and weapon aiming information collimated to infinity, therefore superimposed on the pilot's forward view.
Description	Optical system to project the image focussed to infinity in the pilot's direct vision, connected to the avionic systems to obtain navigation and weapons data.
Safety/integrity aspects	Safety involved – safety critical if used for primary flight information.
Key integration aspects	Human factors integration, cockpit display suite.
Key interfaces	Cockpit installation, must not infringe ejection clearances.
Key design drivers	Combat performance, may also be used as a landing aid.
Modelling	Mission system test rig.
References	Jukes (2003).
Sizing considerations	HUD assembly, cockpit mounting.
Future considerations	Used on civil aircraft types.

Table 12.63 Helmet-mounted display system characteristics.

System title	Helmet-mounted display
Purpose of system	To provide primary flight information and weapon information to the crew whilst allowing freedom of movement of the head.
Description	Display surface mounted to the pilot's helmet, may also contain a sighting mechanism.
Safety/integrity aspects	Mission critical.
Key integration aspects	Integration with mission computing and avionics. Human factors.
Key interfaces	Interface with standard aircrew helmet.
Key design drivers	Combat performance, low workload, health and safety (of user).
Modelling	Mission system test rig.
References	Jukes (2003).
Sizing considerations	Treat as pilot role equipment.
Future considerations	

Table 12.64 Data link system characteristics.

System title	Data link
Purpose of system	To provide transmission and receipt of messages under secure communications using data rather than voice.
Description	Terminal with encoding/decoding facility, mission data uploading capability, and encryption devices.
Safety/integrity aspects	Mission critical.
Key integration aspects	Integration with suitable radio transmitters, data link protocol suitable for co-operative working.
Key interfaces	Communications, mission data loads.
Key design drivers	Security of transmission.
Modelling	Mission system test rig.
References	Schleher (1999).
Sizing considerations	Transmitter/receiver, message workstation, antenna.
Future considerations	

12.5 Sizing and Scoping Systems

There are occasions when it is necessary to obtain a quick estimate of the size and scope of a project. An example of this is a student project in which teams compete to evolve a preliminary design of a project and need to do some trade studies to determine the most cost-effective solution, or at least to understand what their solution will do and to understand the cost and mass at a rudimentary level. This section will give a brief description of a process for doing this, making use of the key characteristics tabulated in the previous sections. The process is illustrated in Figure 12.2.

A) The project requirement will provide key parameters such as target weight, range, endurance, operating altitude etc.
B) An analysis of the requirements will result in one or more solutions that can be considered for comparison.
C) This will enable a top-level architecture to be developed to define the major systems, their sub-systems, and the most likely sources of power.
D) From the architecture it will be possible to list the main components of the individual systems. For this level of analysis this means major components; it is not necessary to include all components, especially if they are of low mass and low energy demand. The output from this stage is an equipment list.
E) The components on the list can now be evaluated to determine the key parameters required for the trade-off. Typically, this will include mass, power requirement, dissipation, cost etc.
F) These can be obtained from a number of sources. The references in the tables in this chapter will provide some information and there are textbooks that quote parameters for equipment (note that textbook information does age and may not be current). The internet is a valuable source of information with searches conducted against components, systems or suppliers (note that not all information can be validated). Suppliers can be helpful and their websites may contain appropriate information. Alternatively, an email or a telephone call to their publicity departments will usually prompt a response.
G) The information on electrical loads can be used to great effect by compiling a load analysis, as illustrated in Figure 12.3. In this example the aircraft mission or aircraft typical flight is divided into sections or phases of flights in order to record the load and an approximation of the time that the load is active, the duty cycle.
H) In a similar manner the rates of flow to the hydraulic system components can be estimated and recorded to provide information about the complete hydraulic system, as shown in Figure 12.4.
J) Each system major component will dissipate heat if it is electrically powered or if it is converting energy from one form to another, such a hydraulic actuator. Pumps, generators, and motors are not 100% efficient and the inefficiency usually results in heat. A thermal load analysis will determine how much energy is being dissipated in the systems and how much of this needs to be cooled. People are a considerable source of heat energy, each passenger and crew member generates typically 200 W, and also require more energy to be expended in their flight entertainment system and the galley.

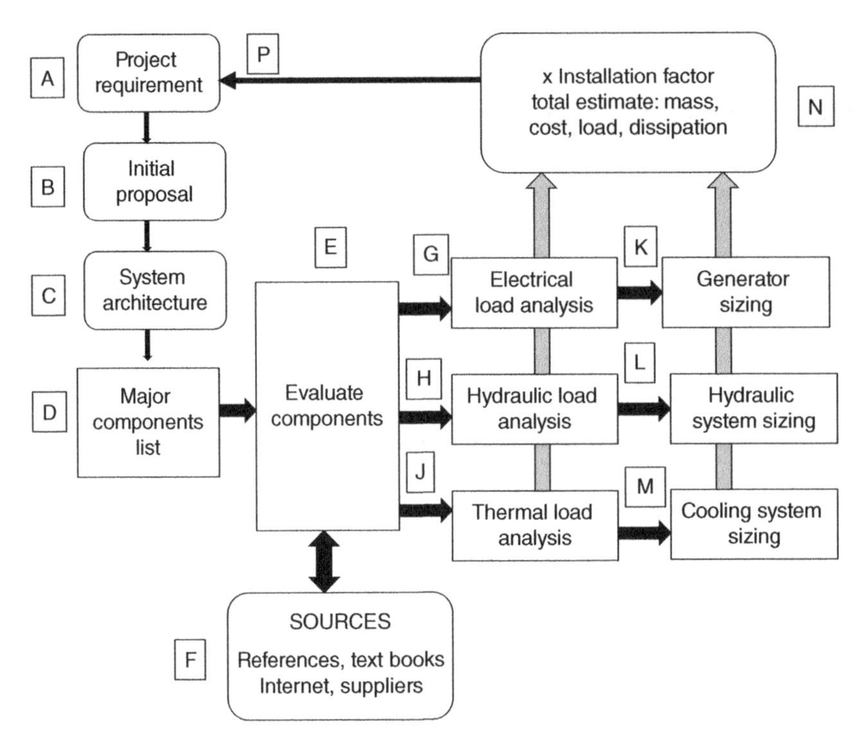

Figure 12.2 Process for producing a project estimate for trade-off.

Load	Service	Start	Taxy	Take-off	Climb	Cruise	Combat	Descent	Land
DC loads									
AC loads									
Emergency									
etc									

Figure 12.3 Example of an electrical load analysis.

Load	Service	Start	Taxy	Take-off	Climb	Cruise	Combat	Descent	Land
Brakes									
Gear Doors									
Gear									
Flaps									
Slats									
Canard									
Ailerons									
Rudder									
Elevator									
Bomb Bay									
etc									

Figure 12.4 Example of a hydraulic load analysis.

K) Knowledge of the duty cycle allows the mean load and peak load on the generator to be estimated – information that is used to select the most appropriate size generator and to determine the most suitable gauge for the main bus bars. Further design analysis will lead to decisions on batteries and auxiliary power unit (APU) requirements.
L) The hydraulic load analysis can be used to determine the most appropriate size of pump, reservoirs, and piping. An analysis of system robustness will indicate what degree of redundancy is required, which will indicate the number of pumps and systems required.
M) The thermal load analysis will lead to the determination of the cooling system for the cabin and flight deck, for avionic equipment cooling, and for special systems such as liquid cooling for specialist sensors. This may also have an impact on drag if ram air primary heat exchangers are needed.
N) This collection of data can be summarised to form a view of the total project. The system mass can be refined by applying an installation factor to account for equipment mounting, connectors, and wiring. The mass obtained from supplier data sheets in usually uninstalled mass and a factor of, say 1.25, brings the mass closer to reality.
P) The collection of information is then compared with the targets defined in the project requirement.

12.6 Analysis of the Fuel Penalties of Aircraft Systems

A knowledge of the mass of the systems and their key components (examples are given in the tables) can be used to estimate fuel penalties, and hence the impact on performance. This section is used at Cranfield University on their Airframe Systems Design short course and is reproduced with kind permission of Dr C.P. Lawson.

12.6.1 Introduction

Airframe systems have a very significant effect on overall aircraft performance. Therefore, as well as designing suitably optimised individual airframe systems, the airframe systems designer should also consider the optimisation of their systems selections on a whole-aircraft level. Airframe systems cause penalties in aircraft fuel consumption directly due to the three following factors:

1) system weight
2) system power off-take requirements (shaft power and/or bleed)
3) system resultant direct aircraft drag increases.

Airframe systems also cause penalties in aircraft fuel consumption due to indirect effects. For example, the extra fuel capacity that may be required for the aircraft to overcome the direct penalties may result in greater capacity fuel tanks, which may require added structure to support them. This will impose further fuel penalties due to the added weight caused by the extra fuel and structure, and greater drag due to the larger fuel tanks, particularly if external tanks are required. Consequently, larger engines may be required to provide more thrust, further increasing aircraft weight and drag. These factors are all linked and therefore the fuel penalties caused by airframe systems can be seen to increase with a snowballing effect.

This section goes on to consider a simplified method for calculating the fuel weight penalty incurred by the direct effects of the addition of a system to an aircraft. Whilst more complex methods exist, the method presented here is easy to apply and provides a good understanding of the parameters used as they can largely be analysed separately.

12.6.2 Basic Formulation of Fuel Weight Penalties of Systems

In this section a basic method is formulated to predict the fuel weight penalties of aircraft systems for a single flight phase, based on the three factors identified in Section 12.6.1. The first step in deriving fuel weight penalties equations is to make an assumption about aircraft drag. Thus, the following relationship is used to represent drag:

$$\text{drag} = \frac{\text{weight}}{\frac{\text{lift}}{\text{drag}}\text{ratio}} \tag{12.1}$$

In fact, Eq. (12.1) is a simplification since it only holds true if the lift/drag ratio is constant.

Consider an aircraft of weight W_A (excluding system weight) flying at Mach number M. Then, the range dR covered by the aircraft over the period dt is given by:

$$\mathrm{d}R = aM\mathrm{d}t \tag{12.2}$$

where a is the speed of sound.

During the period dt, the aircraft mass of fuel may be expressed as:

$$\left(f + \Delta f_{\mathrm{w}} + \Delta f_{\mathrm{p}} + \Delta f_{\mathrm{D}}\right)\mathrm{d}t = -\mathrm{d}\left(M_{\mathrm{F}} = \Delta M_{\mathrm{F}}\right) \tag{12.3}$$

where the negative sign is indicative of the fuel weight decreasing as flight time (t) increases, and:

- f = rate of fuel used by aircraft without system
- Δf_W = rate of fuel used due to system weight
- Δf_P = rate of fuel used due to system power off-take
- Δf_D = rate of fuel used due to system drag
- M_F = mass of fuel used excluding system effect
- ΔM_F = extra mass of fuel used due to system effect.

Eq. (12.2) may be rearranged and substituted into Eq. (12.3) in order to eliminate the dt term, and consequently rearranged to give Eq. (12.4):

$$\mathrm{d}R = \frac{-aM\left[d\left(M_{\mathrm{F}} + \Delta M_{\mathrm{F}}\right)\right]}{f + \Delta f_{\mathrm{w}} + \Delta f_{\mathrm{p}} + \Delta f_{\mathrm{D}}} \tag{12.4}$$

In this case, thrust specific fuel consumption (sfc) (c) is assumed constant and may be expressed as:

$$c = \frac{f + \Delta f_{\mathrm{w}} + \Delta f_{\mathrm{p}} + \Delta f_{\mathrm{D}}}{\text{total drag}} \tag{12.5}$$

Thrust sfc is defined as fuel flow rate per unit thrust, and drag is equal to thrust in this case. Thus, by recalling the assumption that was made to write Eq. (12.1), aircraft drag (excluding the effect of the system) can be written as:

$$\text{drag} = \frac{\text{weight}}{\frac{\text{lift}}{\text{drag}}\text{ratio}} = \frac{W_{\mathrm{A}} + W_{\mathrm{F}}}{r} \tag{12.6}$$

where:

- W_{A} = aircraft empty weight excluding system
- W_{F} = weight of fuel used excluding system effect
- r = lift/drag ratio.

Therefore, aircraft drag including the effect of the system weight may be expressed as:

$$\text{weight drag} = \frac{W_{\mathrm{A}} + \Delta W_{\mathrm{A}} + W_{\mathrm{F}} + \Delta W_{\mathrm{F}}}{r} \tag{12.7}$$

where:

- ΔW_{A} = system weight
- ΔW_{F} = extra weight of fuel used due to system effect.

The total aircraft drag, including the drag increase due to system direct drag increase (ΔD), and the effective increase in drag due to the engine power off-take required by the system ($\Delta f_{\mathrm{p}}/c$), may be expressed as:

$$\text{total drag} = \frac{W_{\mathrm{A}} + \Delta W_{\mathrm{A}} + W_{\mathrm{F}} + \Delta W_{\mathrm{F}}}{r} \Delta D + \frac{\Delta f_{\mathrm{p}}}{c} \tag{12.8}$$

Substituting Eq. (12.8) into Eq. (12.5) gives:

$$c = \frac{f + \Delta f_{\mathrm{W}} + \Delta f_{\mathrm{p}} + \Delta f_{\mathrm{D}}}{\left(W_{\mathrm{A}} + \Delta W_{\mathrm{A}} + W_{\mathrm{F}} + \Delta W_{\mathrm{F}}\right)\frac{1}{r} + \Delta D + \frac{\Delta f_{\mathrm{p}}}{c}} \tag{12.9}$$

Rearranging Eq. (12.9) gives:

$$f + \Delta f_{\mathrm{W}} + \Delta f_{\mathrm{p}} + \Delta f_{\mathrm{D}} = \frac{c}{r}\left(W_{\mathrm{A}} + \Delta W_{\mathrm{A}} + W_{\mathrm{F}} + \Delta W_{\mathrm{F}} + R\Delta D + \frac{r\Delta f_{\mathrm{p}}}{c}\right) \tag{12.10}$$

Substituting Eq. (12.10) into Eq. (12.4) yields:

$$dR = \frac{r}{c} \cdot \frac{-aM\left[d\left(W_F + \Delta W_F\right)\right]}{\left(W_A + \Delta W_A + W_F + \Delta W_F + r\Delta D + \frac{r\Delta f_P}{c}\right)} \tag{12.11}$$

Substituting $M_F = W_F/g$ and $\Delta M_F = \Delta W_F/g$ and integrating Eq. (12.11) gives the aircraft range, R:

$$R = aM\frac{r}{cg} \cdot \ln \frac{W_A + \Delta W_A + W_{FO} + \Delta W_{FO} + r\Delta D + \frac{r\Delta f_P}{c}}{W_A + \Delta W_A + r\Delta_D + \frac{r\Delta f_P}{C}} \tag{12.12}$$

where:

- W_{FO} = weight of fuel used to fly range, R, excluding system
- ΔW_{FO} = extra weight of fuel used to fly range, R, due to system
- g = gravitational constant of acceleration.

It is convenient at this point to define t as the time taken to fly range R ($R = aMt$). Eq. (12.12) can then be simplified to:

$$t\frac{cg}{r} = ln \frac{W_{FO} + \Delta W_{FO}}{W_A + \Delta W_A + r\Delta_D + \frac{r\Delta f_P}{C}} + 1 \tag{12.13}$$

Finally, Eq. (12.13) can be rearranged to give the total weight of fuel used by the aircraft with the system fitted, $W_{FO} + \Delta W_{FO}$:

$$W_{FO} + \Delta W_{FO} = \left(W_A + \Delta W_A + r\Delta D + \frac{r\Delta f_P}{c}\right)\left(e^{\frac{ctg}{r}} - 1\right) \tag{12.14}$$

From Eq. (12.14) it is simple to obtain the weight of fuel used by an aircraft with the system excluded (W_{FO}) by setting $\Delta W_{FO} = \Delta W_A = \Delta D = \Delta f_P = 0$, giving:

$$W_{FO} = W_A\left(e^{\frac{ctg}{r}} - 1\right) \tag{12.15}$$

Equally, from Eq. (12.14) it is simple to obtain the increased weight of fuel used due to the system (ΔW_{FO}) by setting $W_{FO} = W_A = 0$, yielding:

$$W_{FO} = \left(\Delta W_A + r\Delta D + \frac{r\Delta f_P}{c}\right)\left(e^{\frac{ctg}{r}} - 1\right) \tag{12.16}$$

From Eq. (12.16) the fuel weight increase due to the three components identified in Section 12.6.1 can be written in three separate equations as follows:

Fuel weight increase due to system weight:

$$\left(\Delta W_{\mathrm{FO}}\right)_{\Delta W_{\mathrm{A}}} = \Delta W_{\mathrm{A}}\left(e^{\frac{ctg}{r}} - 1\right) \tag{12.17}$$

Fuel weight increase due to system power off-take:

$$\left(\Delta W_{\mathrm{FO}}\right)_{\Delta f_{\mathrm{P}}} = \frac{r}{c}\Delta f_{\mathrm{P}}\left(e^{\frac{ctg}{r}} - 1\right) \tag{12.18}$$

Fuel weight increase due to system drag:

$$\left(\Delta W_{\mathrm{FO}}\right)_{\Delta D} = r\Delta D\left(e^{\frac{ctg}{r}} - 1\right) \tag{12.19}$$

Provided the required data is available Eqs (12.17)–(12.19) can be used to directly calculate the fuel weight penalties of a system.

12.6.3 Application of Fuel Weight Penalties Formulation for Multi-phase Flight

In Section 12.6.2 equations to calculate the fuel weight penalties incurred by a system for a single flight phase were derived. Of course, any real aircraft flight will involve multiple phases of flight at different operating conditions. Therefore, this section goes on to consider how these equations may be applied to a multi-phase flight. First, it is assumed that there is a step change in conditions between flight phases. It is then convenient to define a variable, F, which represents the fuel weight penalty incurred due to the system in all subsequent flight phases. Therefore, F is the sum of the system fuel weight penalties in all flight phases that take place after the flight phase being considered. It is thus clear that $F = 0$ in the final flight phase. Applying this weight penalty to Eq. (12.16) produces an equation for the fuel weight penalty due to the system for a single flight phase, i, of a multi-phase flight.

$$\left(\Delta W_{\mathrm{FO}}\right)_i = \left(\Delta W_{\mathrm{A}} + F_i + r\Delta D + \frac{r\Delta f_{\mathrm{P}}}{c}\right)\left(e^{\frac{ctg}{r}} - 1\right) \tag{12.20}$$

If n is defined as the total number of phases in the flight, then the fuel weight penalty due to the system throughout the entire flight is:

$$\Delta W_{\mathrm{FO}} = \sum_{i=1}^{n}\left(\Delta W_{\mathrm{FO}}\right)_i \tag{12.21}$$

12.6.4 Analysis of Fuel Weight Penalties Formulation for Multi-phase Flight

Differentiation of Eq. (12.20) with respect to time results in an equation for the instantaneous increase in fuel flow rate due to the system, Δf, for a given flight phase, i:

$$\Delta f_i = \left(\Delta W_A + F_i + r\Delta D + \frac{r\Delta f_p}{c} \right) \frac{c}{r} e^{\frac{ctg}{r}} \quad (12.22)$$

Eq. (12.22) can be split into three equations describing the contributions of system weight, off-take power, and direct drag increases, as follows:

$$\left(\Delta f_{\Delta W_A} \right)_i = \left(\Delta W_A + \left(F_{\Delta W_A} \right)_i \right) \frac{c}{r} e^{\frac{ctg}{r}} \quad (12.23)$$

$$\left(\Delta f_{\Delta f_p} \right)_i = \left(\frac{r\Delta f_p}{c} + \left(F_{\Delta f_p} \right)_i \right) \frac{c}{r} e^{\frac{ctg}{r}} \quad (12.24)$$

$$\left(\Delta f_{\Delta D} \right)_i = \left(r\Delta D + \left(F_{\Delta D} \right)_i \right) \frac{c}{r} e^{\frac{ctg}{r}} \quad (12.25)$$

Considering Eq. (12.24) for instantaneous fuel flow rate increase due to system power off-take, it can be shown that at the end of the final flight phase, where $t = 0$, the instantaneous fuel flow rate increase due to the system is only equal to the direct increase due to the system power off-take. At all earlier stages of the flight it is higher than this due to the engine thrust needing to be higher to overcome the extra drag caused by the necessary fuel being carried.

12.6.5 Use of Fuel Weight Penalties to Compare Systems

When comparing systems, the overall system weight penalty should be used, W_T, where W_T is the system weight plus the additional fuel carried due to system effects:

$$W_T = \Delta W_A + \Delta w_{FO} \quad (12.26)$$

Substituting Eq. (12.16) into Eq. (12.26) yields:

$$W_T = \Delta W_A + \left(\Delta W_A + r\Delta D + \frac{r\Delta f_p}{c} \right) \left(e^{\frac{ctg}{r}} - 1 \right) \quad (12.27)$$

It is often the case that a more massive system (greater ΔW_A) will incur a smaller fuel weight penalty (ΔW_{FO}) than a less massive system and thus potentially a lower overall system weight penalty (W_T), particularly for long-range missions. Therefore, the system choice will be dependent upon the range that the subject aircraft most commonly flies. The choice of the optimum system has the potential to deliver better aircraft performance in terms of range and payload capabilities. This is represented schematically in Figure 12.5.

In such trade-off cases secondary effects are often important. The mass of fuel tank and supporting structure, as well as payload support structure and external payload-caused

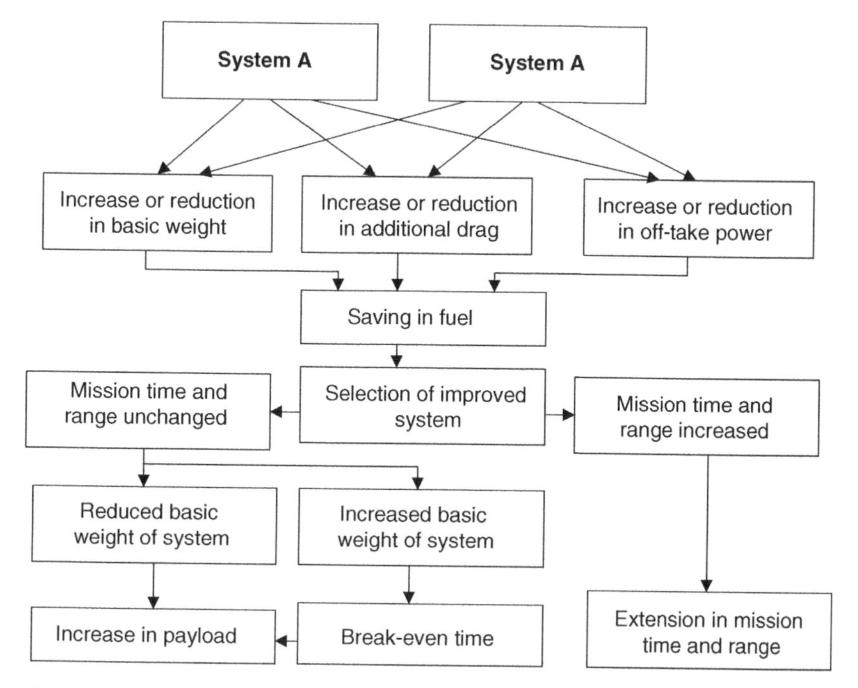

Figure 12.5 Flow chart for suggested system comparison.

drag increases, may have a significant influence on the choice of the best overall system. A more sophisticated level of analysis would take these secondary effects into account. A thorough analysis would also take system costs into account, in terms of both initial cost and lifecycle cost.

12.6.6 Determining Input Data for Systems Weight Penalties Analysis

In order to assess weight penalties, several parameters are required. These can then be used in the equations derived in this section to calculate fuel weight penalties. Ideally, accurate values would be calculated or measured experimentally, thus providing the most accurate results from the fuel weight penalties analysis. However, this is often impossible or impractical, particularly when the systems penalties analysis is being performed at the aircraft design stage. This section presents methods to approximate these values in the absence of accurate data being available.

12.6.6.1 Lift/Drag Ratio

The lift/drag ratio depends on many parameters and will vary significantly between different flight conditions, even for the same aircraft. The most rapid way to get a very rough value for the lift/drag ratio is to look up a table or a chart from the available literature. A value for the lift/drag ratio obtained in this way is adequate to allow a first iteration of a penalties analysis to be performed. For later iterations of a penalties analysis, better accuracy may be obtained by calculating lift and drag. Methods for achieving this in various degrees of complexity are presented in textbooks on aircraft design (Roskam 1990). Measuring lift and drag through wind tunnel testing would provide a still greater level of accuracy in the values of lift/drag ratio obtained, and these may be complemented by aircraft aerodynamic simulations.

12.6.6.2 Specific Fuel Consumption

Specific fuel consumption depends upon the flight condition of the aircraft. The most rapid way to get a rough value for sfc is to look up a table or a chart from the engine manufacturer's data on uninstalled engine performance. A greater level of accuracy may be obtained by using a computer model of an engine. A programme such as Cranfield University's Turbomatch allows models of engines to be built and simulations performed.

12.6.6.3 System Mass

At the early stages of aircraft design, system masses can be estimated using basic methods, such as those presented in Torenbeek (1982) and Roskam (1990). Thus, systems masses may be estimated from aircraft mass and other parameters known early on in the design process by using equations. These methods provide rough mass estimates for conventional systems. Therefore, alternative systems may be compared with these estimated masses, with analysis showing by what percentage the alternative systems would increase or reduce weight compared to the conventional systems. The potential lack of accuracy with the comparisons here can be mitigated somewhat by carrying out a sensitivity study.

12.6.6.4 System Drag Increase

A common source of system-induced drag is ram drag, caused by taking in air, typically for cooling purposes. This may be (pessimistically) estimated by assuming total momentum loss occurs. In the case of systems that impact on the aircraft externally, drag should be estimated using approximated geometries. Methods for calculating fluid-dynamic drag are presented in numerous textbooks on the subject (Hoerner 1965).

12.6.6.5 Increase in sfc Due to Systems Power Off-takes

Increases in sfc due to systems power off-takes are often difficult to obtain accurately, therefore an estimation is usually made. In the case of shaft power off-takes, the increase in sfc varies fairly linearly with shaft power off-take to net thrust ratio for all but improbably large shaft power off-takes. This is illustrated in Figure 12.6, where data are plotted for several civil application turbo-fan engines with by-pass ratios of around five.

Military turbo-fans with lower by-pass ratios also display a similar trend to that shown in Figure 12.6. From this an equation can be written describing the linear trend:

$$\%\text{ increase in sfc} = 0.175\left(\text{kg / sN}\right) \times \text{shaft power off-take/net thrust} \tag{12.28}$$

where power is in units of Watts (Nm/s) and thrust is in units of Newtons. For bleed off-take power, the relationship between sfc increase and bleed flow rate to net thrust ratio is linear at relatively low off-take levels. However, it is non-linear at commonly used higher off-take levels. Therefore, for bleed off-takes a simple equation such as that for shaft off-takes (Eq. (12.28)) cannot generally be used. Plots of percentage sfc increase against bleed to net thrust ratio, such as that in Figure 12.7, are required for a fuel weight penalties analysis.

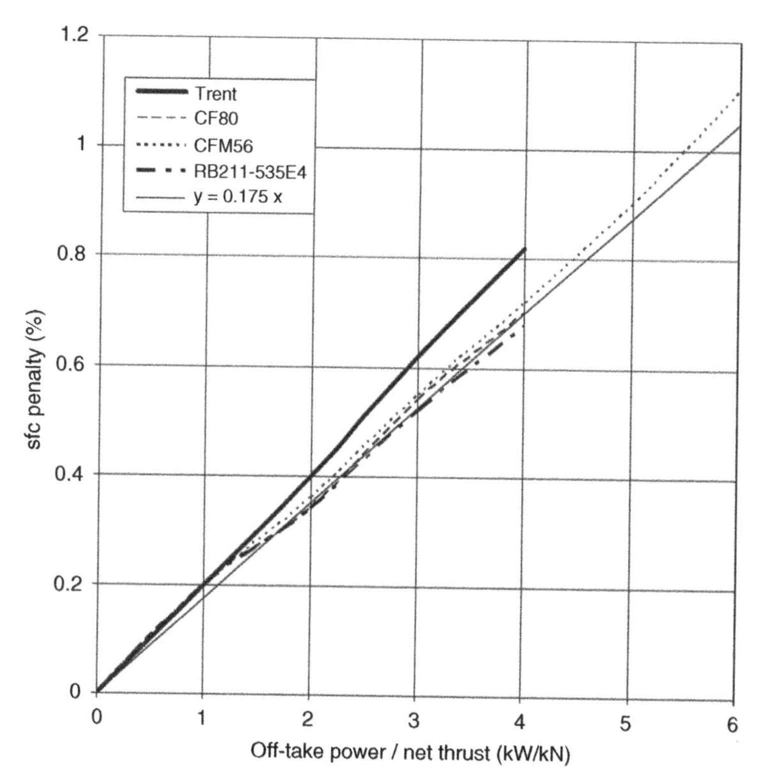

Figure 12.6 Variation in percentage sfc increase with shaft off-take power to net thrust ratio for civil turbo-sfans.

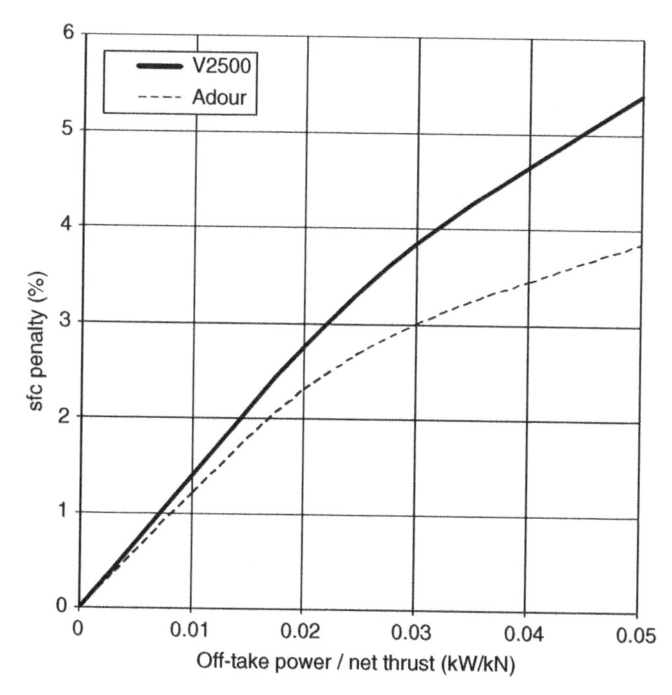

Figure 12.7 Variation in percentage sfc increase with bleed off-take rate to net thrust ratio for turbo-fans engines.

Nomenclature

a	Speed of sound
c	Thrust specific fuel consumption
$\mathrm{d}R$	Range covered by aircraft
$\mathrm{d}t$	Period of time
f	Rate of fuel used by aircraft without system
F	Fuel weight penalty due to system effect for subsequent flight phases
g	Gravitational constant of acceleration
i	Flight phase number
M	Mach number
M_{F}	Mass of fuel used excluding system effect
n	Total number of phases in the flight
r	Lift/drag ratio R range
t	Time taken to fly range
R_{WA}	Aircraft empty weight excluding system
W_{F}	Weight of fuel used excluding system effect
W_{FO}	Weight of fuel used to fly range, R, excluding system effect
W_{T}	Total weight penalty due to system
ΔD	System direct drag increase
Δf	Instantaneous additional fuel flow rate due to system effect
Δf_{D}	Rate of fuel used due to system drag
Δf_{p}	Rate of fuel used due to system power off-take
Δf_{W}	Rate of fuel used due to system weight
ΔM_{F}	Extra mass of fuel used due to system effect
ΔW_{A}	System weight
ΔW_{F}	Extra weight of fuel used due to system effect
ΔW_{FO}	Extra weight of fuel used to fly range, R, due to system

References

Adamy, D.A. (2003). *EW 101 A first Course in Electronic Warfare*. Artech House.

Agarwal, R. (2016). Energy optimization for solar powered aircraft. In: *Encyclopedia of Aircraft Engineering, Green Aviation* (eds. R. Argawal, F. Collier, A. Schäfer and A. Seabridge), 147–163. Wiley.

Airey, T.E. and Berlin, G.L. (1985). *Fundamentals of Remote Sensing and Airphoto Interpretation*. Prentice Hall.

Atkin, E.M. (2010). Aerospace avionics systems. In: *Encyclopedia of Aerospace Engineering*, Vol. 8: Chapter 391 (eds. R.H. Blockley and W. Shyy), 4787–4797. Wiley.

Bamford, J. (2001). *Body of Secrets*. Century.

Bomani, B.M.M. and Hendricks, R.C. (2016). Biofuels for green aviation. In: *Encyclopedia of Aircraft Engineering, Green Aviation* (eds. R. Argawal, F. Collier, A. Schäfer and A. Seabridge), 179–191. Wiley.
Bryson, R.E. Jr. (1994). *Control of Spacecraft and Aircraft*. Princeton University Press.
Burberry, R.A. (1992). *VHF and UHF Antennas*. Peter Pergrinus.
Conway, H.G. (1957). *Landing Gear Design*. Chapman & Hall.
Cramer, M.R., Herndon, A., Steinbach, D., and Mayer, R.H. (2010). Modern aircraft flight management systems. In: *Encyclopedia of Aerospace Engineering*, Vol. 8: Chapter 397 (eds. R.H. Blockley and W. Shyy), 4861–4872. Wiley.
Currey, N.S. (1984). *Landing Gear Design Handbook*. Lockheed Martin.
FAA (2018). *Flight Standards Information Management System (FSIMS) – Master Minimum Equipment List (MMEL)*. Available: fsims.faa.gov/wdocs/mmel (accessed September 2019).
Freeh, J.E. (2016). Hydrogen fuel cells for auxiliary power units. In: *Encyclopedia of Aircraft Engineering, Green Aviation* (eds. R. Argawal, F. Collier, A. Schäfer and A. Seabridge), 193–191. Wiley.
Galotti, V.P. Jr. (1998). *The Future Air Navigation System (FANS)*. Ashgate Publishing.
Gardner, W.J.R. (1996). *Anti-submarine Warfare*. Brassey's.
Gent, R.W. (2010). Ice detection and protection. In: *Encyclopedia of Aerospace Engineering*, Vol. 8: Chapter 408 (eds. R.H. Blockley and W. Shyy), 5005–5015. Wiley.
Giguere, D.'.A. (2010). Aircraft emergency systems. In: *Encyclopedia of Aerospace Engineering*, Vol. 8: Chapter 407 (eds. R.H. Blockley and W. Shyy), 4995–5003. Wiley.
Gradwell, D.P. (2010). Physiology of the flight environment. In: *Encyclopedia of Aerospace Engineering*, Vol. 8: Chapter 382 (eds. R.H. Blockley and W. Shyy), 4693–4702. Wiley.
Hall, M.R.M. and Barclay, L.W. (1980). *Radiowave Propagation*. Peter Pergrinus.
Hoerner, S.F. (1965). *Fluid Dynamic Drag*. Hoerner.
Hunt, T. and Vaughan, N. (1996). *Hydraulic Handbook*, 9e. Elsevier.
Jackson, A.J.B. (2010). Choice and sizing of engines for aircraft. In: *Encyclopedia of Aerospace Engineering*, Vol. 8: Chapter 401 (eds. R.H. Blockley and W. Shyy), 5123–5134. Wiley.
Jukes, M. (2003). *Aircraft Display Systems*. Professional Engineering Publishing.
Kayton, M. and Fried, W.R. (1997). *Avionics Navigation Systems*. Wiley.
Langton, R. (2006). *Stability and Control of Aircraft Systems*. Wiley.
Langton, R. (2010). Gas turbine fuel control system. In: *Encyclopedia of Aerospace Engineering*, Vol. 8: Chapter 405 (eds. R.H. Blockley and W. Shyy), 4973–4984. Wiley.
Langton, R., Clark, C., Hewitt, M., and Richards, L. (2009). *Aircraft Fuel Systems*. Wiley.
Langton, R., Clark, C., Hewitt, M., and Richards, L. (2010). Aircraft fuel systems. In: *Encyclopedia of Aerospace Engineering*, Vol. 8: Chapter 402 (eds. R.H. Blockley and W. Shyy), 4919–4938. Wiley.
Lawson, C.P. (2010). Environmental control systems. In: *Encyclopedia of Aerospace Engineering*, Vol. 8: Chapter 406 (eds. R.H. Blockley and W. Shyy), 4985–4994. Wiley.
Lloyd, E. and Tye, W. (1982). *Systematic Safety*. Taylor Young.
MacIsaac, B. and Langton, R. (2011). *Gas Turbine Propulsion Systems*. Wiley.
Moir, I. and Seabridge, A.G. (2010). Vehicle management systems. In: *Encyclopedia of Aerospace Engineering*, Vol. 8: Chapter 401 (eds. R.H. Blockley and W. Shyy), 4903–4917. Wiley.
Moir, I. (2010). Electrical power generation and distribution. In: *Encyclopedia of Aerospace Engineering*, Vol. 8: Chapter 404 (eds. R.H. Blockley and W. Shyy), 4955–4972. Wiley Ltd.

Moir, I. and Seabridge, A. (2013). *Civil Avionics*, 2e. Wiley.
Moir, I. and Seabridge, A. (2006). *Military Avionics Systems*. Wiley.
Moir, I. and Seabridge, A. (2008). *Aircraft Systems*, 3e. Wiley.
Oxlee, G.J. (1997). *Aerospace Reconnaissance*. Brassey's.
Pallett, E.H.J. (1987). *Aircraft Electrical Systems*. Longmans Group.
Pallett, E.H.J. (1992). *Aircraft Instruments & Integrated Systems*. Longmans Group: EHJ Pallett.
Poisel, R.A. (2003). *Introduction to Communication Electronic Warfare Systems*. Artech House.
Pornet, C. (2016). Electric drives for propulsion of transport aircraft. In: *Encyclopedia of Aircraft Engineering, Green Aviation* (eds. R. Argawal, F. Collier, A. Schäfer and A. Seabridge), 201–219. Wiley.
Pratt, R. (2000). *Flight Control Systems: Practical Issues in Design & Implementation*, 2000. IEE Publishing.
Avionics Communications (1995). *Principles of Avionics Data Buses*. Avionics Communications.
Purdy, S.I. (2010). Probe and drogue aerial refuelling systems. In: *Encyclopedia of Aerospace Engineering*, Vol. 8: Chapter 409 (eds. R.H. Blockley and W. Shyy), 5018–5027. Wiley.
Rankin, J.M. and Matolak, D. (2010). Aircraft communications and networking. In: *Encyclopedia of Aerospace Engineering*, Vol. 8: Chapter 394 (eds. R.H. Blockley and W. Shyy), 4829–4852. Wiley.
Raymond, E.T. and Chenoweth, C.C. (1993). *Aircraft Flight Control Actuation System Design*. Society of Automotive Engineers.
Rigby, K. (2010). Weapons integration. In: *Encyclopedia of Aerospace Engineering*, Vol. 8: Chapter 417 (eds. R.H. Blockley and W. Shyy), 5107–5116. Wiley.
Roskam, J. (1990). *Airplane Design*. University of Kansas.
Schleher, C. (1999). *Electronic Warfare in the Information Age*. Artech House.
Schleher, C.D. (1978). *MTI Radar*. Artech House.
Schutte Jeff, S., Payan, A.P., Briceno Simon, I., and Marvis, D.N. (2016). Hydrogen-powered aircraft. In: *Encyclopedia of Aircraft Engineering, Green Aviation* (eds. R. Argawal, F. Collier, A. Schäfer and A. Seabridge), 165–174. Wiley.
Seabridge, A. (2010). Hydraulic power generation and distribution. In: *Encyclopedia of Aerospace Engineering*, Vol. 8: Chapter 403 (eds. R.H. Blockley and W. Shyy), 4939–4953. Wiley.
Skolnik, M.I. (1980). *Introduction to Radar Systems*. McGraw-Hill.
Spitzer, C. (1993). *Digital Avionics Systems, Principles and Practice*, 2e. McGraw-Hill.
Srivastava, A.N., Meyer, C., and Mah, R.W. (2010). In-flight vehicle health management. In: *Encyclopedia of Aerospace Engineering*, Vol. 8: Chapter 436 (eds. R.H. Blockley and W. Shyy), 5327–5338. Wiley.
Stimson, G.W. (1998). *Introduction to Airborne Radar*, 2e. SciTech Publishing.
Torenbeek, E. (1982). *Synthesis of Subsonic Airplane Design*. Delft University Press.
Urick, R.J. (1982). *Principles of Underwater Sound*. Peninsula Publishers.
Urick, R.J. (1983). *Sound Propagation in the Sea*. Peninsula Publishers.
Van Brunt, L.B. (1995). *Applied ECM*. EW Engineering.
Walton, J.D. (1970). *Radome Engineering Handbook*. Marcel Dekker.
Weller, B. (2018). *A History of the Fly-by-Wire Jaguar*. BAE Systems Heritage Department. (available via Amazon).
Xue, N., Wenbo, D., Martin, J.R.R.A., and Wei, S. (2016). Lithium-ion batteries: thermomechanics, performance and design optimization. In: *Encyclopedia of Aircraft Engineering, Green Aviation* (eds. R. Argawal, F. Collier, A. Schäfer and A. Seabridge), 221–237. Wiley.

13

Conclusions

This book has attempted to portray the design and development of aircraft systems as practised in the aerospace industry. The industry is currently dominated by the need to deliver hard products, many of them complex interactions of airframe, components, human operators, and systems, both hardware and software. These products are provided to customers as a part of their armed forces or airline infrastructure, which in turn may be part of a wider national or international entity. This increasingly complex nature of products has led to an approach to dealing with them as complex systems.

An understanding of what constitutes a system is important. There is an increasing tendency for domain-specific engineers to take a broader view of their system, a state of mind that is stimulated by the increasing integration of systems in the modern aircraft. Thus, individual systems are perceived as being sub-systems of larger integrated systems existing in complex environments. Chapter 2 looked at these wider concepts of system and identified the commonality between them in terms of the form and terminology of systems. There are many books and papers available for further study to build on this understanding and to examine how the field of systems engineering has a large part to play.

Systems engineering is the science, discipline or art that is employed to understand the initial need or requirement for a system and to progress in an orderly manner to the delivery of a completed entity. The skills required for this are many and varied. Some skills are innate, some can be taught, and others are acquired by experience. Again there is plenty to read: the website and the transactions of the International Council in Systems Engineering (INCOSE) provide a wide range of applications and experiences in the wider world of systems engineering. There is an increasing awareness of the need to provide some formal education at all levels from high school to postgraduate to supplement the existing courses. This formal education is complemented in the UK by short courses and continuing professional development courses, especially by aerospace specialist universities like Cranfield.

The concept of an environment in which all systems exist is extremely useful in identifying factors that influence the system. Visualising the environment as a set of nested or intersecting environments enables these factors to be structured or prioritised in order to deal with them as similar groups or as factors of differing impact. Chapter 4 described some examples of factors or design drivers, but this list is by no means exhaustive and time spent identifying all the relevant factors in an individual project is time well spent.

Design and Development of Aircraft Systems, Third Edition. Allan Seabridge and Ian Moir.

Systems architectures are a convenient method of visualising emerging concepts in both functional and physical form. The block diagram is a convenient notation for identifying the form of systems and is used as a medium for brainstorming, debate, and discussion. Chapter 5 introduced this topic and presented an example. The topic is one that is better described in real life by developing architectures from a blank sheet of paper. Read the chapter, but try it for yourself, preferably in a group, to feel the power of pictorial representation and the ideas that it stimulates.

In Chapter 6 the topic of systems integration was introduced in order to explain that the topic has many interpretations. The use of techniques to reduce systems from a top-level visualisation down to smaller and smaller sub-system elements is useful in simplifying any one part of the system for a design solution to be sought. The skill of systems integration allows the products of this reductionist approach to be assembled to fulfil the original top-level requirement. This is achieved by ensuring that the top-level requirements are decomposed down to their elemental levels, and that the corresponding design maps directly on to the requirement. Although the term 'systems integration' has come to mean different things to different people, all these uses are valid in their own application. The aircraft systems engineer shows tolerance and understanding to allow them all to co-exist.

A technique that enables systems engineers to expand the understanding of their concepts at all stages of the lifecycle is that of modelling. Modelling takes many forms, from simple sketches to balsa wood and modelling clay representations through sophisticated mathematical models running on supercomputers to full-scale prototypes. One important aspect of modelling is that of designing and testing the integration of the human into the system – human factors, not merely physically but also cognitively. Each model has its part to play in the evolution of a system and Chapter 7 only scratched the surface of this topic.

Chapter 8 encapsulated the experience of the author and colleagues to illustrate some aspects around the periphery of systems engineering. These aspects make a contribution to the process of system evolution. Again, this is a limited and personal perspective. There should be no barrier to capturing and using best practice, and taking note of poor practice wherever it is encountered – there is no substitute for practical experience. This experience is extended to the installation of systems and the design of wiring harnesses. Chapter 9 introduced the topic of configuration control, which is a vital element to keep order in the dynamic and changing world of a long-term development. It is a discipline that is being undermined by the proliferation of personal devices that enable people to create, disseminate, and store information that is vital to the decision-making process. There is a danger that this information will not be recoverable for analysis and there is a danger in the way we use information that needs to be carefully observed. The world of personal devices and social media means that we exchange information without necessarily recording how it was obtained, or even how valid it is. This can lead to discrepancies and gaps in the paper trail that should not be permitted.

Chapter 10 provided an example of systems development from top-level representation and looked at redundancy considerations in detailed system design. Chapter 11 looked at trends towards increasing complexity in systems that should create a slight nervousness about the certainty with which systems can be understood and exhaustively tested. This will only become more serious as the aircraft becomes more dependent on integration with external systems. The likelihood of chaotic behaviours in these extended system environments must be seriously considered.

Chapter 12 summarised the systems that have been the subject of this book and defined some key integration and interfacing aspects of those systems. A process was introduced to enable students to attempt to put some scale to their designs by sizing the system in terms of mass, power demand, and dissipation. A process for estimating the impact of mass on fuel requirements was also presented.

13.1 What's Next?

Science will not stand still and new opportunities will arise from new scientific discoveries. These will be turned into new technologies by an industry that has great enthusiasm for new technology. Technology advance will continue to provide a focus for implementations to gradually improve the performance of aircraft or to provide solutions for specific problems. A brief scan through the current literature reveals the following topics that will be developed over the next decade.

Computing has been at the heart of most advances in aircraft systems since the emergence of microprocessors suitable for use in aircraft applications in the 1970s. Changes in technology have resulted in faster computers and denser memories, and the application of these technologies in the commercial field of home computing, games, and internet applications has been put to good use in aircraft systems. Artificial intelligence algorithms in design and in real-time on-board applications will benefit from neuromorphic computing for fast and energy-efficient processing. This will have an impact on real-time pattern recognition, speech processing, and image classification. This will affect the design of intelligence gathering, sensor processing, human factors design, and navigation systems in future aircraft and may pave the way towards unmanned passenger aircraft with real autonomy. The main benefit of neuromorphic devices is their lower energy demand and the potential to incorporate more processing power into smaller volumes. This is predicted to improve over the next three decades, exceeding the predictions of Moore's law.

One use of this advance in computing will be to move towards synthetic vision in cockpits and flight decks. To some extent this has been applied in head-up displays and in cameras mounted to provide all-round vision from within the aircraft – a perceived spherical view by the pilot. Artificial intelligence (AI) and learning software techniques will lead to the provision of such facilities as object recognition and avoidance, techniques being developed for autonomous road vehicles. There is a danger that this can lead to an overload of information in the cockpit unless robust human factors approaches are applied. Techniques such as only displaying the required contextual information to an operator at a point in time and managing certain tasks on the operator's behalf will be applied to control the workload. Emerging analysis models such as trusted reasoning and trusted AI will be used to increase the certainty with which systems can be understood and tested.

There will always be a need for high-energy systems to control the attitude of air vehicles, and there will continue to be a move towards electrical rather than hydraulic or pneumatic effectors. This will stimulate a demand for electrical actuation and electro-hydrostatic systems with a need for electric motors with materials capable of operating at higher temperatures, higher magnetic field strengths, and higher rotational speeds using in-built failure prediction software which will greatly improve flight and engine control systems. This is expected to improve the power and rate performance of control surfaces at reduced mass.

It has been shown that power demand on successive generations of aircraft is growing. Despite the promises of lower power requirements of succeeding generations of semiconductors, this always off-set by an increasing demand for more processing power. To reduce generated power and large heavy generators connected to the main engines it is expected that next-generation thermo-electric generators will be used to scavenge waste and rejected heat sources to convert thermal energy directly into electrical energy. This can be used to capitalise on waste engine heat from engines, auxiliary power units, brakes, and avionics sources. There have been studies on how to use heat generated by the brakes to contribute to the energy needed to taxi an aircraft. A slight dilemma arises here, in that one of the major heat-generating systems on an aircraft is the hydraulic system, and of course its replacement by electrical systems is already being predicted.

On the subject of electrical power, it is expected that electrical generation components are likely to be embedded in the shafts of future engines. Engine manufacturers have been pursuing this for some time and it is now very likely to happen.

Composite materials have been changing the approach to design and manufacturing over the past 30 years to great advantage in commercial and military aircraft types. New materials such as graphene will pave the way for novel approaches to structure design and the design of electronic equipment housings which may require alternative approaches to screening and bonding.

A serious interest in environmental issues is at last moving away from simply banning the use of materials and substances and towards a creative use of technology and techniques to reduce carbon emissions. This has resulted in studies of different ways of operating aircraft and different shapes and propulsions solutions. This, in turn, has led to consideration of novel sources of fuel being sought for cleaner combusting engines.

In the avionic systems of the aircraft the electrical wiring, generally copper with shielding and insulation, is a contributor to weight, it needs a large volume of the fuselage interior for harnesses or bundles of wire (many tens of kilometres), connectors are always a contributor to faults, and the whole assembly is costly. Testing and repair are further costs. Despite the widespread use of data buses, including some fibre-optics, there is still a lot of wiring in an aircraft. In the military data-centric world the introduction of new mission computers, real-time video capture, high-definition cockpit displays, and advanced electronic warfare systems means extensive re-wiring for updates. Future systems will need to put greater emphasis on optical networks and signal multiplexing, novel mechanisms for incorporating wiring or optical fibres into the airframe structure and using transceivers. The commercial aircraft industry is leading the way with fibre optical backbones in the A380 and B787.

In the last edition of this book it looked as though the future in aviation was going to be in very large passenger-carrying aircraft and unmanned autonomous aircraft. There are indications in 2019 that this trend may have reached its limit with the run-down in A380 sales and strong indications that the next generation of military aircraft will be manned (Allen and Turner-Blow 2016–2018).

13.2 A Historical Footnote

In his description of the life and work of Isambard Kingdom Brunel the British pioneering engineer of the Victorian age, Angus Buchanan (Buchanan 2002), provides evidence that the great engineer had distinct tendencies towards a systems engineering discipline. With regard to the Great Western Railway (GWR) for example, his vision was for an integrated system:

> *Once Brunel had become the Engineer for the GWR, his vast resources of energy ... were turned to realising the vision of an integrated railway system.*

This, of course was what the GWR and many lines in the west of England became. Buchanan also draws attention to the vision and process adopted:

> *Brunel's vision for the GWR departed from what was then the standard view of railway construction in two important respects: he envisaged a system, first, which was primarily devoted to the movement of passengers and secondly ... one which would achieve high speeds in order to reduce journey times. These criteria determined his integrated approach to the design of the railway, which he saw as a system on interdependent parts, the efficiency of each being essential to the operation of the whole. The creation of such a system required a series of stages. The first stage was the promotion of the project, in which support was canvassed, and necessary legislation acquired. Secondly, there was the survey, to secure the best possible rout for the railway. Thirdly, came the construction stage, in which the vital civil engineering works were performed. Fourthly, the operation of the system and required provision of the necessary locomotives and rolling stock, stations and signalling to be in place. Fifthly, measures of consolidation were needed to prepare the sub-structure of workshops, offices and accommodation which would guarantee the permanence of the enterprise. And, sixthly, there was the further development of the railways to be considered, whereby long-term modifications and extensions could be introduced, and viable relationships established with neighbouring railways.*

Clearly this is a modern viewpoint; it cannot be confirmed that Brunel believed himself to be a 'systems engineer'. However, it is clear from a study of his work that he embodied some of the characteristics of a systems thinker. He was an innovator who took a holistic view of the subject, he took into account a consideration of the wider aspects of rail transport such as the infrastructure needed to make such a system successful, and the need to test his concepts thoroughly.

He was also refreshingly human, and displayed some characteristics that are present in today's engineers – he was stubborn, he tried to retain a system that was clearly out of step with a national standardisation trend, and he tried to achieve perfection, almost at all costs. This may be evidence of a 'not invented here' syndrome, although Brunel was able to show quantitatively that his broad-gauge system was more efficient than the narrower gauge. It is an example of modern engineering behaviour that continues to surface in modern complex product developments.

The modern view of systems engineering seeks to engender a balance of skill, experience, and judgement that removes personal bias in order to achieve a system that meets the needs of the customer and is considered fit for its intended purpose. The creative aspects of systems engineering must not be overlooked. There are links between creativity in art and creativity in engineering (MacDonald 1998) and this should not be underestimated. The discipline of systems engineering is about the creation and realisation of elegant systems solutions in a complex world.

References

Allen, Mark; Turner-Blow, James. (2016–2018). Prescient – a collection of technology trends and impacts. Unpublished Work. BAE Systems (Operations) Ltd.

Buchanan, A. (2002). *Brunel: The Life and Times of Isambard Kingdom Brunel*. London: Hambledon and London.

MacDonald, John S. (1998). Keynote speech at the 1998 INCOSE Symposium.

Index

Design and Development of Aircraft Systems, Third Edition. Allan Seabridge and Ian Moir.

g

h

i

j

l

m

n